Rubber Boots Methods for the Anthropocene

T0395884

Rubber Boots Methods for the Anthropocene

Doing Fieldwork in Multispecies Worlds

Nils Bubandt

Astrid Oberborbeck Andersen

Rachel Cypher

Editors

UNIVERSITY OF MINNESOTA PRESS

MINNEAPOLIS • LONDON

Excerpts from Kamau Brathwaite, "Coral," in *The Arrivants: A New World Trilogy* (Oxford University Press, 1973), are reproduced with permission of the Licensor through PLS Clear.

Published by the University of Minnesota Press
111 Third Avenue South, Suite 290
Minneapolis, MN 55401-2520
http://www.upress.umn.edu

ISBN 978-1-5179-1164-5 (hc)
ISBN 978-1-5179-1165-2 (pb)

A Cataloging-in-Publication record for this book is available from the Library of Congress.

Printed in the United States of America on acid-free paper

The University of Minnesota is an equal-opportunity educator and employer.

31 30 29 28 27 26 25 24 23 22   10 9 8 7 6 5 4 3 2 1

Contents

Acknowledgments

The conference on which this book is based took place on November 26–27, 2018. It had the form of a "node conference," designed to experiment with the format for academic conferences, presentations, and knowledge circulation in ways that both stimulated thinking and collaboration but also countered the carbon cost of the conventional, physical conference. The Anthropocene requires, so we thought in 2018, such double consciousness. Since then, of course, the Covid-19 pandemic has accelerated and overtaken our experiment, forcing all of us to convene in such digitally connected nodal formats as a norm rather than an exception. Ironically, and very aptly for the multispecies historical perspective at the heart of *Rubber Boots Methods for the Anthropocene*, the viral zoonosis of a global pandemic pushed humans and technology to align in novel ways, reminding us of the urgency of rethinking the methods and means through which we know and connect in the world. The 2018 node conference was in that sense an unintended harbinger of things to come. We would like to thank all the participants in the node conference, who were distributed across eight nodes in Europe, Asia, Australia, and the United States. Apart from those who have contributed to this anthology, we would like to thank Caitlin Morgan and Donna Haraway in particular for generously sharing their insights and observations. We also thank Christine Fentz from the arts collective Secret Hotel, who led us through an artistic and participatory silence dialogue called "Dark Walks." Mia Korsbæk was indispensable with her technical and logistical support. The conference was the final event in the collaborative and interdisciplinary research project Aarhus University Research on the Anthropocene (AURA). AURA was headed innovatively, caringly, brilliantly, by Anna Tsing and funded generously by the Danish National Research Foundation. We owe a huge debt of gratitude to both and to the many collaborators from biology, science studies, history of ideas, and other disciplines who have been part of, visited, and shared their thoughts and methods with AURA over the years since 2013. Christian Buono Kolborg helped with layout and proofreading of the final manuscript of this anthology, and the research program of anthropology at Aarhus University provided financial support for the process. The research time of Nils Bubandt in the final frantic period was generously funded by the Carlsberg Foundation. The editors would, finally, like to thank the anonymous reviewers for wonderful and thought-provoking comments, as well as Doug Armato, director of the University of Minnesota Press, for his unwavering support of this anthology during the tumultuous and economically uncertain period of the pandemic.

Rubber Boots Methods

Outline for a Multispecies Study of the Anthropocene

NILS BUBANDT, ASTRID OBERBORBECK ANDERSEN,
and RACHEL CYPHER

The former brown coal mines near the tiny Danish town of Søby are an unlikely microcosm of the Anthropocene. Covering an area of only eleven hundred hectares, the former mines are easily dwarfed by the megamines and tar sand pits elsewhere in the world. What they lack in size, they also lack in drama. Set in the flat and rather drab heaths of western Denmark, the former mines are now a wooded, rolling landscape that has been declared a cultural heritage site as testimony to a bygone chapter of resource exploitation. And located as they are within a welfare state on the northern rim of Europe, the mines are part of a multispecies landscape that has been relatively cushioned by geographic fortitude and economic history from the much more clearly violent historical dramas that have become the hallmarks of a planetary climate and environment going awry. No monstrous storms, no unprecedented forest fires, no king tides, and no mass extinctions plague these landscapes. No draconian forms of government intervention, genocidal evictions of Indigenous populations, or resource curses haunt these landscapes and their human and nonhuman inhabitants. Søby is, in both spatial scale and historical drama, an unremarkable patch of the Anthropocene. And yet, it is exactly the undramatic mundaneness of this version of the Anthropocene that has attracted us, the members of the research project Aarhus University Research on the Anthropocene (AURA), to its simplified landscape since 2013. Headed by Anna Tsing and funded by the Danish National Research Foundation, AURA made the former brown coal mines in Søby a methods lab, an experimental site that anthropologists, biologists, ecological historians, and philosophers of science could visit to test their own fieldwork-based methods and learn from those of others as we all tried to make sense of the landscape. The former brown coal mines in Søby became a playground for methods experimentation across disciplines that shared a tradition of empirical fieldwork attuned to multispecies socialities and histories, as well as a test site for how best to retool these fieldwork traditions for the Anthropocene.

At a time when both biology and anthropology have come to realize that the conventional objects of their disciplines—natural ecosystems and human sociocultural worlds, respectively—can no longer be studied in isolation, it seemed to us particularly timely to try to regain the lost common ground between the disparate fieldwork traditions of these two disciplines. Anthropology, after all, borrows its tradition of modern fieldwork from biology (Stocking 1983a) and developed from this its own robust methods for studying human worlds. But these methods have primarily been geared toward humans. They have until recently been anthropocentric in the best sense of the term: driven by an ambition to describe human ingenuity and injustice, to understand how human lives are lived differently, and to analyze how human worlds might be constructed otherwise. Anthropological fieldwork methods were born out of attempts to work against Occidental prejudice, colonial oppression, and modern myopia. But even though the worlds otherwise that anthropologists sought to paint through their discipline-specific fieldwork and writing methods often included cows and other animals, plants and forests, spirits and materials, the worlds anthropology depicted were first and last *human* worlds. In recent years, however, the unwavering faith in a world in which humans are the only protagonists has crumbled, challenged

FIGURE I.1. A view of the former brown coal mining area in Søby. Photograph by Nils Bubandt.

from within academia as well as by the more-than-human environmental and climatic crisis sometimes called the Anthropocene. In the Anthropocene, the issue of social justice is still and more than ever a matter of addressing the social, racial, and gender inequalities and forms of historical violence that affect human lives. And it indexes the intimate entanglement of those inequalities and forms of violence in broader landscapes of multispecies livability.

The Anthropocene is a contested concept that means many things to many people (Swanson, Bubandt, and Tsing 2015). The conventional claim is that the Anthropocene describes the Great Acceleration of an increasingly globally connected world after World War II in which humans have become a force of nature on a geological scale (McNeill and Engelke 2014; Steffen et al. 2015; Zalasiewicz et al. 2017). This claim has been the object of important critiques from a variety of scholars in the social and human sciences who point out that the concept of the Anthropocene is driven by a Western worldview, a universal and gender-blind idea of "the human," and a politically liberal notion of history that ignores the dynamics and history of capitalism (Danowski, Viveiros de Castro, and Latour 2014; Demos 2017; Grusin 2017; Malm and Hornborg 2014; Moore 2016; Yusoff 2018). This critique notwithstanding, the Anthropocene is also a fundamental provocation to the academic division of labor and modern accounts of the world that for some scholars activate concerns to rethink "life" and to tell new stories about the positions and responsibilities of humans in the world (Zylinska 2014); to shift the way that the past, present, and future are envisioned, breaking with the self-evidence of modern progress (Chakrabarty 2009); and to speculate about novel foundations for future coexistence (Latour 2018; Morton 2016). We use the notion of the Anthropocene heuristically here, exploring it both critically and with an eye for the provocations it raises. Most importantly, we insist that the Anthropocene is emphatically not an epoch of the human (the *anthropos*) but rather a multispecies phenomenon—and a multispecies crisis—through and through (Haraway et al. 2016). Understood in this way, Anthropocene scholarship aligns with the recent "multispecies turn" within anthropology, geography, philosophy, and related disciplines (DeMello 2012; Kirksey and Helmreich 2010; Lorimer 2015; Wolfe 2013), a turn that itself be can be traced to a broader "nonhuman turn" and a critical posthumanism (e.g., Wolfe 2009; Braidotti 2013) as well as an array of intellectual and theoretical developments including but not limited to actor network theory and the politics and ecology of things (Latour 2004, 2005; Ingold 2012); feminist scholarship and the ontological turn (Viveiros de Castro 1998; Henare, Holbraad, and Wastell 2006; Grusin 2017; Kohn 2013); new approaches to materiality (Bennett 2010; Harvey, Krohn-Hansen, and Nustad 2019); affect theory, animal studies, and assemblage theory (Deleuze and Guattari 1987); new media theory, speculative realism, and systems theory (Grusin 2015); as well as postcolonial and decolonial approaches to human–environment relations (Gómez-Barris 2017; Green 2015; de la Cadena 2015).

It is out of this theoretical and analytical trajectory that the methodological experiments of this book grow. We offer a multiplicity of methodological approaches, attuned to a variety of particular landscapes and the specific socioecological histories tied to them. And throughout it all, the former brown coal mines in Søby infuse the methods, remaining a place from which to think and a place to think with, a reminder to always ground ourselves in the concrete. This is because the landscapes of Søby were the practical common testing ground for our first efforts in retooling and reconceptualization. In this miniature landscape of a mere eleven square kilometers, we experimented with the methods needed to study on a planetary scale the more-than-human coordinations and perturbations of the Anthropocene in a critically comparative and curiously transdisciplinary mode. It was by thinking and fieldworking with others within this site that it became clear that "more-than-human" does not mean "without humans"; it does not mean ignoring unequal economic history and (post)colonial political history or indeed human sociality and experience. On the contrary, it means taking the uneven complexity and multispecies entanglement of human existence seriously by not reducing it to the modern conceit of (Western) Man, a peculiarly racialized species in splendid isolation from and yet somehow also in full control over other species. Highlighting the multispecies nature of the Anthropocene is not a denial of human agency, injustice, and suffering. In fact, it is the opposite: a refocusing of diverse histories of human agency, injustice, and suffering through their complicity and synchrony with relations to nonhuman beings and forces in a multispecies, historical landscape shaped by modernity and colonialism. As the chapters that follow will illustrate, more-than-human methods for the Anthropocene seek to attend to the ways that human injustice and ecological disruption are entangled in the same treacherous histories (Baker et al. 2020). The chapters explore how we might retool anthropological and aligned disciplinary fieldwork methods and analytics to better deal with multispecies forms of life and death after human exceptionalism. The chapters also address how different disciplinary approaches might be activated together to be attuned to particular landscapes and the socioecological histories tied to them. They show that more-than-human methods are a way of attending to issues of human existence, social justice, colonial history, and geopolitics by studying them in their multispecies entanglements (Tsing, Mathews, and Bubandt 2019).

LEARNING TO TRACE MULTISPECIES TRACKS

So how can we learn to track the histories of more-than-human socialities (Tsing 2013)? Answering this question means experimenting with the ways in which human-centered fieldwork methods might be augmented by the fieldwork methods of biology, ecology, and landscape history without giving up their critical edge and ambitions to be open to serendipity and wonder (Helmreich 2009; Tsing 2015; Myers 2015). It also means rethinking, along the lines already pursued by critical political ecology (Johannes 1981; Pauly 1995; Robbins 2007; Swyngedouw and Kaika 2014; Escobar 2018), how ecological fieldwork practices could learn from ethnographic fieldwork. We offer this book as testimony to the many kinds of method experiments necessary to study multispecies livability. It is our claim and point of departure that the changing conditions for human and nonhuman life on Earth—for which the concept Anthropocene is a contested and admittedly imperfect moniker (Haraway 2016; Yusoff 2018; de la Cadena 2015)—demand new modes of sensing and seeing, new forms of collaboration, and new foci for description in which landscape histories matter. Studies of more-than-human life in the Anthropocene, in short, require new kinds of multispecies fieldwork methods.

Søby was a place to cultivate multispecies curiosities, transdisciplinary collaborations, and landscape histories that such multispecies methods require. In Søby, we experimented with multispecies methods by following the tracks left by humans and nonhumans alike. This meant finding traces inscribed on the landscape, such as in the sand or in the muddy lake sediments. It also meant locating the traces left in the local archives and scientific reports. In Søby, multispecies methods involved studying the historical forms of deer grazing patterns and forest regrowth areas as much as the memory of Søby's human inhabitants (see chapter 11). In addition to practicing the classical methods of anthropological fieldwork—participant observation, interviewing, and "deep hanging out"; collecting oral histories and undertaking archival studies to get a historical perspective; following pipes and technologies and documents as science and technology studies (STS) scholars might do—multispecies methods in Søby entailed cultivating landscape observations; paying attention to animal etiologies and plant forms; attempting field identification of plants and animals; experimenting with the use of microscopes and wildlife cameras; and seeking additional information from DNA analysis, sediment sampling, and transect studies. It meant reading works from patch ecology and landscape history as well as from anthropology, STS, and history (see chapter 12). It meant developing new forms of "slow collaboration" across disciplines (see chapter 9), and it meant experimenting with new ways of doing-fieldwork-with-others, humans as well as nonhumans (see chapters 6 and 8).

"Rubber boots methods" is the playful name we have given to these more-than-human method experiments across disciplines. The name reflects our practical realization that waterproof footwear was essential in the cold, waterlogged landscapes of the former brown coal mines. The term also grew out of a wish to defend "rubber boots biology," a derogatory term used in Danish biological circles to denote fieldwork-based biological observations of ecosystemic relations. Rubber boots biology (Danish: *gummistøvlebiologi*) has in recent years been marginalized from university education and research, pressured by more prestigious branches of biology, such as laboratory-based microbiology, molecular biology, computational biology, ecosystems modeling, and biotechnology. This marginalization of field-based approaches is part of an overall institutional trend (see chapter 13). The natural sciences have been moving toward more computational methods and technologies for decades and have become dependent on big data and mathematical modeling for their knowledge production. This is not in itself a problem. We do not categorically suggest that remote sensing and other advanced technologies and practices for data generation, storage, and analysis are not useful; they are! It was, after all, computer modeling and big data that allowed climate change, species extinction, ocean acidification, and planetary boundaries (Rockström et al. 2009) to emerge as scientific objects, political concerns, and moral problems (Barnosky et al. 2011; Edwards 2010). We wonder, however, what is lost when scientific knowledge production about changing ecosystems is dominated by sensing from afar and by modeling. What realities and stories, what kinds of critical connections, never make it out of the dust, mud, or water of lived lives and fieldwork observations into research results? What sensibilities do we need to cultivate to tell the multispecies stories of a changing world that remote sensing and big data cannot deliver? Rubber boots are the metaphorical device through which we call for and invoke the sensitivities, interactions, and histories necessary to bring out these realities and stories. We do so by noting that these realities and stories are the stuff that makes up fieldwork and that one only gets access to tell such stories by wearing the methodological footwear most appropriate for the terrain and sensibilities of the particular field. Rubber boots methods are the tools and perspectives needed to get into the patchy landscapes of the Anthropocene (Tsing, Mathews, and Bubandt 2019), to notice and generate empirical material for the critical descriptions and analyses required in and for our times. Once walked through and sensed "from below," environmental and more-than-human histories may well be enriched by the perspectives produced by ecological models or microscopes. And, so we argue, ecological models may well be complemented by multispecies stories (see chapters 1 and 2).

Rubber Boots Methods in the Muddle of Our Time

Lowly rubber boots seemed a fitting metaphor for the AURA attempt to restore the broken bonds of empirical, anecdotal, and fieldwork-based research between the human and natural sciences, a type of research that is as denigrated by the inclination toward "high" (often French) theory and speculation in the human sciences as it is by the big data revolution and "evidence-based" styles in the natural sciences. It was "rubber boots biological" skills—detailed attention to landscape change and co-species coordinations—that allowed James Estes and Robert Paine, rubber boots biologists if ever there were, to observe the complex ecosystemic relations between sea urchins, sea otters, and kelp forests in the seas off the Aleutian Islands in Alaska, for instance (Estes 2016). Their careful field observations were critical to the development of the ecological notions of "keystone species" and "trophic cascades," ideas that have since become central to extinction studies and debates about planetary limits in the Anthropocene (Estes et al. 2011). Our call for scholars to "put on rubber boots" and experiment with multispecies, transdisciplinary, and landscape-historical field methods should in this light also be seen as an invitation to develop forms of knowledge production that grow from a meandering, curious, and peripatetic imagination even as they are empirically grounded, based on fieldwork. It is a call to cultivate the multiple forms of slow, bodily, and intimate (as opposed to "remote") sensing implied in fieldwork that for many decades has characterized anthropology as well as biology and other field-based sciences (see chapter 10). Anthropology's tradition of grounded research and insistence on the value of long-term field-based experiences, observations, relations, and anecdotes as traces of larger patterns seemed, to us, to be a suitable experimental ground on which to restage a transdisciplinary encounter between fieldwork-based methods in a world of science otherwise addicted to models, big data, and "large theory." At the same time, we are acutely aware that if big data methods are epistemologically tainted, so, too, are fieldwork methods.

Fieldwork as a research practice emerged, after all, at the beginning of the twentieth century as part and parcel of a colonial legacy of exploration, penetration, and political control (Asad 1973; Clifford 1997; Fabian 2000; Stocking 1983b). Indeed, it is hard to imagine a more colonial piece of footwear than rubber boots. Andrew Mathews repeatedly reminded us of this colonial legacy of rubber boots during the conference on which this anthology is based when he playfully and critically referred to the conference title ("Rubber Boots Methods for the Anthropocene") as "Boots on the Ground" (see also the discussion about the term in chapter 14). Boots give access, they extend thinking and perception, but they also isolate and trample. And here they are not alone. What can be said of boots methods can be said of all methods. The term *rubber boots methods* attempts to make explicit the historicities of war and colonialism that are latent within all scientific methods and seeks to make awareness of these violences part of its application.

Wellington boots, the British word for rubber boots, are named after Arthur Wellesley (1769–1852), the first Duke of Wellington and arguably the most famous British general, known primarily as commander of the Anglo-Dutch-German army that defeated Napoleon Bonaparte at Waterloo in 1815. Before becoming the Duke of Wellington, however, Arthur Wellesley spent seven years, from 1797 to 1805, in India. Here he led a series of military campaigns against local sultans and rajas from which he not only amassed a considerable private fortune but also—in close collaboration with his older brother Richard Wellesley, the governor-general of India—became the "architect" of the British colonial empire in India (Severn 2007). Propelled by the fame following his victory at Waterloo, the Duke of Wellington went on to become a key political figure, serving twice as British prime minister. It was as such that he in the public mind in England became synonymous with the particularly high leather boots that he wore. The rubber version of the Wellington boot emerged when Hiram Hutchinson in 1852 bought the patent to apply sulfur vulcanization, a technique invented some years earlier by Charles Goodyear, to rubber footwear. It took the First World War to make rubber boots a global commodity. This happened when the British War Office commissioned the rubber boot company "Hunter" to produce rubber boots for standard issue to its soldiers in the trenches of World War I. The start of the twentieth century also saw rubber becoming a key colonial commodity on plantations from Asia to Latin America, where entirely new arrangements of rubber trees, human labor, and colonial control reshaped landscapes and human lives on a global scale (Aso 2018; Garfield 2014; Harp 2016; Stoler 1995; Tully 2011). As the Brazilian rubber plant *(Hevea brasiliensis)* was planted in straight lines on ever larger plantations in Africa, Latin America, and Southeast Asia, rubber joined the ranks of coffee and sugar as a key cash crop in a global constellation linking landscape simplification to regimented labor control and global relations of production and consumption (Mintz 1985; Scott 1998). The Plantationocene, the term suggested for this global constellation as the shadow figure of the Anthropocene (Haraway et al. 2016), is in that sense very much the story of rubber entangled in a history of racial inequality, social suffering, and ecological destruction. Rubber is also the story of the Capitalocene, another contender for the title of our time: the transformation of "cheap" nature into profit by "cheap" labor in the production of goods for "cheap" lives (Patel and Moore 2017). And rubber is the story of the Great Acceleration of human industrial production and consumption after World War II that now stretches the carrying capacity of the planet beyond its limits (Zalasiewicz et al. 2017). Despite the invention of synthetic rubber, which now accounts for two-thirds of the global use of rubber for the production of tires, belts, and flooring, the production of natural rubber has continued to grow exponentially in the last half decade, fed mainly from megaplantations in Southeast Asia (Kenney-Lazar and Ishikawa 2019). In 1951, the annual global production of natural rubber was around two thousand metric tons; today, it is ten times as much (Umar et al. 2011).

Rubber boots are intimately embroiled in these histories of war, patriarchy, colonization, racism, resource extraction, and ecological change in the same landscapes that we propose to study using rubber boots as a methodological figure. This contradiction is intentional. It serves not as a rubber boot confession, absolving the authors of this volume from their obligation to analyze structures of violence and privilege that have shaped both their fields and their field methods (Smith 2014; Fortier 2017). Quite the opposite, it highlights the tainted nature of all method. The name *rubber boots methods* springs from an attempt to acknowledge that scientific methods are always historically embroiled in the phenomena they seek to study. Rubber boots methods seek to interpellate these histories, not to bury them. Let us offer a concrete example: in the regional dialect of eastern Indonesia, where Bubandt has conducted fieldwork, rubber boots are referred to as *jenggelbot.* The word is an Indonesian adaptation of the English word *jungle boot,* a type of military combat boot developed by the United States Rubber Company and adapted for tropical warfare during the Vietnam War. The word probably came to eastern Indonesia from Borneo, where British special forces used their own adaptations of the U.S. jungle boot—with a higher, Wellington-like cut—during the Malayan Emergency from 1948 to 1960, a guerilla war between Commonwealth forces and the Malayan National Liberation Army that in turn very much grew from the grievances of rubber plantation workers in an export economy in ruins following World War II. Today, people in eastern Indonesia use the word *jenggelbot* to refer to the heavy-duty boots that male workers are issued by mining companies. For the contributors to *Rubber Boots Methods for the Anthropocene,* the fact that rubber boots evoke histories of war, gender, resource exploitation, and the multispecies violence of ecological change *is* the point: methods are never innocent or pure.

Acutely aware of its etymological problems—the association between rubber boots and war, landed aristocracy, and the colonial legacy of rubber being among them—the chapters seek to "stay with the trouble" (Haraway 2016) of rubber boots methods, making integral the reflexive study of the colonial legacies of scientific methods to their deployment in the field.

Staying with the troubled legacies of our methods, however, also opens up unexpected twists. For as it turns out, rubber boots are more-than-colonial, more-than-patriarchal, and more-than-Western. They are ambivalent and full of tensions—like all colonial phenomena (Cooper and Stoler 1997). Wellingtons may be the footwear of choice for the British landed aristocracy, but they are also highly popular among French smallholders struggling against landed elites. Similarly, "gumboots" may be an icon of colonial endurance for white Pakeha farmers in New Zealand, but settler colonialists are not the only ones who use rubber boots in projects of cultural self-making. As Pierre du Plessis reminded us during the conference on which this book is based, in the 1920s, black stevedores and mining workers in apartheid South Africa developed the "gumboot dance" (called *isicathulo* in the Zulu language) as a nonverbal form of protest against racial inequality and outrageous labor conditions (Erlmann 1989; Mills 2016). Today, in postapartheid South Africa, gumboot dancing, performed in straight lines in mimicry of military parades, has become a major tourist attraction. If colonialism is full of ambivalence, so, too, are protests against it.

As Tsing (2005, 6) points out, the history of industrial rubber is also the rich and ambivalent history of multispecies vicissitudes rather than merely human hegemony or progress: industrial rubber is made in a complex process that involves the savagery of European conquest, the confusion of war and technoscience, but also the resistance strategies of peasants, Brazilian rubber plants, insect pests, and more. It is this very vicissitude, tension, and friction that we emphasize in "rubber boots methods." Rubber boots are our "figuration" for fieldwork methods that come with genealogies of colonialism but that also open up to the collaborative study of multispecies relations that combine anthropological fieldwork with the denigrated tradition of "rubber boots biology" and ask us to pay "troubled" attention to the entangled human and nonhuman lifeworlds subjugated by colonialism. This attention is indexed by the three sections that make up the anthology: critical description, curiosity, and collaboration. In that sense, rubber boots methods express an effort to decolonize fieldwork methods across the biological and anthropological traditions and to move beyond narratives about human exceptionalism by suggesting that such efforts require not only critical attention to histories and concepts but also cautious and curious experiments with both natural science methods and Indigenous knowledges.

We invited the contributors to take rubber boots methods figuratively, not literally. "Rubber boots methods" is not a Foot Locker advertisement pitch for a particular kind of methodological footwear; rubber boots methods are a "figuration" for situated multispecies methods (Haraway 2004), choosing whatever methods (metaphorical and actual footwear) that are most appropriate and that allow for most sensibility in the field studied. In the chapters that follow, the authors take us to fields where different rubber boots extensions, such as rubber boots, horses' hooves, rubber fins, dogsleds, boats, or bare feet become methodological prosthetics that help extend how more-than-human worlds are experienced, thought, felt, and known. Rubber boots methods in their varied forms share an attention to the feet and other means of movement. They emphasize how movement depends on the affordances and histories of the landscapes and seascapes in which it occurs. Rubber boots methods seek to highlight both the sensibilities enabled by the physical contact to the ground (or water) and the tracks and disturbances that all movement leaves behind. Extending calls to see ethnography as a method of walking (Vergunst and Ingold 2016; Kusenbach 2003), rubber boots methods focus on the possibilities of peripatetic thinking and sensing while also suggesting the need to critically historicize our scientific prosthetics and modes of locomotion.

In the broadest sense, rubber boots methods are an injunction to engage with methodology in a new way: an invitation to get out to follow the more-than-human tracks that crisscross the landscape of any field; an invitation also to get outside of one's own disciplinary comfort zone by experimenting with other methods; and an invitation, finally, to "get out of" the field itself by seeing the landscapes of one's field as particular landscapes-in-the-making connected historically, politically, and ecologically to the other multispecies landscapes that came before and that exist beside it. Rubber boots methods are an invitation to the critical, curious, and collaborative study of multispecies landscapes.

MULTISPECIES METHODS FOR THE ANTHROPOCENE

We imagine this anthology as a methodological companion piece to *Arts of Living on a Damaged Planet* (Tsing et al. 2017), a collection of analyses by scholars from anthropology, ecology, science studies, art, literature, and bioinformatics that sought, like this volume, to understand the Anthropocene by beginning in landscapes, "the overlaid arrangements of human and nonhuman living spaces" (Gan et al. 2017, G1). *Arts of Living* sought to understand the varied and ruined landscapes of the Anthropocene analytically through the double figure of monsters and ghosts: the entangled bodily forms of multispecies symbiosis through which nonhuman life proliferates "monstrously" in ruined landscapes that are also haunted by the ghostly histories of more-than-human socialities and destructive processes that have made them what they are. *Rubber Boots Methods for the Anthropocene* is similarly focused on landscapes in their more-than-human spatiality and historicity but takes a methodological rather than an analytical approach. To flesh out what this means by way of example, let us return briefly to the landscape of Søby.

In one sense, the Søby landscape is unremarkable. In another, it is as extraordinary as any landscape in the Anthropocene. Indeed, the drabness of the Søby landscape is an effect of a dramatic history. And make no mistake, the landscape around Søby has changed radically in recent centuries. From the start of the twentieth century, the area became an important source of brown coal, and during the Second World War, the mines supplied up to one-third of the national energy needs in a Danish version of the Great Acceleration. Trenches and holes up to thirty-five meters deep, dug first by hand and eventually by machines, cut through the topsoil, exposing the sand and groundwater underneath. Busy but long since abandoned rail lines transported the brown coal to urban hubs. When mining finally ceased in the early 1970s, the flat heathlands of Søby had disappeared. In their place had become what locals describe as a barren and hilly landscape of sand dunes. Between the dunes, potholes filled with acidic water turned red from the oxidization of pyrite in the soil. It was a lunar landscape made by humans, a polluted wasteland on which even the most insistent reforestation efforts appeared to fail. It was also an unstable landscape that exceeded human agency: the waterlogged, sandy soils made landslides a common occurrence and agriculture untenable as a result (Højrup and Swanson 2018). So, while the rest of Denmark after the 1970s busied itself building up a mechanized agricultural industry that would finance and feed the growing urban population of this Nordic welfare state increasingly hooked on fossil fuels much more efficient than brown coal (Brichet and Hastrup 2018), the Søby mines were left to their own devices. And these devices, as it turned out, were very much of a multispecies kind. While most humans looked the other way, species of conifer, abandoned survivors from the failed reforestation efforts of the 1960s and 1970s, began to proliferate on their own in the mining spoils. In particular, the lodgepole pine *(Pinus contorta)*—a tree common to western North America—thrived in the mining ruins, aided by its symbiotic alliance with ectomycorrhizal fungi that entangled themselves in its roots (Gan, Tsing, and Sullivan 2018). Pine forests sprouted, as did thistles and other nitrate-philic and weedy plants, in a landscape that by Danish standards was "wild," unstable, and uninhabited. Roe deer *(Capreolus capreolus)* cautiously moved in, and by the late 1990s, the roe deer were followed by red deer *(Cervus elaphus),* the most prized of all game animals in northern Europe (Forssman and Root-Bernstein 2018). This got the attention of humans. Well-padded hunters soon followed, shooting their first red deer in the area in 2002. Almost overnight, land in the ruined and rewilded Søby mines went from worthless to priceless as hunters paid premium rates for the red deer hunting grounds. In 2007, this former wasteland was declared a national heritage site, owing to both its cultural history and its undomesticated natural beauty (Hoag, Bertoni, and Bubandt 2018).

The former brown coal mines in Søby are a Danish version of the Anthropocene, one might say. They are a patchy landscape of unintended effects, at once specific in its political ecology and yet intimately connected to the simplified agricultural and industrial landscapes of the rest of Europe. In that sense, Søby sits in a patchy landscape that is full of the same ambiguities that characterize the Anthropocene in general. It is a landscape of overturned geological layers, fundamentally remade by the anthroturbations of the Great Acceleration (Zalasiewicz, Waters, and Williams 2014); a banal landscape shaped by dramatic changes that now, like the structural violence that shaped it, lie hidden from view (Nixon 2011; Swanson 2017b); a multispecies landscape made not just by humans but by cascades of more-than-human action and reaction that shaped what players entered the scene and how they could interact (Bubandt and Tsing 2018b; Haraway et al. 2016).

The ecologist James Estes argues that it was the very simplicity of the Aleutian seascapes that allowed him to identify the trophic relations and cascades between otters, kelp, and urchins—relations that would have been hidden from view by the sheer complexity of trophically more complex ecosystems, such as a tropical forest or a coral reef (Estes 2016, 23). The former coal mines in Søby are a simplified landscape in a similar way, a landscape of fungi, pines, red deer, and humans. Animals, plants, fungi, bacteria, and other nonhumans clearly made history in this landscape. At the same time, it is obvious that these nonhumans did not make history under conditions of their own choosing. Nonhuman history-making, like human history, always happens, as Karl Marx ([1852] 2005, 7) put it, "under circumstances directly encountered, given and transmitted from the past." The landscape in Søby remade by fungi, pines, red deer, and others was a landscape already made simple by a history of fossil fuel extraction, world war, wastelanding, and state building. But even this history also involved nonhumans. Take brown coal itself. Brown coal, a low-grade fossil fuel, is sedimented rock of compressed peat that in Søby is up to fifteen million years old. Brown coal is partly fossilized nonhuman life. The Great Acceleration of Western modernity, with its dreams of human freedom and control, was powered by the same ironic relation to fossil fuels: it was petrified forms of nonhuman life that fueled human dreams and combustion engines. These are the human and nonhuman histories of Anthropocene landscapes that rubber boots methods seek to draw out, by tracing the braided processes of industrial simplification and more-than-human proliferation (Tsing, Mathews, and Bubandt 2019; Tsing et al. 2017).

While Søby was the first landscape in which we tried to hone rubber boots methods by attending to the histories that can be read in landscape forms and multispecies coordinations as much as in the words and practices of human interlocutors or archival records (Mathews 2018; Bubandt and Tsing 2018a), *Rubber Boots Methods for the Anthropocene* asks its contributors to take the next comparative step: to take the rubber boots methods from Søby to their own fields and to adapt or exchange their methodological rubber boots for whatever methodological footwear is best suited to their particular patch of the Anthropocene. From this provocation, as the chapters will show, a range of transformed rubber boot methods emerged: walking the hills of the Tuscan forests (chapter 1), exploring shrubs in the grasslands of Lesotho (chapter 2), tracking in the Kalahari desert (chapter 3), following changing settler crops in South Asian farmlands (chapter 4), firing Australian eucalyptus forests (chapter 5), snorkeling in the Indian Ocean (chapter 6), figuring and thinking in the New Delhi monsoon (chapter 7), riding across the South American Pampas (chapter 8), following nonhuman hitchhikers in the global aquarium trade (chapter 9), dogsledding and sailing through the High Arctic (chapter 10), padding the Jutland heathlands (chapter 11), dust balling in a Danish museum (chapter 12), and teaching students to pay attention to landscapes in outdoor university courses (chapter 13). Understanding rubber boots methods as the field-based tools, techniques, and practices that each contributor engages with and develops in doing research, and seeing the resulting methodology as a system of knowledge practices guided by a certain set of "rubber boots–inspired" theoretical underpinnings and principles, the contributions that make up this anthology seek to identify the sensitivities and devices that point toward a multispecies Anthropocene methodology and—so we hope—serve to inspire future empirical studies. The possibilities and limits of rubber boots methods are discussed in a roundtable between Kirsten Hastrup, Ursula Münster, Anna Tsing, and Nils Bubandt in the concluding chapter of the anthology (chapter 14).

Methods for the Contemporary Moment

Rubber Boots Methods for the Anthropocene joins other attempts to develop methodological approaches capable of "figuring out" and intervening in the contemporary moment. Anthropology, the humanities, and the social sciences engaged in studying environmental crises and Anthropocenic realities have been experimenting with methods for decades. This has included experiments with new modes of doing fieldwork that are critical and self-reflexive (Faubion and Marcus 2009), global in outlook (Burawoy 2000), mobile or multisited by design (Büscher and Urry 2009; Marcus 1998), anticipatory in spirit (Pandian 2019), decolonial in kind (Smith 2012), or driven by analytical curiosity (Fischer 2018) and a multimodal design imaginary (Dattatreyan and Marrero-Guillamón 2019). These varied experiments in the methods and analytical imagination of fieldwork attuned to the contemporary world grow from an attempt to make anthropology public and the discipline "matter" in new ways to the crises of our time (Marcus and

Pisarro 2008). They are also attempts to respond to the changing conditions of life that characterize the twenty-first century—a globalized, mediated, technoscientific, biosocial, crisis-ridden world (Faubion and Marcus 2009)—even as they question what crisis might mean anthropologically in the first place (Roitman 2013). Such novel methodological reflection and innovation is needed because the forces of a globalized world implicate the very conditions of doing fieldwork, necessitating a decoupling of fieldwork from its conventional theoretical fixtures of cultures, peoples, communities, societies, and villages. Situated fieldwork requires new kinds of reflexivity and storytelling in a deterritorialized world that problematizes what the "field" is in the first place (Gupta and Ferguson 1997; Strathern 2004). It also entails a critical distance to scientific authority and truth without giving up on knowledge, shifting epistemological ambitions from grand narratives to "big-enough stories" (Clifford 2013; Comaroff 2010).

The Anthropocene has arguably introduced a whole new set of methodological challenges on top of those that attend to a globalized world. After all, the Anthropocene is the dark planetary ecological twin of globalization and ideas of modern progress. In the wake of the cosmopolitan dreams and fears of globalization of the 1980s and 1990s, the Anthropocene opens up to a very different world: a more uncanny space, planetary in scale but also fragmented; a fractal world populated with growing multitudes of more-than-human actors and agents; a moment in time that grows from a nonlinear form of acceleration as well as from innumerable ambivalent histories; a landscape of queer socialities and more complex relations and interlinkages across species and matter than the concept of "globalization" would ever allow. As Timothy Morton (2016, 10) puts it, "now that the globalization dust has settled and the global warming data is in, we humans find ourselves on a very specific planet with a specific biosphere." The Anthropocene, as the heuristic concept we take it to be, is in that sense an interruption of global modernity's narrative about progress, nature, globalization, and the human. The recognition of the power that climate change, ocean acidification, biodiversity loss, and chemical imbalances of nitrogen and phosphorous will have in the future fundamentally challenges modernity's own account of humans and human society as somehow aloof and in control of their own global fate. It is this failure of modern and secular accounts of human history, Bruno Latour (2017, 84) points out, that the term Anthropocene sums up, however disputed its date of origin and its definition may be: "The earth system reacts henceforth to your action in such a way that you no longer have a stable and indifferent framework in which to lodge your desires for modernization."

The study of power in the Anthropocene—meaning power in the classical sense of having an ability to affect the actions and conditions of life of others—has to allow multiple new actors into its accounts. Beyond humans and their institutions, a study of power also has to include documents, laws, technologies, and other material actors to whose power STS has alerted us for some time (Latour 1996; Riles 2006) as well as the profusion of living and chemical beings (animals, plants, bacteria, fungi, vira, protists, CO_2, phosphorous, nitrogen, radiation, and water) on which an emergent and rapidly expanding more-than-human social science focuses (e.g., Tsing 2015; Brown 2013; Ballestero 2019). And ought a study of power in the Anthropocene not also include spirits and ghosts? After all, it was a Hegelian idea of spirit that drove modern ideas of progress for two centuries. Today, the Anthropocene is a time full of new monsters, ghosts, and spirits (Tsing et al. 2017). Indeed, climate change has come to take the form of a monstrous and unpredictable spirit, conjured up by computer models and scientific fact. This anthology, then, asks what kinds of methods are needed to address this proliferation of protagonists in their multiple human and nonhuman forms. The Anthropocene, we suggest, is a multispecies phenomenon, an uncanny valley (Bubandt 2018, 2022), and a patchwork of landscapes (Tsing, Mathews, and Bubandt 2019; Stengers 2018) that require methods and forms of fieldwork that are more-than-human, that are nonsecular, and that take landscapes as their key unit (Swanson 2017a; Bubandt 2018; Tsing 2015).

If "mobile," "multisited," "experimental," and "anticipatory" methods were the ways in which anthropology (and other human and social sciences) sought to retool itself to understand human life in a global world, we suggest that multispecies and "more-than-secular" methods comprise the methodological retooling necessary to understand more-than-human life in the political ecology of the Anthropocene. Within anthropology, the discipline with which the three coeditors and the majority of our coauthors are most familiar, many "environmental methods" exist, and there are numerous examples of fieldworkers combining tools to explore human–environment relations in ways that bring forth perspectives beyond the human. One might think of community ecology methods, ethnobotanical observations and vegetation inventories, GIS, topographic mapping, biodiversity indices, terrestrial scanning, plant use recall, and a number of other, more traditional methods for research design and observation, such as participant observation, natural experiments, ethological observation, and household surveys (see Bernard, Ryan, and Wutich 2016; Vaccaro, Smith, and Aswani 2010), as well as methods that include collaborative modalities and postcolonial forms of learning about landscapes (see Strang 2010; Flora and Andersen 2017).

And yet, despite the breadth of methodological experimentation, strikingly few publications are dedicated to multispecies methods. An exception is the recent and excellent volume by Lisa Hamilton and Nik Taylor (2017) titled *Ethnography after Humanism*. Writing from the perspective of critical animal studies, the book outlines the breadth of visual, multisensorial, and "monstrous" methods that might be applied to include animals in research (6). Hamilton and Taylor call for methods that allow us to see, understand, and analyze social relations and worlds that are more-than-human. This is an ambition we share. We also share Hamilton and Taylor's commitment to ethnography-as-method as well as their critique of ethnography's historical tendency to obliquely reproduce speciesism and anthropocentric hierarchies of power by its exclusion of nonhumans. However, our exploration of multispecies methods differs from and complements that of Hamilton and Taylor in two ways. First, where Hamilton and Taylor's multispecies methods rest on an "ethnography for human-animal relations" (53) that grows from critical animal studies and takes the relationships between humans and specific animal species as its focus, *Rubber Boots Methods for the Anthropocene* takes landscapes and landscape history, with their multiplicity of species relations, as its starting point. Following Bubandt, Mathews, and Tsing (2019), we adopt the notion of "patch" from ecology. Any particular "patch" of landscape is unique and yet its uniqueness is generated by its historical-ecological relations to other patches (Pickett and White 1985): a forest is made through its ecological and earth systemic connections to the neighboring grassland, the adjacent sea, and the distant city. Human landscapes in a globalized world are patchy in similar ways. Rubber boots methods seek to study the patchiness of landscapes in their overlapping and mutually conditioning histories of ecology and political economy.

It is out of these entangled histories that specific lifeworlds, human and nonhuman, emerge. Second, and following from this, *Rubber Boots Methods for the Anthropocene* has as its focus the history of landscapes as they are made by multiple kinds of humans and nonhumans over time rather than the more "microsociological" focus on relations between humans and the lifeworlds of another specific species that characterizes the chapters of Hamilton and Taylor's anthology. *Rubber Boots Methods for the Anthropocene* is in that sense an experiment with the anthropological tradition of holism, with all of its problems and baggage (Otto and Bubandt 2010), by seeing landscapes as a multispecies "total fact" that is social and natural, political and ecological, in historical ways that can be studied empirically.

Landscapes and seascapes are the analytical lenses through which the chapters in this volume tell their multispecies stories in ways that are both particular and general. The Anthropocene of a rubber boots methods approach is a multispecies landscape or seascape that is historically and spatially patchy in its more-than-human makeup. The method experiments explored and narrated in the chapters of this book range across many "more-than-human" scales, from bacteria and fungi to holobionts, humans, colonial and capitalist histories, and spirits. The chapters show how the landscapes of the Anthropocene grow out of unstable ecologies, open-ended and in-the-making through multispecies histories of many kinds, linked historically to other landscapes. These ecologies—their instability and their uniqueness shaped by their historical relations to other landscapes in ways we call "patchy"—demand a commitment to curious, collaborative, and critically descriptive method experiments through which both the historical uniqueness and structural linkages of these multispecies landscapes can be captured.

Anthropos and the Material (Harvey, Krohn-Hansen, and Nustad 2019) is another recent book that seeks to broaden ethnographic methods to embrace and work with objects of study beyond the human. The focus here is on the ways in which ethnography can engage with more-than-human and material relationalities and on the political forms and spaces that emerge in such engagements. As such, the approach to method is mainly analytical rather than empirical. Some chapters of *Anthropos and the Material* address the methodological challenges that multispecies relations and world making present to ethnography (Tsing 2019), but emphasis is mainly on the conceptual and analytical repertories required to collapse existing boundaries between natural and human histories (Harvey, Krohn-Hansen, and Nustad 2019, 4). Of course, the boundaries between method and analysis are fuzzy. Methods are always effects of analytical attention and the form of the particular empirical object engaged with. Neither methods nor the analytical objects that one appreciates through them are ever isolated phenomena (see chapter 12). They cohere to each other and to the histories of the scholarly engagement out of which they grow. Still, unlike those of *Anthropos and the Material,* the majority of the chapters of this book begin in empirical experiments with method rather than in analytical deliberation. We also differ in the confidence we have in the ethnographic method being able to retool itself in disciplinary isolation. *Anthropos and the Material* suggests that a novel approach to the material should be specifically anthropological. We very much agree and sympathize with this ambition to put ethnography at the methodological center of a novel approach to more-than-human worlds. At the same time, the chapters in *Rubber Boots Methods for the Anthropocene* take as a shared starting point that fieldwork in Anthropocene landscapes demands transdisciplinary experimentation and that anthropology does not have all the necessary tools to open up for the required understanding, sensitivities, and critical engagements with multispecies relationalities in ruined ecologies. Several chapters explore how different scientific, shamanic, and other modes of noticing and knowing can be brought together for such studies, and although anthropology plays a central part in this knowledge production, the discipline of anthropology as such is not our only methodological anchor. We believe that a retooling of anthropology's methods requires venturing beyond analytical reflection into the empirical muck of our times as well as an experimentation with methods beyond those of ethnography. This is why critical description, curiosity, and collaboration are the pillars of our common exploration.

Critical Description, Curiosity, and Collaboration

Three characteristics unite the methodological experiments in *Rubber Boots Methods for the Anthropocene*: an empirical, descriptive approach that insists on being *critical*; *curiosity* about multispecies worlds; and a *collaborative* ambition that seeks to cross disciplinary and epistemic divides. Rubber boots methods are critical, insisting on empirical descriptions that are historically founded and that challenge the limits of the real. Rubber boots methods practice curiosity about the social relationships that all beings, whether human or nonhuman, establish with each other. And rubber boots methods are collaborative because we need to employ all modes of knowing and efforts of engagement based on empirical and fieldwork-based curiosity to find pathways for coexistence. The method experiments presented in this book thus share a critical (that is, empirical, but not empiricist), curious, and collaborative description of the way landscapes come into being through historical relations and coordinations, human as well as nonhuman, in what might be called a "patchy Anthropocene" (Bubandt, Mathews, and Tsing 2019). The current and evolving patchiness of the Anthropocene is very much formed by colonial and capitalist histories, but these do not take the form of a hegemonic and exhaustive logic (Bear et al. 2015). Because it is more-than-human, patchiness is also more-than-capitalist and more-than-Western. A patchy Anthropocene, we suggest, following scholars of STS (Law 2004), is a particular kind of mess that requires a methodological rather than a theoretical approach.

Starting with methods rather than analytics, ontology, or epistemology, this collection gathers thirteen empirical studies of the multispecies messiness of contemporary life that ask some of the big questions of our time. How, for instance, can the shared tradition of empirical fieldwork be retooled to study co-species socialities in a grounded way when the ground upon which we stand, the landscapes we inhabit, the oceans we navigate, and the air we breathe are themselves changing in dramatic ways? Can our methods for the Anthropocene be designed to capture *both* the ontological specificity of each patchy case *and* the planetary scale of our current mess? What does it take to make methods transdisciplinary and collaborative as well as self-reflexive and critical? Might we study co-species relations in all of their ecological, historical, and political complexity while also challenging the secular logic that informs most Western accounts of this complexity? How might we, in other words, design methods for the Anthropocene that can deal with the more-than-human in terms that are also more-than-Western? The chapters in this book address these questions through distinctive accounts of practical methodological experimentation as well as through dialogical "slow science" that has an eye for failures and mistakes as well as achievements.

Working across disciplines is often hard, and it takes time: time to learn each other's languages and to understand (and accept/respect) different epistemological logics and practices. Borrowing the words of Marilyn Strathern (2004), the ground on which such cross-disciplinary collaborations take place is often not common. Therefore, in this book, we celebrate the possibilities offered by this uncommon ground, make space for slow and multiple experimentation and reflexivity across disciplines, and pay attention to the spaces of possibility (and impossibility) that reveal themselves in engagements across disciplines. Celebrating such possibilities does not mean denying that they may lead down many a blind alley. Each contribution is united by the shared understanding that methods are experiments; they are about trial and error, and each researcher or research team must learn from failures and successes alike. For this reason, in this book, we do not present a how-to guide that will lead readers step by step through a multispecies fieldwork or a systematic approach to analyzing multispecies data. Neither is this anthology a guide to collaboration, because every team collaboration is different, depending on the object of study and who is joining around the table (Swanson 2015). Combining critical description, curiosity, and collaboration in different ways, the authors of the thirteen chapters that follow engage with their fieldwork to show how this methodological premise shifts the very terms of analysis.

The book is divided into three parts mirroring the three characteristics that unite rubber boot methods. Part I, "Critical Description," demonstrates how empirical description emerges from critical engagement with historical and more-than-human landscape change. This also means a historical awareness of the rubber boots methods deployed. In his chapter, Andrew S. Mathews walks through the formerly cultivated chestnut forests of the Monte Pisano in central Italy, seeing wherever he goes patches and landscape structures. Combining landscape observation, map making, and archival work, Mathews provides five scaling stories about plant–human relationships through the methods of walking, noticing, and drawing: morphology, landscape structure, species distribution models, and stories about coordination. Colin Hoag follows dwarf shrub encroachment on grasslands in the high-altitude mountains of Lesotho in his chapter, arguing for attention to landscape "forms" as a specific dimension of rubber boots methods. Critical description of landscape forms draws history and power into thinking about ecological processes and thus combines—or rather works across—approaches and methods from anthropology and ecology. Pierre du Plessis takes us to Botswana's Kalahari Desert, where he demonstrates how tracking can be made into a method for critical landscape description, a form of multispecies noticing that is empirical and imaginative and that, in the words of his teacher, "is good because it tells you that you are here and that you are not alone." Through the experience of tracking a grumpy aardvark, du Plessis shows how tracking as method cultivates multispecies empathy and perspectival sensibility that is situated and collaborative. At the same time, du Pleissis makes tracking a rubber boots method through a historical awareness of the colonial legacies of tracking in the Kalahari. In his chapter, Daniel Münster is on the track of a spate of farmer suicides in Kerala, South India, at the turn of the twenty-first century, arguing that a critical description of the historical landscape changes associated with a succession of settler crops gives a vital multispecies clue to the deteriorating environmental living conditions under which suicides occur in Kerala. Maybe we need to pay critical attention to the landscape changes wrought by cassava and lemongrass, pepper, ginger, and bananas, to understand why life became unbearable for so many Kerala settler farmers in the new millennium? Perhaps, Münster's chapter implies, we need to broaden classical social determinist ideas of "anomie" to understand human suffering and suicide (Durkheim [1897] 1997). Münster also demonstrates that cassava, recently highlighted as the quintessential crop deployed to escape state power and settler colonialism (Scott 2009), was, in Kerala, the first settler crop. Rubber boots methods show that plants and power may align in multiple ways.

Part II is titled "Curiosity." We argue that it is necessary first and foremost to be curious, to wonder, and, through curiosity, to allow new questions to emerge about the world. This curiosity is, as the following chapters show, always grounded in questions that emerge from a direct engagement with the field site. Such a materially based engagement upends even basic categories, such as fire, water, air, and earth. Jon Rasmus Nyquist takes us to the forests of Western Australia, where he engages with the materiality of prescribed burning to propose a fire method for the Anthropocene. He follows the forests that burn to see how fire feeds off and synthesizes its surroundings without being reducible to them. From the pulse of fire to the plants that grow back after a burn, he outlines a method of burning for knowing a warming world. Nils Bubandt, in his chapter, snorkels the tidal zones of West Papua, arguing for a multispecies tidalectic methodology that attunes to the shifting zone of critical overlap between land and sea. Tides go in and out, rhythmic and rising, and Bubandt's rubber flipper methods follow such rhythm to frame this critical zone within histories of colonialism and capitalism, challenging terracentrism to argue for multispecies amphibian methods that take us beyond the "mainland of the real" and allow us to notice not only the fish and invertebrates but also the spirits that inhabit coral reefs. Harshavardhan Bhat, in contrast, engages with the lively and pulsing quality of monsoonal winds to propose a method and methodology of stickiness. Troubling the term *monsoon* as a colonially imposed hegemonic weather formation, Bhat proposes a sticky method for knowing atmospheres in which one feels first through the body the anticipation of summer rains gathering. In an Anthropocenic world where atmospheric science imposes its knowing on climate change, Bhat uses his South Asian fieldwork to ask what it might mean to employ other methods for knowing air, winds, and storms, in which "times of the rains" are conceived of in conversation with each other. In her chapter, Rachel Cypher rides horses through the pampas of Argentina on the margins of soybean landscapes, showing how riding may be a method for landscape attention. Critical to this attention is the awareness that riding itself has become a marginal activity on the pampas as industrial, mechanized soybean farming has colonized the landscapes of cattle ranching. Riding, Cypher argues, is an inherently affective multispecies method of landscape attention, a specific form of a more generalized commitment to what Cypher calls "affective empiricism"—curious, critical, and affective method/ologies for a multispecies Anthropocene.

Part III is titled "Collaboration." As the chapters of this part make clear, the arts of working across disciplinary difference are varied, difficult, disturbing, and unsettling. But the arts of methodological collaboration are also, precisely because of this, immensely important. Following Donna Haraway's mandate to "stay with the trouble," the authors of the chapters in Part III exploit the methodological trouble in which they find themselves, giving themselves over to it even as they often feel fatigue and despair. Working across disciplinary difference, these chapters show, is like working across difference in anthropology, rewarding precisely because it disturbs, challenging taken-for-granted *ways of knowing the world* and, in so doing, upending anthropology's epistemological and ontological methods and opening up new insights into multispecies worlds.

Joseph Klein, Stine Vestbo, Peter Funch, and Anna Tsing bring us on a wild rubber boots journey that has as its center the question of collaboration across disciplines. Their objects of study are aquarium hitchhikers, organisms that come along for the ride in the live coral trade. Through their shared struggle to develop a project across ecology, biology, and multispecies anthropology, the authors develop three key insights: that the living world is composed of nested holobionts, that human disturbance transforms these, and that supply chains generate and redistribute this human disturbance. At the end, they include a model research proposal for future collaboration between biologists and anthropologists, asking whether the aquarium trade unintentionally acts as a global vehicle for weedy sponge invaders. In their chapter, Kirsten Hastrup, Janne Flora, and Astrid Oberborbeck Andersen take us to the Avanersuaq (Thule) region of Northwest Greenland to follow the tracks, trails, and traces of humans and animals in an interdisciplinary project that combines anthropologists, archaeologists, and biologists. Along these paths, they tease out the challenging but productive frictions between different disciplinary approaches and epistemological positions. They find that, despite sharing the same landscape as field site, distinct understandings of the "empirical" may be at play. *Sensations*—the acknowledgment of the interplay of forces at work in shaping the world—become the methodological pathway to understanding that fields are always more-than-anthropological, constituted by a plurality of materialities, temporalities, and viewpoints, human and more-than-human. Meredith Root-Bernstein, Filippo Bertoni, Natalie Forssman, and Katy Overstreet, in their chapter, put on rubber boots and draw us through the muddy terrain of the Søby brown coal mines in Denmark. In the process, they explore how disturbance might be not merely an object of study but also itself a method—a way to experiment, play with, and learn about the affordances and historical structures of a multispecies world. This chapter, itself a coauthored collaboration between an ecologist, two STS scholars, and an anthropologist, studies ecologically disturbed patches—deer meadows, the waterlogged pits of abandoned mines, and a pig farm—and allows these disturbed landscapes to disturb in turn the rubber boots metaphor by exploring how zigzagging between fields with different methods can be productive of new insights about emotional and affective registers of humans and more-than-human actors in Anthropocene landscapes. In Nathalia Brichet's chapter, we find ourselves inside the milky white fluid of a cholera bottle at a Danish museum. Starting with this bottle, with its unknown but likely lethal content, Brichet seeks to find "common ground" from which collaboration between an anthropologist and a museum research project, with epidemiologists, microbiologists, and cholera researchers—and the opening of the bottle as an object of research—might proceed. Such common ground is never solely defined by the fieldworker. Through a series of fieldwork encounters—from research on pandemics in a Philadelphia museum to virulent strains of cholera emerging from horizontal gene transfer—Brichet argues for an unsettling approach in the making of a shared and collaborative object of interest, and of ecologies, in the Anthropocene.

While many of these chapters trouble the conception of the field, Heather Anne Swanson makes the university her "field" and asks what it takes for the university to be relevant in the Anthropocene. Focusing on the teaching of rubber boots methods in her chapter, Swanson highlights the strong institutional and neoliberal forces that prevent the teaching of curious, collaborative, and critical methods and takes us on fieldwork in a methods course in which she and Pierre du Plessis attempted to put the lessons from AURA's "slow seminars" into didactic practice. "Rubber boots teaching," Swanson argues, needs to be informed by the same feminist scholarship and "slow science" that inform rubber boots methods. This commitment to situated, empirical knowledge implies a blurring of the distinction between the classroom and the field. It means not merely teaching students the abstract theories of the Anthropocene or debating the epistemologies and ontologies of multispecies ethnography but also taking students into the landscape, allowing them to practice the modes of multispecies curiosity, the transdisciplinary methods, and the critical historical attention that rubber boots methods require.

Together, the thirteen chapters chart the outline of a rubber boots anthropology, a multispecies methods of the otherwise in which methods are rethought in direct response to the more-than-human disruptions of a patchy and multiple Anthropocene. In its various forms, this means affectively tracking, walking, riding, snorkeling, dogsledding, sailing, sampling, teaching, and getting lost in the disturbed landscapes of our time with curious, collaborative, and critical attention to its changing multispecies assemblies. The chapters argue for methods that are wide-ranging and that emerge in a reflexive engagement with the material and historical conditions of fieldwork on a disturbed planet. This means that rubber boots methods have to be themselves disturbed.

The afterword to the volume, "Troubling Methods in the Anthropocene," is a transcribed and edited version of a roundtable discussion between scholars Kirsten Hastrup, Ursula Münster, Anna Tsing, and Nils Bubandt. It touches on many of the issues that we have outlined in this introduction by tracing four lines of inquiry: towards critical description, mixing methods, decolonizing methods, and methods for a nonsecular Anthropocene. The conversation addresses the possibilities as well as the challenges and limits of transdisciplinary experiments with methods in the Anthropocene and offers a general snapshot of how AURA sought to push anthropological responses to the concept of the Anthropocene from conceptual critique to critical description. In the spirit of the anthology as a whole, the roundtable discussion seeks to be a curious and collaborative opening that is critical rather than consensual. While the other chapters of the anthology provide empirical examples of how researchers learn and experiment with curiosity, collaboration, and critical description, the roundtable—in lieu of a conclusion—situates these chapters in a larger conversation and suggests ways forward for future research engagements.

REFERENCES

Asad, Talal, ed. 1973. *Anthropology and the Colonial Encounter.* London: Ithaca Press.

Aso, Michitake. 2018. *Rubber and the Making of Vietnam.* Chapel Hill: University of North Carolina Press.

Baker, Janelle, Paula Ebron, Rosa Ficek, Karen Ho, Reyna Ramirez, Zoe Todd, Anna Tsing, and Sarah Vaughn. 2020. "The Snarled Lines of Justice: Women Ecowarriors Map a New History of the Anthropocene." *Orion Magazine.* https://orion magazine.org/article/the-snarled-lines-of-justice/.

Ballestero, Andrea. 2019. *A Future History of Water.* Durham, N.C.: Duke University Press.

Barnosky, A. D., N. Matzke, S. Tomiya, Guinevere O. U. Wogan, Brian Swartz, Tiago B. Quental, Charles Marshall et al. 2011. "Has the Earth's Sixth Mass Extinction Already Arrived?" *Nature* 471, no. 7336: 51–57.

Bear, Laura, Karen Ho, Anna Tsing, and Sylvia Yanagisako. 2015. "Gens: A Feminist Manifesto for the Study of Capitalism." https://culanth.org/fieldsights/652-gens-a-feminist-manifesto-for-the-study-of-capitalism.

Bennett, Jane. 2010. *Vibrant Matter: A Political Ecology of Things.* Durham, N.C.: Duke University Press.

Bernard, H. Russell, Gery W. Ryan, and Amber Y. Wutich. 2016. *Analyzing Qualitative Data: Systematic Approaches.* 2nd ed. Los Angeles, Calif.: SAGE.

Braidotti, Rosa. 2013. *The Posthuman.* Cambridge: Polity.

Brichet, Nathalia, and Frida Hastrup. 2018. "Industrious Landscaping: The Making and Managing of Natural Resources at Søby Brown Coal Beds." *Journal of Ethnobiology* 38, no. 1: 8–23.

Brown, Kate. 2013. *Plutopia: Nuclear Families, Atomic Cities, and the Great Soviet and American Plutonium Disasters.* Oxford: Oxford University Press.

Bubandt, Nils. 2018. "Anthropocene Uncanny: Nonsecular Approaches to Environmental Change." In *A Non-secular Anthropocene: Spirits, Specters and Other Nonhumans in a Time of Environmental Change,* edited by Nils Bubandt, 2–19. AURA More-Than-Human Working Paper 3. Aarhus, Denmark: Aarhus University Research on the Anthropocene (AURA). http://anthropocene.au.dk/working-papers-series/.

Bubandt, Nils. 2022. "The Uncanny Valley of the Anthropocene: Short Stories about the Undead under the Brightest of Lights." In *The Anthropocene and the Undead: Cultural Anxieties in Contemporary Narratives,* edited by Simon Bacon, 67–84. Washington, D.C.: Lexington Books.

Bubandt, Nils, Andrew Mathews, and Anna Tsing, eds. 2019. "Patchy Anthropocene: The Frenzies and Afterlives of Violent Simplifications." Special issue of *Current Anthropology* 60, no. S20.

Bubandt, Nils, and Anna Tsing. 2018a. "An Ethnoecology for the Anthropocene: How a Former Brown-Coal Mine in Denmark Shows Us the Feral Dynamics of Post-industrial Ruin." *Journal of Ethnobiology* 38, no. 1 (online suppl.): 1–13. https://bioone.org/journals/journal-of-ethnobiology/volume-38/issue-1/0278-0771-38.1.001/Feral-Dynamics-of-Post-Industrial-Ruin-An-Introduction/10.2993/0278-0771-38.1.001.short.

Bubandt, Nils, and Anna Tsing. 2018b. "Feral Dynamics of Post-industrial Ruin: An Introduction." *Journal of Ethnobiology* 38, no. 1: 1–7.

Burawoy, Michael, ed. 2000. *Global Ethnography. Forces, Connections and Imaginations in a Postmodern World.* Berkeley: University of California Press.

Büscher, Monica, and John Urry. 2009. "Mobile Methods and the Empirical." *European Journal of Social Theory* 12, no. 1: 99–116.

Chakrabarty, Dispesh. 2009. "The Climate of History: Four Theses." *Critical Inquiry* 35, no. 2: 197–222.

Clifford, James. 1997. *Routes: Travel and Translation in the Late Twentieth Century.* Cambridge, Mass.: Harvard University Press.

Clifford, James. 2013. *Returns: Becoming Indigenous in the Twenty-First Century.* Cambridge, Mass.: Harvard University Press.

Comaroff, John. 2010. "The End of Anthropology, Again: On the Future of an In/Disciplne." *American Anthropologist* 112, no. 4: 524–38.

Cooper, Frederick, and Ann Laura Stoler, eds. 1997. *Tensions of Empire: Colonial Cultures in a Bourgeois World*. Berkeley: University of California Press.

Danowski, Deborah, Eduardo Viveiros de Castro, and Bruno Latour. 2014. "The Thousand Names of Gaia: From the Anthropocene to the Age of the Earth." https://thethousandnamesofgaia.files.wordpress.com/2014/07/position-paper -ingl-para-site.pdf.

Dattatreyan, Ethiraj G., and Isaac Marrero-Guillamón. 2019. "Introduction: Multi-modal Anthropology and the Politics of Invention." *American Anthropologist* 121, no. 1: 220–28.

de la Cadena, Marisol. 2015. *Earth Beings: Ecologies of Practice across Andean Worlds*. Durham, N.C.: Duke University Press.

Deleuze, Gilles, and Félix Guattari. 1987. *A Thousand Plateaus: Capitalism and Schizophrenia*. London: Athlone.

DeMello, Margo. 2012. *Animals and Society: An Introduction to Human-Animal Studies*. New York: Columbia University Press.

Demos, T. J. 2017. *Against the Anthropocene: Visual Culture and Environment Today*. Berlin: Sternberg Press.

Durkheim, Emile. (1897) 1997. *Suicide: A Study in Sociology*. London: Routledge.

Edwards, Paul N. 2010. *A Vast Machine: Computer Models, Climate Data, and the Politics of Global Warming*. Cambridge, Mass.: MIT Press.

Erlmann, Veit. 1989. "'Horses in the Race Course': The Domestication of Ingoma Dancing in South Africa, 1929–39." *Popular Music* 8, no. 3: 259–73.

Escobar, Arturo. 2018. *Designs for the Pluriverse: Radical Interdependence, Autonomy, and the Making of Worlds*. Durham, N.C.: Duke University Press.

Estes, James. 2016. *Serendipity: An Ecologist's Quest to Understand Nature*. Berkeley: University of California Press.

Estes, James, John Terborgh, Justin Brashares, Mary E. Power, Joel Berger, William J. Bond, Stephen R. Carpenter et al. 2011. "Trophic Downgrading of Planet Earth." *Science* 333, no. 6040: 301–6.

Fabian, Johannes. 2000. *Out of Our Minds: Reasons and Madness in the Exploration of Central Africa*. Berkeley: University of California Press.

Faubion, James D., and George Marcus, eds. 2009. *Fieldwork Is Not What It Used to Be: Learning Anthropology's Method in a Time of Transition*. Ithaca, N.Y.: Cornell University Press.

Fischer, Michael J. 2018. *Anthropology in the Meantime: Experimental Ethnography, Theory, and Method for the Twenty-First Century*. Durham, N.C.: Duke University Press.

Flora, Janne, and Astrod O. Andersen. 2017. "Whose Track Is It Anyway? An Anthropological Perspective on Collaboration with Biologists and Hunters in Thule, Northwest Greenland." *Collaborative Anthropologies* 9, no. 1–2: 79–116.

Forssman, Natalie, and Meredith Root-Bernstein. 2018. "Landscapes of Anticipation of the Other: Ethno-ethology in a Deer Hunting Landscape." *Journal of Ethnobiology* 38, no. 1: 71–87.

Fortier, Craig. 2017. "Unsettling Methodologies/Decolonizing Movements." *Journal of Indigenous Social Development* 6, no. 1: 20–36.

Gan, Elaine, Anna Tsing, and Daniel Sullivan. 2018. "Using Natural History in the Study of Industrial Ruins." *Journal of Ethnobiology* 38, no. 1: 39–54.

Gan, Elaine, Anna Tsing, Heather Swanson, and Nils Bubandt. 2017. "Introduction: Haunted Landscapes of the Anthropocene." In *Arts of Living on a Damaged Planet: Ghosts and Monsters of the Anthropocene,* edited by Anna Lowenhaupt Tsing, Heather Anne Swanson, Elaine Gan, and Nils Bubandt, G1–14. Minneapolis: University of Minnesota Press.

Garfield, Seth. 2014. *In Search of the Amazon: Brazil, the United States, and the Nature of a Region.* Durham, N.C.: Duke University Press.

Gómez-Barris, Macarena. 2017. *The Extractive Zone: Social Ecologies and Decolonial Perspectives.* Durham, N.C.: Duke University Press.

Green, Lesley. 2015. "The Changing of the Gods of Reason: Cecil John Rhodes, Karoo Fracking, and the Decolonizing of the Anthropocene." *e-flux journal,* June 9. http://supercommunity.e-flux.com/texts/the-changing-of-the-gods-of-reason/.

Grusin, Richard. 2015. *The Nonhuman Turn.* Minneapolis: University of Minnesota Press.

Grusin, Richard, ed. 2017. *Anthropocene Feminism.* Minneapolis: University of Minnesota Press.

Gupta, Akhil, and James Ferguson, eds. 1997. *Anthropological Locations: Boundaries and the Grounds of a Field Science.* Berkeley: University of California Press.

Hamilton, Lindsay, and Nik Taylor. 2017. *Ethnography after Humanism: Power, Politics and Method in Multi-species Research.* New York: Palgrave Macmillan.

Haraway, Donna. 2004. *The Haraway Reader.* New York: Routledge.

Haraway, Donna. 2016. *Staying with the Trouble: Making Kin in the Chthulucene.* Durham, N.C.: Duke University Press.

Haraway, Donna, Noboru Ishikawa, Scott F. Gilbert, Kenneth Olwig, Anna Tsing, and Nils Bubandt. 2016. "Anthropologists Are Talking—about the Anthropocene." *Ethnos* 81, no. 3: 535–64.

Harp, Stephen L. 2016. *A World History of Rubber: Empire, Industry, and the Everyday.* Chichester, U.K.: Wiley.

Harvey, Penny, Christian Krohn-Hansen, and Knut G. Nustad, eds. 2019. *Anthropos and the Material.* Durham, N.C.: Duke University Press.

Helmreich, Stefan. 2009. *Alien Ocean: Anthropological Voyages in Microbial Seas.* Berkeley: University of California Press.

Henare, Amira, Martin Holbraad, and Sari Wastell, eds. 2006. *Thinking through Things: Theorising Artefacts Ethnographically.* London: Routledge.

Hoag, Colin, Filippo Bertoni, and Nils Bubandt. 2018. "Wasteland Ecologies: Undomestication and Multispecies Gains on an Anthropocene Dumping Ground." *Journal of Ethnobiology* 38, no. 1: 88–104.

Højrup, Marie, and Heather Swanson. 2018. "The Making of Unstable Ground: The Anthropogenic Geologies of Søby, Denmark." *Journal of Ethnobiology* 38, no. 1: 24–38.

Ingold, Tim. 2012. "Toward an Ecology of Materials." *Annual Review of Anthropology* 41: 427–42.

Johannes, Robert E. 1981. *Words of the Lagoon: Fishing and Marine Lore in the Palau District of Micronesia*. Berkeley: University of California Press.

Kenney-Lazar, Miles, and Noburo Ishikawa. 2019. "Mega-plantations in Southeast Asia: Landscapes of Displacement." *Environment and Society: Advances in Research* 10: 63–82.

Kirksey, Eben, and Stefan Helmreich. 2010. "The Emergence of Multispecies Ethnography." *Cultural Anthropology* 25, no. 4: 545–76.

Kohn, Eduardo. 2013. *How Forests Think: Towards an Anthropology beyond the Human*. Berkeley: University of California Press.

Kusenbach, Margarethe. 2003. "Street Phenomenology: The Go-along as Ethnographic Research Tool." *Ethnography* 4, no. 3: 455–85.

Latour, Bruno. 1996. *Aramis, or the Love of Technology*. Cambridge, Mass.: Harvard University Press.

Latour, Bruno. 2004. *Politics of Nature: How to Bring the Sciences into Democracy*. Cambridge, Mass.: Harvard University Press.

Latour, Bruno. 2005. "From Realpolitik to Dingpolitik—an Introduction." In *Making Things Public: Atmospheres of Democracy*, 14–41. Cambridge, Mass.: MIT Press.

Latour, Bruno. 2017. *Facing Gaia: Eight Lectures on the New Climatic Regime*. London: Polity Press.

Latour, Bruno. 2018. *Down to Earth: Politics in the New Climatic Regime*. Oxford: Polity Press.

Law, John. 2004. *After Method: Mess in Social Science Research*. New York: Routledge.

Lorimer, Jamie. 2015. *Wildlife in the Anthropocene*. Minneapolis: University of Minnesota Press.

Malm, Andreas, and Alf Hornborg. 2014. "The Geology of Mankind? A Critique of the Anthropocene Narrative." *Anthropocene Review* 1, no. 1: 62–69.

Marcus, George E. 1998. *Ethnography through Thick and Thin*. Princeton, N.J.: Princeton University Press.

Marcus, George E., and Marcello Pisarro. 2008. "The End(s) of Ethnography: Social/Cultural Anthropology's Signature Form of Producing Knowledge in Transition." *Cultural Anthropology* 23, no. 1: 1–14.

Marx, Karl. (1852) 2005. *The Eighteenth Brumaire of Louis Bonaparte*. New York: Mondial.

Mathews, Andrew. 2018. "Landscapes and Throughscapes in Italian Forest Worlds: Thinking Dramatically about the Anthropocene." *Cultural Anthropology* 33, no. 3: 386–414.

McNeill, John R., and Peter Engelke. 2014. *The Great Acceleration: An Environmental History of the Anthropocene since 1945*. Cambridge, Mass.: Belknap Press of Harvard University Press.

Mills, Dana. 2016. *Dance and Politics*. Manchester, U.K.: Manchester University Press.

Mintz, Sidney. 1985. *Sweetness and Power: The Place of Sugar in Modern World History*. New York: Penguin.

Moore, James W. 2016. *Anthropocene or Capitalocene? Nature, History, and the Crisis of Capitalism*. Oakland, Calif.: PM Press.

Morton, Timothy. 2016. *Dark Ecology: For a Logic of Future Coexistence*. New York: Columbia University Press.

Myers, Natasha. 2015. "Conversations on Plant Sensing: Notes from the Field." *Nature-Culture* 3: 35–66.

Nixon, Rob. 2011. *Slow Violence and the Environmentalism of the Poor*. Cambridge, Mass.: Harvard University Press.

Otto, Ton, and Nils Bubandt, eds. 2010. *Experiments in Holism: Theory and Practice in Contemporary Anthropology*. Oxford: Wiley-Blackwell.

Pandian, Anand. 2019. "Introduction: An Ethnographer among Anthropologists." In *A Possible Anthropology: Methods for Uneasy Times*, 1–14. Durham, N.C.: Duke University Press.

Patel, Raj, and James Moore. 2017. *History of the World in Seven Cheap Things*. Berkeley: University of California Press.

Pauly, Daniel. 1995. "Anecdotes and the Shifting Baseline Syndrome of Fisheries." *Trends in Ecology and Evolution* 10, no. 10: 430.

Pickett, S. T. A., and P. S. White, eds. 1985. *The Ecology of Natural Disturbance and Patch Dynamics*. Amsterdam: Academic Press.

Riles, Annelies, ed. 2006. *Documents: Artefacts of Modern Knowledge*. Ann Arbor: University of Michigan Press.

Robbins, Paul. 2007. *Lawn People: How Grasses, Weeds, and Chemicals Make Us Who We Are*. Philadelphia: Temple University Press.

Rockström, Johan, Will Steffen, Kevin Noone, Åsa Persson, F. Stuart Chapin, Eric Lambin, Timothy M. Lenton et al. 2009. "Planetary Boundaries: Exploring the Safe Operating Space for Humanity." *Ecology and Society* 14, no. 2: 32.

Roitman, Janet. 2013. *Anti-crisis*. Durham, N.C.: Duke University Press.

Scott, James. 1998. *Seeing Like a State: How Certain Schemes to Improve the Human Condition Have Failed*. New Haven, Conn.: Yale University Press.

Scott, James. 2009. *The Art of Not Being Governed: An Anarchist History of Upland Southeast Asia*. New Haven, Conn.: Yale University Press.

Severn, John. 2007. *Architects of Empire: The Duke of Wellington and His Brothers*. Norman: University of Oklahoma Press.

Smith, Andrea. 2014. "Native Studies at the Horizon of Death: Theorizing Ethnographic Entrapment and Settler Self-Reflexivity." In *Theorizing Native Studies*, edited by Audra Simpson and Andrea Smith, 207–34. Durham, N.C.: Duke University Press.

Smith, Linda T. 2012. *Decolonizing Methodologies: Research and Indigenous Peoples*. New York: Zed Books.

Steffen, Will, Wendy Broadgate, Lisa Deutsch, Owen Gaffney, and Cornelia Ludwig. 2015. "The Trajectory of the Anthropocene: The Great Acceleration." *Anthropocene Review* 2, no. 1: 81–98.

Stengers, Isabelle. 2018. *Another Science Is Possible: A Manifesto for Slow Science*. Cambridge: Polity.

Stocking, George W. 1983a. "The Ethnographer's Magic: Fieldwork in British Anthropology from Tylor to Malinowski." In *Observers Observed*, edited by G. W. Stocking, 70–120. Madison: University of Wisconsin Press.

Stocking, George W., ed. 1983b. *Observers Observed: Essays on Ethnographic Fieldwork*. Madison: University of Wisconsin Press.

Stoler, Ann L. 1995. *Capitalism and Confrontation in Sumatra's Plantation Belt, 1870–1979*. Ann Arbor: University of Michigan Press.

Strang, Vera. 2010. "Mapping Histories: Cultural Landscapes and Walkabout Methods." In *Environmental Social Sciences: Methods and Research Design*, edited by Ismael Vaccaro, Eric Alden Smith, and Shankar Aswani, 132–56. Cambridge: Cambridge University Press.

Strathern, Marilyn. 2004. *Commons and Borderlands: Working Papers on Interdisciplinarity, Accountability, and the Flow of Knowledge*. Oxford: Sean Kingston.

Swanson, Heather. 2015. "Who's in the Room? The Importance of Multidisciplinary Spaces for Anthropology and STS." *HAU: Journal of Ethnographic Theory* 5, no. 1: 445–48.

Swanson, Heather. 2017a. "The Banality of the Anthropocene." In *Dispatches: Cultural Anthropology* (blog), February 22. https://culanth.org/fieldsights/1074-the-banality-of-the-anthropocene.

Swanson, Heather. 2017b. "Methods for Multispecies Anthropology: Analysis of Salmon Otoliths and Scales." *Social Analysis* 61, no. 2: 81–99.

Swanson, Heather, Nils Bubandt, and Anna Tsing. 2015. "Less than One but More Than Many: Anthropocene as Science Fiction and Scholarship-in-the-Making." *Environment and Society: Advances in Research* 6: 149–66.

Swyngedouw, Erik, and Maria Kaika. 2014. "Urban Political Ecology: Great Promises, Deadlock . . . and New Beginnings?" *Documents d'Anàlisi Geogràfica* 60, no. 3: 459–81.

Tsing, Anna Lowenhaupt. 2005. *Friction: An Ethnography of Global Connections*. Princeton, N.J.: Princeton University Press.

Tsing, Anna Lowenhaupt. 2013. "More-Than-Human Sociality: A Call for Critical Description." In *Anthropology and Nature*, edited by Kirsten Hastrup, 27–42. New York: Routledge.

Tsing, Anna Lowenhaupt. 2015. *The Mushroom at the End of the World: On the Possibility of Life in Capitalist Ruins*. Princeton, N.J.: Princeton University Press.

Tsing, Anna Lowenhaupt. 2019. "When the Things We Study Respond to Each Other: Tools for Unpacking 'the Material.'" In *Anthropos and the Material*, edited by Penny Harvey, Christian Krohn-Hansen, and Knut G. Nustad, 221–44. Durham, N.C.: Duke University Press.

Tsing, Anna, Andrew Mathews, and Nils Bubandt. 2019. "Patchy Anthropocene: Landscape Morphology, Multispecies History and the Retooling of Anthropology." *Current Anthropology* 60: S186–97.

Tsing, Anna Lowenhaupt, Heather Anne Swanson, Elaine Gan, and Nils Bubandt, eds. 2017. *Arts of Living on a Damaged Planet: Ghosts and Monsters of the Anthropocene*. Minneapolis: University of Minnesota Press.

Tully, John. 2011. *The Devil's Milk: A Social History of Rubber.* New York: Monthly Review Press.

Umar, H. Y., D. Y. Giroh, N. B. Agbonkpolor, and C. S. Mesike. 2011. "An Overview of World Natural Rubber Production and Consumption: An Implication for Economic Empowerment and Poverty Alleviation in Nigeria." *Journal of Human Ecology* 33, no. 1: 53–59.

Vaccaro, Ishmael, Eric A. Smith, and Shanka Aswani. 2010. *Environmental Social Sciences: Methods and Research Design.* Cambridge: Cambridge University Press.

Vergunst, Jo Lee, and Tim Ingold, eds. 2016. *Ways of Walking: Ethnography and Practice on Foot.* London: Routledge.

Viveiros de Castro, Eduardo. 1998. "Cosmological Deixis and Amerindian Perspectivism." *Journal of the Royal Anthropological Institute* 4: 469–88.

Wolfe, Cary. 2009. *What Is Posthumanism?* Minneapolis: University of Minnesota Press.

Wolfe, Cary. 2013. *Before the Law: Humans and Other Animals in a Biopolitical Frame.* Chicago: University of Chicago Press.

Yusoff, Kathryn. 2018. *A Billion Black Anthropocenes or None.* Minneapolis: University of Minnesota Press.

Zalasiewicz, Jan, Colin Waters, and Mark Williams. 2014. "Human Bioturbation, and the Subterranean Landscape of the Anthropocene." *Anthropocene* 6: 3–9.

Zalasiewicz, Jan, Colin Waters, Alexander Wolfe, Anthony Barnosky, Alejandro Cearreta, Matt Edgeworth, Erle C. Ellis et al. 2017. "Making the Case for a Formal Anthropocene Epoch: An Analysis of Ongoing Critiques." *Newsletters on Stratigraphy* 50, no. 2: 205–26.

Zylinska, Joanna. 2014. *Minimal Ethics for the Anthropocene.* Ann Arbor, Mich.: Open Humanities Press.

PART I

CRITICAL
DESCRIPTION

Walking in Italian Forests and Telling Stories about Global Environmental Change

ANDREW S. MATHEWS

Over the last few years, I have been learning to tell stories about people, trees, and landscapes in the Monte Pisano of central Italy, between the cities of Lucca and Pisa. Like most Mediterranean landscapes, this area has been cultivated for several thousand years, and traces of human presence are everywhere. The traces of encounters between people, plants, and soils are visible in the morphologies of trees, of terracing and drainage systems, and of ruined houses. By walking, looking, and wondering, I have learned to tell the biographies of trees, terracing systems, and buildings. By walking through the landscape and by drawing trees and landscapes, I have learned to see larger landscape patches and structures (Tsing, Mathews, and Bubandt 2019), each of which partially describes intertwined social and ecological processes. Walking allows me to see, to touch, to smell, and to taste, but it also allows me to speculate, to imagine, and to wonder. Walking is both sensory engagement and reflective practice. Drawing allows me to slow down and notice the shape-changing capacities of trees. By wondering about the histories that shaped trees and brought landscape patches and structures into being, I come up with causal accounts for the transformations that destroyed formerly cultivated chestnut forests and left fire-blasted landscapes. My landscape walks are made possible by histories of peasant and pastoral mobility, which offer me long-distance footpaths through the landscape. Walking is also made possible by histories of Fascist rule, which have made the countryside more accessible than in most countries. The story goes that the dictator Benito Mussolini wished to make Italian people more martial and that he therefore allowed hunters and mushroom gatherers onto otherwise private property. I wear a bright red backpack during hunting season so as to be visible to the hunters who I hear banging away on autumn mornings.

Rubber boots methods, such as landscape walking, can produce accounts of global environmental change that complement climate and ecological models. This is only possible, however, if we are bold enough to move from our sensory experiences of plants, animals, soils, and people to larger scales in time and space. We can draw examples from the "ontological anarchism" of the world (Viveiros de Castro 2019), but examples are not enough if they remain unique conjunctures. We risk remaining like Borges's (1964) Funes the Memorious, who is alert to the uniqueness of the world but unable to communicate it to others. This chapter is about how different kinds of forms can help us tell larger-scale stories about relationships between plants, soils, people, and diseases; changing climates; and changing economies (for a consideration of form, see also chapter 2). The first type of form is morphology, the shapes of trees, terracing, and drainage systems that tell me of biographies of encounter among people, plants, animals, and soils. Drawing and sketching are the methods through which I slow down and notice the responsiveness of plants and the forms of drainage systems. Drawing returns me to the landscape with further questions and speculations. A second type of form comprises the landscape structures that I learn to notice during my walks and the diagrams through which I show these structures. Another kind of form is the species distribution model that emerged from my collaboration with the botanist and ecological modeler Gabriele Casazza and the climate scientist Michele Brunetti. As we shall see, landscape structures and models project our phenomenological experience of the world into larger-scale stories in very different ways, with different strengths and weaknesses. A final type of form is the persuasive narrative or story, with which anthropologists and historians are familiar (Cronon 1992). This chapter tells a story about the productive coordinations between people, animals, and plants that persisted in this region for many centuries and about the kinds of disasters that can cause a loss of coordination (Tsing and Gan 2018). Productive agroecosystems have given way to the fire- and disease-blasted landscapes of the Anthropocene.

Each of these types of form emerges from different methods of walking, noticing, drawing, diagramming, modeling, and narrating. Each form is a

FIGURE 1.1. Location map, Monte Pisano. Map by Fabio Malfatti, 2020.

limited and partial account of a world of process and movement, of encounter and transformation. None is definitive; each allows me to critically describe the intertwined social and ecological changes in the disturbed landscapes of the Anthropocene; methodological experimentation and collaboration with natural scientists allow me to tell large-scale stories about the effects of plant disease and social change (see the Introduction to this volume).

Walking, Noticing, and Drawing

One of my most important research methods is to walk slowly across a watershed, looking closely at details of tree and landform. These walks are a structured curiosity. People and animals care a lot about plants, and plants care about climate, soil depth, and fertility. This makes walking across a watershed a good way to explore the range of responses between beings in a particular region. In the Monte Pisano, villas and gardens, olive trees and grape vines on the deep and well-drained soils of the valley floor, give way to fire-blasted pine forests on shallow soils higher up the mountain.

As I walk, I look closely at trees and terraces, looking for details of form that can tell me something about their biographies. Noticing gives rise to wondering, as I imagine the encounters with forest fires, peasant firewood cutters, or diseases that gave each tree its particular morphology. I wonder why each terrace was built, whether a stone wall near a villa, a more informal earthen bank in an abandoned olive grove choked by young pines, or the rough stone walls that skirt the roots of chestnut trees still higher up the mountain. This kind of speculative noticing and wondering is a way of looking at the effects of people, plants, fires, diseases, and soils upon each other.

Noticing details does not mean that I look only at things that are close to me. I look also up to the larger landscape, and my attention moves back and forth between near and far. As my attention shifts, I begin to see patterns in what at first looked like a relatively amorphous mélange of evergreen pines and miscellaneous oaks, chestnuts, and other broadleaves. As I walk, I take notes and photographs and sketch particularly interesting trees.

FIGURE 1.2. Field note and photograph of ancient chestnut tree with botanist Francesco Roma Marzio. Photograph by the author.

Pausing to sketch and photograph forces me to focus closely on the forms of puzzling individuals. Trees respond to fire, firewood cutting, grafting, and disease through morphological inventiveness. Tree branches grow toward light, roots grow in response to soil moisture, stumps sprout new limbs in response to pruning and firewood cutting. As sedentary organisms with exquisitely acute senses, plants record their biographies in their forms. Practices of sketching and of looking closely force me to slow down and attend to the morphological inventiveness of plants, shape-changing beings that move across the landscape too slowly for us to notice their responsiveness. My diagrams are empirical descriptions, but they are also a gesture toward the shape-changing capacities and slow mobilities of plants across the landscape. The botanist Francis Hallé (2018, 8) argues that drawings are better than photographs for learning about plants: "It is as if we were visiting a distant planet and encountered a form of extraterrestrial life with which we share no language—a form of life based upon principles that are not our own. If we wish to understand this creature, it is best not to rush." Slowing down can emerge from the moment when a particularly strange tree demanded that I stop, look, photograph, and sketch, but often I set pen to paper months or years later as I decide which of many details to highlight.

Drawings and diagrams have a long tradition in anthropology, from the "diagrammatic representations" of cattle colorations by Evans-Pritchard (1940, 40–41) to Tim Ingold's (2011) use of diagrams that emerge from phenomenological descriptions of travel and movement. My use of drawings is most immediately inspired by the drawings of the heterodox historical ecologist Oliver Rackham (Grove and Rackham 2001) and by traditions of field natural history (Canfield 2011). Whereas anthropologist Michael Taussig's (2011) notebook drawings avoid realism and are "fragments that are suggestive of a world beyond," my drawings attend to morphologies that emerge from slow processes of plant growth and soil movement while also gesturing toward what escapes description. Even as I learn from my visits to the archive and my conversations with farmers and officials, drawings and diagrams emerge from my movement through the landscape, partially describing relationships between people, soil, plants and disease, climate change, and capitalism.

Walking and noticing, speculating and sketching, give me a chance to be pulled into telling stories about the pasts and futures of landscapes. Similar kinds of speculative noticing and wondering are how many natural scientists come up with the questions that drive their research. Ecologists think hard about the landscapes they study, and they entertain many speculative possibilities. Even as they gradually narrow their focus to the most urgent and answerable questions, ecologists, like other scientists, are always aware of the many speculations that cannot be answered. There is common ground between scientists, peasants, and Indigenous people, each of whom engages in their own forms of noticing and principled speculation. The farmers with whom I spoke in Italy and the Zapotec Indigenous people with whom I formerly worked in Mexico notice and wonder at the ecosystems in which they live and work (Mathews 2018, 2011).

Through my practices of observation, wondering, and drawing, I gradually learn to notice larger-scale landscape structures, such as patterns of fire-scarred forest, of dying trees, or of tended chestnut groves. Tacking back and forth between sensory engagement, oral histories and ethnographic interviews, land use records and maps, my perceptions as to which landscape structures matter shift and come into focus. My travels through the landscape have helped me learn of surprising relationships between long-lived trees, changing climates, unstable landscapes, and changing societies.

Histories of regional social and environmental change that are linked to landscape structures emerge from my attunement to the specificities of encounters between plants, people, soils, and diseases, but they also reach toward regional or global scales. World-spanning political and economic changes brought the diseases and agricultural abandonment that have reshaped the Monte Pisano. A regional account, such as I provide in this chapter, is open to surprise, to the possibility that the next tree or the next terrace might tell of some unexpected conjuncture of human care and abandonment. An unruly flourishing of plant and soil morphology might reshape my understanding of what plants are capable of. Regional examples that build up from noticing the ontological anarchy of the world stand in helpful contrast to the systems modeling that has helped us notice global climate change. The systems models that have allowed natural scientists to detect climate change necessarily stabilize relations between plants, people, fire, disease, and soils in order to tell larger-scale stories about the world.

COMING INTO NOTICING

Learning to tell stories about a landscape is the result of a process of sensory attunement (see chapter 5). Let me show you how I came to tell stories that emerged from my noticing of tree stumps in the forest. First, consider this rather anonymous tree stump. It is large, certainly, but what is there about it that might compel storytelling? Allow yourself to be puzzled and look closely at this image before I draw out the traces of human and plant coordination that are preserved in the morphology of this stump. This tree stump, almost two meters in diameter, was an ancient chestnut tree that was converted from food production to firewood. Chestnut trees were formerly a principal source of carbohydrate food to millions of people across the Apennine Mountains of Italy and mountains across the Mediterranean (Giannini and Gabbrielli 2013; Squatriti 2013). Successive pathogen epidemics and the disappearance of peasant agriculture have caused the abandonment of most chestnut cultivation. This is the survivor of another civilization.

The simplifications of this diagram (Figure 1.4) show the details that matter for telling this story. The artist Hannah Caisse and I discussed what I wanted to show and what to leave out, what would work in a drawing and what would not. Each set of smaller stumps records encounters between the tree and firewood cutters over the last seventy years. This practice of repeated cutting, known as *ceduo* (coppice), has been practiced by firewood cutters in Europe and around the world for several thousand years. From walks and encounters with other trees, and from conversations with firewood cutters and farmers, I was primed to notice the significance of this stump, with its pattern of older eroded stems, more recent stems, and new shoots. Diagrams reflect choices about which relations matter and should be highlighted, but they also require that other things be left out. This diagram shows evidence of firewood cutting and abandonment. A different diagram could show traces of disease, wood decay, or still other details. The shoots on the upper left are ambiguous. They might belong to this tree and be a sign of an ongoing growth and care. If I decided that they were from another plant, I could leave them out, and this would be a diagram of ruin and abandonment. The status of this tree is indeterminate, somewhere between living and dying. Only time will tell.

From my conversations with farmers, I could imagine that this unpromising stump had formerly been an enormous ancient chestnut tree that had

FIGURE 1.3. Ancient chestnut stump, Pizzorna, Lucca, Italy, 2014. Photograph by the author.

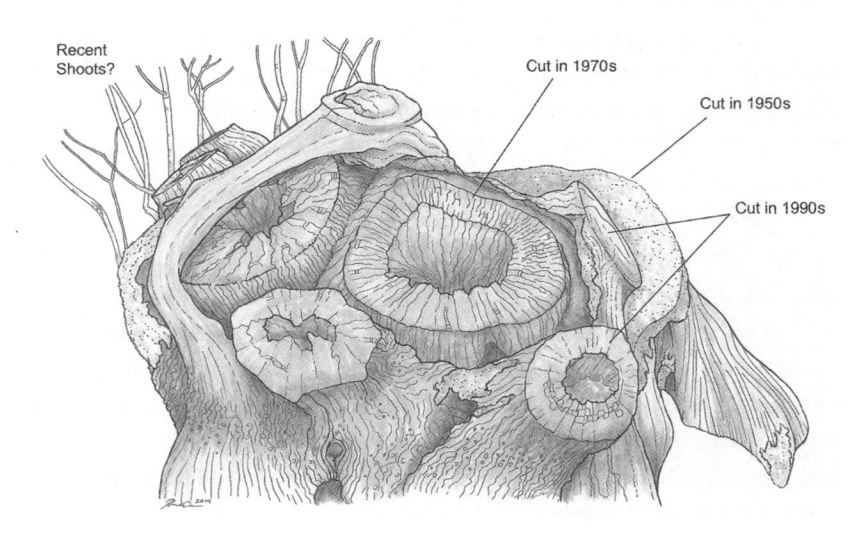

FIGURE 1.4. Diagram of stump response to cutting. Drawing by Hannah Caisse, *Chestnut Stump*, 2019.

been grafted by peasant farmers centuries ago. I could imagine that it had formerly grown in a carefully tended chestnut orchard, as in Figure 1.5.

The trees in this orchard are spaced out to favor fruit production and are perhaps two hundred years old. They are well cared for: none have shoots at the base, and all have well-pruned crowns. The low stone retaining walls, known as *lunette,* keep soil around the roots of the trees. These walls demonstrate peasant farmers' willingness to work hard to attend to chestnut trees' need for moisture. Finally, the orchard is clear and relatively grassy. This requires the work of sheep or goats who graze and of a farmer to rake up leaves and burrs. What cannot be easily seen in this picture is the presence of graft scars. These enormous trees are the long-range echo of a centuries-old event, when a peasant farmer nudged a graft onto a wild rootstock, feeling with skill and imagination for the precise alignment of cambium layers in two small shoots. Trees, fire, people, soils, grass, and sheep have affected each other, in a coordination that required continual attention from peasant farmers.

FIGURE 1.5. Ancient chestnut trees with *lunette* (retaining walls), Fosciandora, Lucca, Italy, 2013. Photograph by the author.

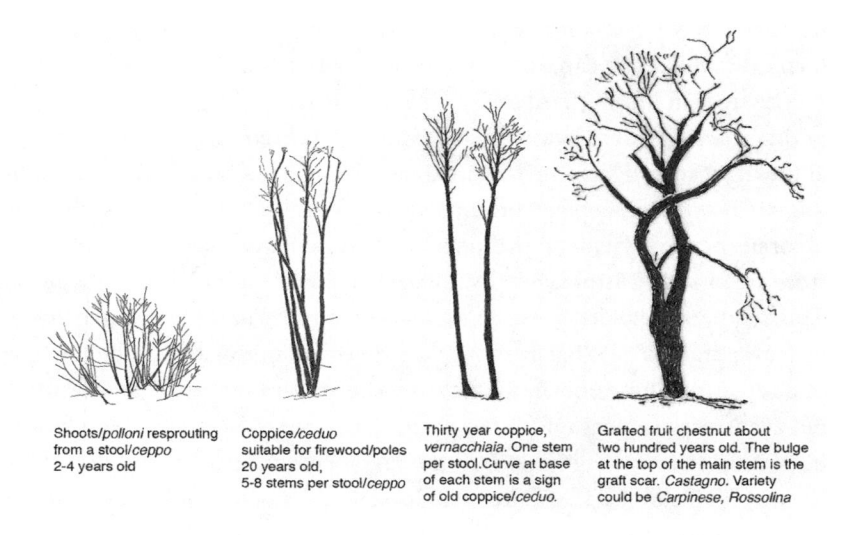

Shoots/*polloni* resprouting from a stool/*ceppo* 2-4 years old

Coppice/*ceduo* suitable for firewood/poles 20 years old, 5-8 stems per stool/*ceppo*

Thirty year coppice, *vernacchiaia*. One stem per stool.Curve at base of each stem is a sign of old coppice/*ceduo*.

Grafted fruit chestnut about two hundred years old. The bulge at the top of the main stem is the graft scar. *Castagno*. Variety could be *Carpinese*, *Rossolina*

FIGURE 1.6. A chestnut bestiary. Drawings by the author.

Figure 1.6 shows what I call my chestnut bestiary, a menagerie of plant forms. All are drawings of chestnut trees, but each has a different name, a different history, and a different shape. The names come from my conversations with smallholders. The drawings are the result of my days of walking across the landscape and my choice of which trees to draw and what features to highlight. I hope that these examples persuade you that trees are wild creatures, stumps that might sprout, gnarly ancient fruit trees that might become firewood, or other shapes entirely. To notice plant morphology is to open yourself to the slow, sly, and strange shape-changing practices of trees. By attending to the morphologies of individual trees, we can learn to imagine dramatic histories that have transformed entire landscapes.

On Rewiring Your Senses: From Morphologies to Landscape Structures

Plant morphologies and the shapes of terracing and drainage systems tell me about histories of human care in landscape patches. Larger-scale landscape structures that are the result of world-transforming processes of industrialization and agrarian change emerge through my practice of moving across the landscape and of shifting my attention between near and far. This is the case with the fire-blasted landscapes of the Monte Pisano (see also chapter 5). One morning, walking up the mountain, I was struck by the silvery color of fire-killed chestnut shoots and by the charcoal at their bases (see Figure 1.7a). I usually scratch what might be charcoal with my thumbnail and put it in my mouth. The crunchy sensation is similar to what you might experience if you try the charcoal from your grill at home. A few minutes later, higher up the mountain, I looked back and saw a larger-scale landscape structure, a silvery area of fire-killed chestnut with a darker green of pine in the background (see Figure 1.7b). Encountering a particular tree had changed my sense of landscape structure. It is only by much travel through the landscape that larger-scale structures settle, always somewhat tentatively. I do my best to continue to be alert to details that might unsettle my sense of larger structures. A diagram of this scene (Figure 1.7c) highlights the ecological relationships between fire, pine, and chestnut that are reshaping this landscape. Fire has opened up space for the fire-adapted *pino marittimo (Pinus pinaster)* in the background to colonize former chestnut coppice. In the foreground, young pine trees are crowding the image. Fire is a rare but powerful visitor that I can see in fire scars and taste in charcoal, that I can see in the presence of fire-adapted pine trees. As I continued to move across the Monte Pisano that morning, I saw the larger landscape structure of fire-killed chestnut.

The landscape structure of fire-killed chestnut made me wonder about what might have caused the fires that had killed these formerly coppiced chestnut trees. What histories might have produced this landscape structure, these patches of fire-killed chestnut, and the morphology of this fire-killed tree with charcoal at its base? What processes might link chestnut forest, rapidly growing pines, and the city of Lucca in the background? To answer these questions, I turned to oral histories and interviews. Elderly peasant farmers and shepherds told me of the practices of landscape care that had formerly made this mountain range much less prone to burning. I learned about the almost forgotten practice of leaf litter raking by talking to Fabio Casella, a municipal planning official who grew up in a peasant household in these mountains. I learned how people used to travel up the mountains into forest to rake leaf litter and carry it down to their fields in heavy baskets. This material would be composted with animal waste and spread across olive terraces and fields, providing essential fertilizer.[1] The flammability of

a

b

c

FIGURE 1.7. (a) Fire-killed chestnut, Monte Pisano, 2013. Photograph by the author. (b) Patch of fire-killed chestnut forest, pines, and city of Lucca in the background. Photograph by the author. (c) Diagram of fire-killed chestnut and flourishing pines. Drawing by Hannah Caise, 2019.

the landscape was due to not only the presence of fire-adapted pines but the abandonment of a way of making a living. The pastoralists who grazed sheep and goats in these mountains and the farmers who raked leaf litter from forest floors so as to produce fertilizer for their olive trees made forests much less flammable. When chemical fertilizers arrived in this part of the world in the 1960s, litter raking for fertilizer was abandoned. Accumulated dry vegetation burns fiercely when fires come. After fires, *pino marittimo*[2] reproduces prolifically on burned sites. Here, as across the Mediterranean, the agricultural and pastoral abandonment caused by the mid-twentieth-century Great Acceleration has produced ever larger and more dangerous fires (Palahi et al. 2008).

From oral history and interview, I moved back to the landscape. My senses had changed. Touch the forest floor in many places on the Monte Pisano and you will feel a layer of pine needles five or ten centimeters thick. It is bouncy and pleasant to walk on, fragrant with the smell of resin, but it is also ideal kindling for fires. This is the smell and the feeling of danger. Old people told me how differently the landscape looked from the past. Formerly, the color of bare soil was visible from across a valley, pastures were trimmed short by sheep and goats, and fallen branches were picked up instantly. The retired peasant Maria Lenzarini told me that forests were kept *puliti* (clean) by daily labor: "one cleaned, one went in the broadleaf woods and in the pine woods, you gathered up everything beneath to make a bed for the sheep" (interview, 2019). Children, women, and men carried huge baskets of leaf litter from forests to sheep sheds. After a few days, this mixture of dung, urine, and leaf litter would be piled up to make compost and eventually dug into the roots of olive trees as fertilizer *(sugo)*. The forest was almost a garden, where every scrap of land was divided by nearly invisible lines of small stones that demarcated woodlots and pastures. "You could walk almost everywhere, whether there was a path or not" and "whether because of litter raking, or . . . if a tree fell someone immediately made it into firewood." Maria went on: "when central heating was installed in most houses, everything, everything was abandoned, [now] a tree falls and stays in the woods, you can't pass any more and it is harder to put fires out." Many shepherds retired or died in the mid-1980s, and sheep have almost disappeared. This landscape is haunted by the absence of the people and animals who formerly grazed, raked, or burned excess vegetation, preventing large wildfires from spreading. A loss of coordination has resulted in disaster.

DIAGRAMMING DISASTER

The kinds of observational evidence I have described so far can be used to produce diagrams of social and ecological change across the landscape. By comparing my observations with early nineteenth-century tax records, I learned of the devastating impacts of plant diseases, a shocking finding that was beyond the reach of oral history or direct observation.

Ecologists often use line transects to study the relative abundance of organisms across an ecological gradient. By walking across the mountain range from one side to the other, I similarly tried to experience the broadest possible range of social and ecological relationships and temporal processes, following footpaths rather than a straight line. Rather than counting plants whose species were already known, I looked for the relationships and responses between people, plants, soils, diseases, and fires. I recorded the presence of trees, shrubs, and understory herbs, but I also described tree and soil morphologies through notes, drawings, and georeferenced photographs. Notebooks, sketches, and photographs allowed me to imagine the encounters that had produced the morphologies of trees or of terracing systems. This practice of walking and noticing was informed by walks across their land with smallholders and by oral histories with peasants and shepherds, but it ultimately depended upon my senses and my capacity to remain alert to surprise.

Oral histories and interviews gave me some sense of the changing cultivation practices and economies that had produced present-day landscape structures. I looked for longer-term histories in early nineteenth-century tax records in the Archivio di Stato of the City of Lucca. Nineteenth-century agronomical treatises and descriptions of cultivation practices told me of changing land uses and of fires and plant disease epidemics. Sometimes an old document would alert me to surprising details of landscape change and rewire my senses. When I returned to the countryside, things sometimes looked different.

By comparing land use from nineteenth-century tax maps of the former Republic of Lucca with my observations from landscape walks, I was able to build diagrams of past and present people–plant–soil relations across the landscape. In 2014 I walked from the village of Vorno at the base of the Monte Pisano to the top of the mountains, at Campo di Croce. Overlaying my landscape observations, field notes, and drawings from 2014 with geographically referenced tax maps from 1843, I built a diagram that showed dramatic landscape transformation (see Figure 1.8).[3] This comparison allowed me to project my capacity to notice a vast range of ecological relationships into a simpler story about long-term landscape change across the whole transect. Overlaying past maps and present observations allowed comparison, but it also silenced the uneven ways that fires, diseases, and people move through the landscape. Humans walk more slowly uphill, and footpaths follow the easiest ways to move across the landscape. Fires rush uphill and pause on

ridgelines. The spores of plant diseases travel with the winds or on the feet
of birds, sometimes traveling hundreds of miles, sometimes pausing at the
edge of a field. A spatial overlay made it harder to imagine these kinds of
mobilities. The resulting diagram confronted the bureaucratic simplifications of the tax map with my own, more open-ended curiosity. Out of the
many kinds of trees and plants that I might notice, only the categories of
land use that mattered to Lucchese elites and their peasant interlocutors could
be shown. Ecological complexity was boiled down to chestnut groves and
firewood forests; the bland term *pasture* at the crest of the mountain con-
cealed fraught struggles over access to grazing.

With all their limitations, bureaucratic simplifications allowed me to project my imagination to a larger spatial scale. As you can see, there has been a
wholesale change in the landscape across the transect between the base and
the crest of the mountain. In 1843, the landscape was dominated by cultivated *castagneto* (chestnut groves), with some *pineta* (pine), a smaller area
of *ceduo* (coppice) managed for firewood, and a communal pasture at the
top of the mountain. In 2014, the *castagneto* is almost entirely gone, *pino
marittimo* occupies many areas, and chestnut *ceduo* covers most of the rest.
This diagram pushed me to look for the history of a disease epidemic that
has left little trace in contemporary memory. This figure is a diagram of a
ghost; it records the cultivated chestnut forests that were destroyed by the
ink disease, *male del inchiostro,* which arrived in this area in the 1840s.

Contemporary accounts told me that the Monte Pisano was ground zero
for a largely forgotten disease epidemic that devastated low-elevation chestnut groves on the Monte Pisano and across Italy in the second half of the
nineteenth century (Gibelli 1876). The Oomycete water mold *Phytopthora
cambivora,* known as *male del inchiostro,* arrived in Italy in this area in the
late 1840s, likely as an unintended effect of the international trade in plants
(Mathews 2021). Affected trees lost their foliage and died within months.
Many chestnut groves were simply abandoned. In the Monte Pisano, chestnut
groves and firewood forests had become fire-prone pine forests by the 1880s
(*L'Agricoltore,* 1884). The economic and social dislocation caused by this
pathogen epidemic likely affected millions of rural people across the Mediterranean, but these events had little resonance in contemporary accounts.
Projects of industrialization and modernization, of expanding wheat cultivation, and of land reclamation saw chestnut cultivation as a backward and
doomed tradition, unworthy of notice. Another disease, the chestnut canker

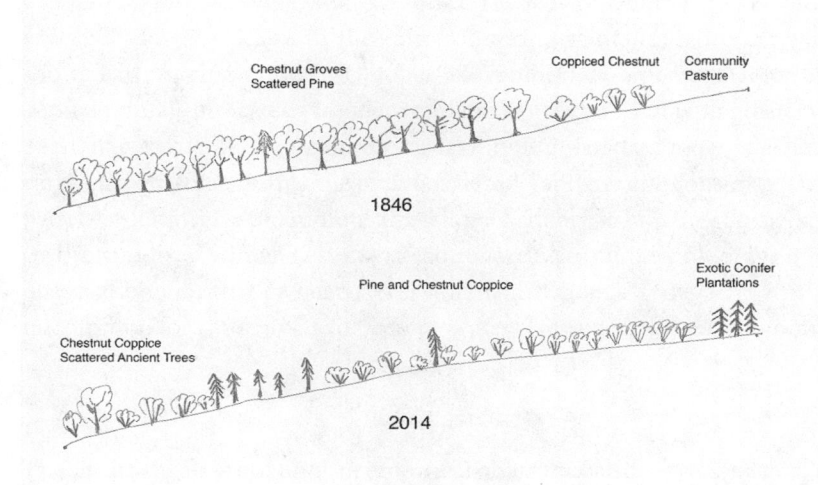

Chestnut Groves
Scattered Pine

Coppiced Chestnut

Community
Pasture

1846

Pine and Chestnut Coppice

Exotic Conifer
Plantations

Chestnut Coppice
Scattered Ancient Trees

2014

FIGURE 1.8. Transect from Cima di Vorno to the top of Monte Faeta, Monte Pisano, comparing dominant tree forms in 1843 and 2014. Drawing by Andrew Mathews.

Cryphonectria parasitica, struck a devastating second blow. Chestnut canker was reported in 1938 near Genoa and in forest nurseries near Lucca in 1946, and it spread rapidly across Italy (Borelli and Pettina 1958; Pavari 1949). Many remaining chestnut orchards were liquidated for tannin factories. Both diseases were likely introduced by live plants from East Asia, and it is likely that chestnut canker was introduced and spread by government plant nurseries in the 1930s (Mathews 2021).

While this transect diagram opened my eyes to the impacts of epidemic disease, it was interviews with present-day chestnut farmers that helped me make sense of ruins that I encountered deep in what was now pine forest. These were small, two-story drying sheds known as *metati*, where chestnuts were formerly dried. Farmers told me that *metati* had to be near or in chestnut groves because fresh chestnuts are heavy and because chestnut groves supply fuel for drying sheds. In Figure 1.9, Fabio Malfatti and I combined past and present, chestnut-drying sheds from an 1846 tax map with vegetation from a vegetation survey in 2000 (Bertacchi, Sani, and Tomei 2004). A view of the Monte Pisano from the northeast (Figure 1.9) shows numerous drying sheds (recorded on nineteenth-century tax maps) deep in present-day

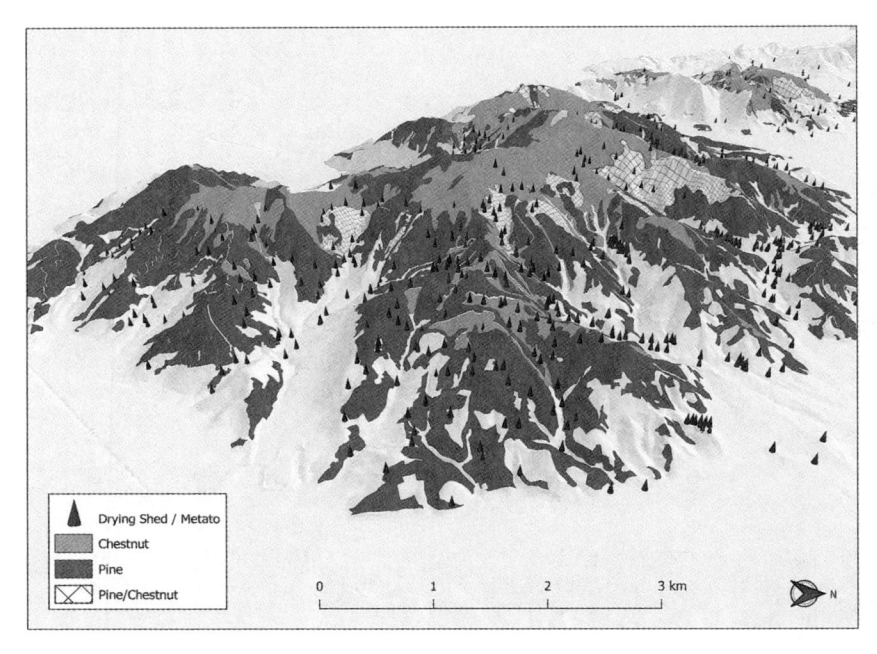

FIGURE 1.9. Present-day vegetation combined with nineteenth-century smoking sheds, Monte Pisano, Lucca, Italy. From Fabio Malfatti, 2020.

pine forest. Drying sheds are a trace of human–plant–animal coordinations that took place in cultivated chestnut groves before the arrival of ink disease. This map is a diagram that links past and present, my sensory practice of walking and looking, my interviews with farmers and foresters, and my visits to archives. Anthropologist Fabio Malfatti has accompanied me on many interviews and visits to the forest, and he has linked nineteenth-century land use and twentieth-century land cover through GIS analysis. The systematic properties of the cadastral map allowed us to make visible a large-scale and dramatic social and ecological transformation even as other details and coordinations disappeared.

DIAGRAMS, EXAMPLES, AND SYSTEMS MODELS

Maps and transect drawings are diagrams that emerge from my movement between the details of plant morphology, landscape structure, and history.

Diagrams can link different kinds of entities and processes, from nineteenth-century drying sheds to diagrams of tree form. I became confident that maps and diagrams had some grasp upon the world only through a gradual process of sensory attunement as I moved back and forth between archive and field, between map and interview. The open-ended curiosity of walking and of looking for evidence of continuously unfolding social and ecological change is in tension with the limited and structured curiosity of the archive. Nineteenth-century tax maps limited my stories to categories shared by early nineteenth-century tax officials and peasant farmers, but their systematicity allowed me to reach larger spatial scales.

The kind of simplified information found in tax maps can be used to tell global-scale stories about the amount of carbon absorbed by forests across the world over the last three centuries (Van Minnen et al. 2009). The systematic simplification of tax records resonates with the requirements of systems modelers who wish to simulate past and future landscapes. Models of the carbon stored in European forests in the seventeenth and eighteenth centuries ignore the vast diversity of cultivation practices and their successive transformations. Large-scale models cannot distinguish between pollarded poplars with grape vines, chestnut groves grazed by sheep, and coppice forests cut for firewood. The infrastructures of the tax archive and the systems model freeze entities and relationships, sustaining large-scale stories while also silencing a multitude of agrosilvopastoral mixtures. Systems accounts such as these find it difficult or impossible to follow changing relationships between beings or to detect rapid transformations caused by disease epidemics or agrarian change.

Systems modeling imposes necessary simplifications, freezing the entities that will be counted and the relationships between them. Systems modelers who tell stories about global-scale carbon absorption by forests have to keep the relationships between people, plants, and diseases fixed. New forms of cultivation or new diseases would make it impossible to calculate the carbon stored in forests and fields. Similarly, the infrastructure of historical weather stations limits the types of data and the relationships that climate change modelers used to detect global climate change (Edwards 2006). The giant computer models that helped us detect global climate change are necessarily unable to contemplate intertwined histories of economic and ecological exploitation and the surprising ways that people and plants respond to weather and to each other. The emergence of new social and ecological relationships and the transformation of landscapes are largely beyond the reach of systems models.

We can see the awkward relationship between global-scale climate models and human experiences of landscape change if we look at species distribution models that predict the locations of key plant species. Species distribution models use climate data to predict whether a species is likely to be found at a particular location. Species distribution models translate the typical one-hundred- by one-hundred-kilometer grid scale of global climate models down to a one- or two-kilometer scale. Such grids assume uniformity within each cell of the grid and make ecological and social details invisible. Although it is difficult to bring historical, ethnographic, and ecological accounts of landscape change in conversation with systems models, such awkward encounters can be informative. In a collaboration between two anthropologists, a historian, a climate modeler, and a botanist, we linked historical ecological evidence, historical climate reconstructions, and species distribution models of pine and chestnut across the twenty thousand hectares of the Monte Pisano (Casazza et al. 2020). Such difficult collaborations take time and goodwill (see the Introduction to this volume). In a preliminary conversation in 2014, Gabriele Casazza and I discussed a species distribution model that would require a one-kilometer grid of land use data across the entire Monte Pisano. Anthropologist Fabio Malfatti and I had to find a way of simplifying and sampling early nineteenth-century land use records. Intensive archival work by Fabio in 2017 sampled the entire Monte Pisano, and in 2019, we verified the locations of chestnut drying sheds in walks through the forest. Complex topography, landscape structure, plant morphology, and the myriad property holdings across the whole mountain range disappeared into a simple checker grid (see Figure 1.10).

The coarse scale of the grid was a productive simplification that allowed the botanist Gabriele Casazza to model the effect of climate change on plant distributions, drawing on historical climate modeling by Michele Brunetti (Brunetti et al. 2006). This large-scale story was qualitatively different from the histories of capitalism and international trade that I had previously linked to this landscape. Simplification allowed us to show the impacts of plant diseases and changed cultivation practices and the relatively less powerful role of climate change. In 1850, chestnut was found in many areas that were

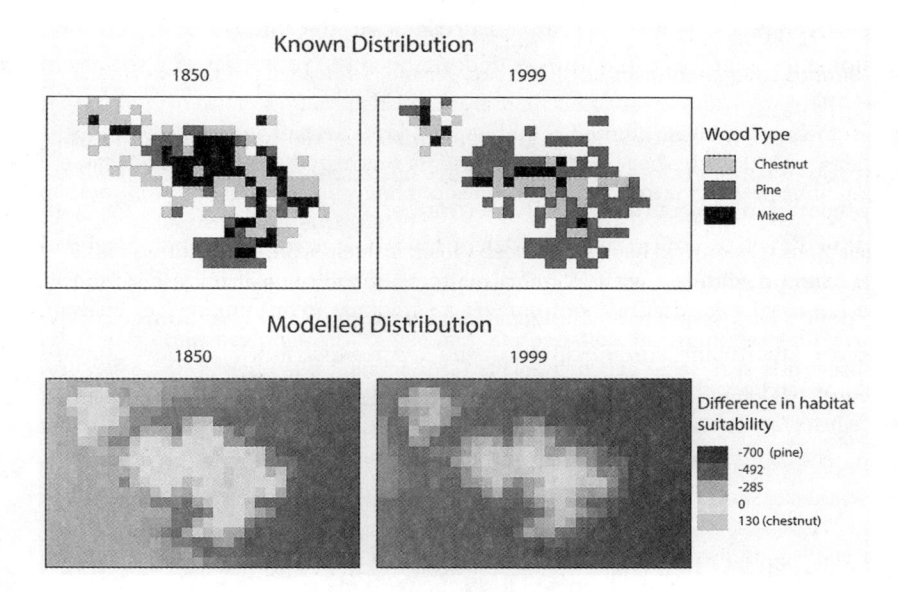

FIGURE 1.10. Known and modeled distributions of chestnut (light gray), pine (medium gray), and mixed (black) forest in 1850 and 1999. From Casazza et al. (2020). By permission Gabriele Casazza.

predicted by the model to be more suitable for pine (dark gray) than for chestnut (light gray). It was only after a century of climate change, land use change, and plant disease that chestnut was found where it was supposed to be according to the model—on the crests of the mountains.

Collaborations do not require agreement. Casazza feels that our results show that peasant cultivators formerly grew climatically stressed chestnut trees at low elevations. I prefer to think that pine and chestnut trees changed their behaviors as a result of relationships with people, diseases, and forest fires. The open-ended curiosity of interviews, drawings, and landscape walks had showed me how peasants could grow chestnut at low elevations by building retaining walls and raking leaf litter. The unintended arrival of *P. cambivora* in the 1850s was a new relationship that changed chestnut's response to climate. Only high-elevation trees that experienced cold winters were protected from the disease. The socioecological change of landscape abandonment in the 1960s caused the unintended accumulation of leaf litter and increase in fires. Expanding pine was the result of interactions between pines, people, and climate. In each case, the emergence of new relationships, many of them unintended, changed the responses of a species to climate.

This example shows how ecological and social details that we can notice through rubber boots methods can have large-scale consequences if we are prepared to project these details up to larger-scale simplifications. Through difficult collaborations with modelers, I was able to link morphological and landscape evidence with global climate change to tell a story about the unintended consequences of international trade and land abandonment. The systematicity and simplification of the species distribution model allowed us to see how climate change alone could not explain changing species distributions. It was through the work of noticing morphologies, landscape patches, and landscape structures that I learned that changing forms of landscape care and disease epidemics had the largest effect on the ranges of key plant species.

The finding that changing cultivation practices and diseases can drastically change species distributions has implications for how we might respond to climate change. Much climate change science and policy focuses on climate change impacts and pays little attention to the ways that people might care for the landscapes in which they live or for the possibility of emerging diseases and ecological relations. In contrast, the example of the Monte Pisano suggests that people can care for landscapes and make them less vulnerable to climate change and that emerging diseases and ecological relationships can be more damaging than climate change alone. Alternatively, we might at least imagine that plants could find ways of adapting to changing climates and new diseases that models cannot predict. The landscape could be made less fire prone by introducing fire breaks, by bringing back controlled grazing, or by cutting back vegetation near buildings. These potential activities involve political institutions that people can call to account, and they focus on their experiences of environmental change. Rather than addressing the national and global scales of climate change science and policy, local residents can think about what municipal leaders and local groups might do. The example of the Monte Pisano is a resource for thinking about alternative climate change policies. Controlling the international trade in live plants could be reimagined as a climate change mitigation policy. Reducing the frequency with which pests and pathogens impact ecosystems will make them more resilient in the face of climate change. This example is also a warning. It suggests that rapid social and ecological change can produce unanticipated landscape transformations and that the carbon accounts and budgets through which policy makers try to slow climate change may not be so accurate. Fires, epidemic diseases, and social change could transform landscapes in unexpected ways. Such examples remind us that the world is less predictable than planners and policy makers often imagine. The Covid-19 epidemic abundantly demonstrates the impacts of emerging diseases and the changing social relations that they bring into being.

Conclusion

In this chapter, I have described how landscape walks allowed me to become attuned to the strange forms of long-lived trees and how we can move from the curiosity of such sensory engagements to larger-scale landscape structures. By moving back and forth between tree morphology, archive, and interview, between map and landscape walk, I gradually became attuned to the larger-scale landscape structures of slowly expanding pine and chestnut forests and of mobile soils and unstable hillsides. Close attention to intertwined social and ecological transformations allowed me to think about the biographies of trees and the histories of economic exploitation and care that had brought larger landscape structures into being. My capacity to tell histories of social and environmental change departs from noticing the shape-changing capacities of plants and the geomorphological transformations produced by peasant cultivators who reshaped hillsides and drainage systems. This form of critical description and analysis builds from curiosity about the transformations caused by relations between beings, from wondering about what encounter caused the morphology of this tree or what history produced that patch of fire-blasted scrub. Each comparison across patches can make me see a different landscape structure and could help me imagine a different history. Speculative noticing supports robust but always provisional accounts of regional environmental change.

The systems models that allow scientists to detect large-scale environmental change are enormously helpful. It was global climate models that helped us detect global climate change in the first place. Systems models, however, impose limitations on the kinds of things that can be noticed, and they struggle with the sudden changes and transformations that we can notice though rubber boots methods of sensory curiosity and historical ecological methods of reconstructing landscape histories. In my brief sketch of species distribution modeling, I outlined the ways that systems models project human sensory curiosity to larger scales in time and space. I explored some of the differences between systems models and arts of noticing, landscape analysis, and history. This tension between systems models and landscape analysis is both productive and problematic. Ecological modeling necessarily freezes the entities that will be counted and the relationships between them, but it allows a bold and simple story to be told. At the cost of freezing our analytic frame to simple land use categories, my collaborators and I were able to show that disease and land use change could be more important than climate change in affecting landscapes, and we were able to extend my observational evidence to a larger spatial scale.

Systems models and landscape analysis meet at a deeper level. Both are concerned with the relationship between detail and form. Landscape structures, transect diagrams, maps, and species distribution models are all forms that are projected from human experience. Speculative noticing is ultimately at the heart of species distribution models, which rely on vegetation surveys; of climate models, which rely on meteorological data; and of landscape analysis, which builds on curiosity about morphology. These types of analysis differ in *how* noticing is projected into larger forms. As I have shown, landscape analysis builds on openness to surprise, to the possibility that the next tree or the next conversation might teach us something different. Such surprises can be extended to larger scales through diagrams, maps, and historical accounts. These larger forms allow us to make partial sense of a world of process, as long as we are willing to project our imagination from our senses to larger scales in time and space. We need both systems models and rubber boots methods of speculative noticing to make sense of our contemporary environmental catastrophe. We need many ways of projecting sensory curiosity and speculation into larger analytic forms that partially grasp the liveliness of the world.

I offer you a final image that may stay with you as a persuasive story, a form that might haunt your imagination. Consider this ancient chestnut tree high up in mountains, grafted perhaps six hundred years ago by peasant cultivators (Figure 1.11). In the mountains of the Mediterranean, the

FIGURE 1.11. The ancient chestnut tree of Pratofosco. Photograph by the author.

most ancient trees are often grafted chestnut trees like this one. Without human care, this tree would not be here. The bulges, loops, and whorls on the trunk are the echo of the moment half a millennium ago when a peasant farmer skillfully aligned the cambium layers of scion and rootstock into an unstable graft. This ancient tree shows not only past cultivation but ongoing care. Without repeated pruning of new shoots from the rootstock, the "wild" shoots would take over, and the cultivated variety would die. I recount this story and show you this image not only for their beauty and strangeness but as an example and a warning. Slow and long-lived beings like this chestnut tree have demonstrably been able to compel humans to take care of them for a very long time. The morphologies of trees can place a call upon us to respond, to be responsible, and to care for them. Perhaps we can plan for the long term by attending to the shapes of trees and by caring for our surroundings. Care is not an easy relationship: pruning and grafting are violent processes whose consequences are never fully known. This tree is also a warning, a refugee from the unintended disease introductions of the last two centuries, a survivor of forests that were liquidated for tannin in the 1950s. If the example of long-term human care for ancient trees shows that humans can care for their environments over the long term, it also warns us that care is an ambivalent and provisional process. This is a humble kind of hope.

Notes

1. Interviews with sixteen elderly former farmers and shepherds in 2016 and 2019 confirmed that leaf litter raking was a ubiquitous practice around the Monte Pisano.

2. *Pino marittimo (Pinus pinaster)* is native to Portugal and North Africa and was introduced in this area in the sixteenth century for timber and to stabilize coastal dunes.

3. The cadastral tax maps of the Duchy of Lucca (Catasto Nuovo Lucchese) and the Grand Duchy of Tuscany (Antico Catasto Toscano; also Catasto Leopoldino) required landowners to report land use for every tax parcel. Land use for the valley of Vorno was recorded in 1843.

References

Bertacchi, Andrea, Alessandra Sani, and Paolo Emilio Tomei. 2004. *La vegetazione del Monte Pisano.* Pisa: Felici.

Borelli, O., and A. Pettina. 1958. "Le Malattie del Castagno." In *Studio monograficao sul castagno nella provincia di Lucca,* edited by O. Borelli, N. Breviglieri, C. A. Cecconi, C. Ciampi, R. Morandini, and A. Pettina, 169–73. Florence, Italy: Centro di Studio Sul Castagno.

Borges, Jorge Luis. 1964. "Funes the Memorious." In *Labyrinths: Selected Stories and Other Writings,* 148–54. New York: New Directions.

Brunetti, Michele, Maurizio Maugeri, Fabio Monti, and Teresa Nanni. 2006. "Temperature and Precipitation Variability in Italy in the Last Two Centuries from Homogenised Instrumental Time Series." *International Journal of Climatology* 26, no. 3: 345–81.

Canfield, Michael R., ed. 2011. *Fieldnotes on Science and Nature.* Cambridge, Mass.: Harvard University Press.

Casazza, Gabriele, Fabio Malfatti, Michele Brunetti, Valentina Simonetti, and Andrew S. Mathews. 2020. "Interactions between Land Use, Pathogens, and Climate Change in the Monte Pisano, Italy 1850–2000." *Landscape Ecology* 36: 601–16.

Cronon, William. 1992. "A Place for Stories: Nature, History and Narrative." *Journal of American History* 78, no. 4: 1347–76.

Edwards, Paul N. 2006. "Meteorology as Infrastructural Globalism." *Osiris* 21, no. 1: 229–50.

Evans-Pritchard, E. E. 1940. *The Nuer: A Description of the Modes of Livelihood and Political Institutions of a Nilotic People.* New York: Oxford University Press.

Giannini, Raffaello, and Antonio Gabbrielli. 2013. "Evolution of Multifunctional Land-Use Systems in Mountain Areas in Italy." *Italian Journal of Forest and Mountain Environments* 68, no. 5: 259–68.

Gibelli, Giuseppe. 1876. "Malattia del castagno." *Gazzetta ufficiale del regno d'Italia,* 258–64.

Grove, A. T., and A. T. Rackham. 2001. *The Nature of Mediterranean Europe: An Ecological History.* New Haven, Conn.: Yale University Press.

Hallé, Francis. 2018. *Atlas of Poetic Botany.* Cambridge, Mass.: MIT Press.

Ingold, Tim. 2011. *Being Alive: Essays on Movement, Knowledge and Description.* Vol. 1, online resource. Hoboken, N.J.: Taylor and Francis.

L'Agricoltore: periodico mensuale del comizio agrario Lucchese. 1884. "Sulla e Ginestra." Anno XX: 173–78.

Mathews, Andrew S. 2011. *Instituting Nature: Authority, Expertise, and Power in Mexican Forests.* Politics, Science, and the Environment. Cambridge, Mass.: MIT Press.

Mathews, Andrew S. 2018. "Landscapes and Throughscapes in Italian Forest Worlds: Thinking Dramatically about the Anthropocene." *Cultural Anthropology* 33: 386–414.

Mathews, Andrew S. 2021. "The Echoes of Exotic Diseases Are Visible in Italian Forests, If We Know How to Look for Them." In *Feral Atlas: The More-Than-Human Anthropocene,* edited by Anna Tsing, Jennifer Deger, Alder Keleman, and Feifei Zhou. Stanford, Calif.: Stanford University Press. http://www.feralatlas.org/.

Palahi, M., R. Mavsar, C. Gracia, and Y. Birot. 2008. "Mediterranean Forests under Focus." *International Forestry Review* 10, no. 4: 676–88.

Pavari, Aldo. 1949. "Chestnut Blight in Europe." *Unasylva* 3, no. 1.

Squatriti, Paolo. 2013. *Landscape and Change in Early Medieval Italy: Chestnuts, Economy, and Culture*. Cambridge: Cambridge University Press.

Taussig, Michael T. 2011. *I Swear I Saw This: Drawings in Fieldwork Notebooks, Namely My Own*. Chicago: University of Chicago Press.

Tsing, Anna, and Elaine Gan. 2018. "How Things Hold: A Diagram of Coordination in a Satoyama Forest." *Social Analysis* 62, no. 4: 102–45.

Tsing, Anna Lowenhaupt, Andrew S. Mathews, and Nils Bubandt. 2019. "Patchy Anthropocene: Landscape Structure, Multispecies History, and the Retooling of Anthropology: An Introduction to Supplement 20." *Current Anthropology* 60, no. S20: S186–97.

Van Minnen, Jelle G., Kees Klein Goldewijk, Elke Stehfest, Bas Eickhout, Gerard van Drecht, and Rik Leemans. 2009. "The Importance of Three Centuries of Land-Use Change for the Global and Regional Terrestrial Carbon Cycle." *Climatic Change* 97, no. 1: 123.

Viveiros de Castro, Eduardo. 2019. "On Models and Examples: Engineers and Bricoleurs in the Anthropocene." *Current Anthropology* 60, no. S20: S296–308.

CHAPTER 2

Interpreting Dwarf Shrub Patterns
in the Lesotho Highlands

COLIN HOAG

A hillslope in the highlands of Lesotho is dominated by a few species of dwarf shrubs, with grasses and other herbs growing in the interstices. Some time ago, the situation was reversed, with dwarf shrubs in the minority. What is the cause of this shift in ecological pattern? And is that question a humanistic one, or strictly a scientific one? Let's consider its substance.

The pattern has political stakes for both plants and humans. The dominance of dwarf shrubs typically diminishes opportunities for herbs, diminishing the total richness and diversity of plant species there. The abundance of shrubs also diminishes the amount of forage available to livestock who graze in the area, and livestock production is hugely important to families in the highlands, where people raise cattle for plowing agricultural plots, as well as merino sheep and angora goats for their wool, mohair, and meat.

The pattern of shrubs and grasses arrayed on the hillslope also has semiotic stakes, in that its meaning is a matter of debate among people with whom I spoke during field research, including livestock owners and government conservation bureaucrats. The former emphasized the role of changes in the rains in promoting shrub encroachment, while the latter emphasized livestock grazing. In their accounts, the two groups used different parts of the shrub's body to theorize different causes of landscape change. In doing so, they indicated that I needed to understand plant ecological relations to assess the veracity or implications of their accounts. The two interpretations not only offered a landscape history but also implied different solutions to the problem of shrub encroachment. Clearly this work of critical description and landscape historiography is interpretive—that is, it involves the careful scrutiny of traces and signs for possible meanings and the curation of those possibilities in a story. Even in the most icy-cold phrasing of positivist natural science, as I will show, the gap between sign and referent, data and exegesis, yawns.

The pattern is most definitely "ecological"—concerned with nonhuman nature and therefore the purview of ecological science. But it also bears some relation to human practices in that the Lesotho highlands are sites of significant human disturbance through livestock production, road construction, pasture burning, and more. Human disturbance suffuses plant community dynamics, suggesting that plant ecology can at times demand attention to human ideas and practices. This does not mean that natural environments are mere reflections of human practice (see Caple 2017). Interspecific relations between nonhuman organisms matter, and the vast majority of these interactions have nothing to do with humans. In the case of Lesotho's highland pastures, however, human disturbance is certainly in play, muddying the distinction between cultural and natural histories.

The pattern is therefore clearly of humanist concern. But to what extent can it be interpreted by humanist techniques? The shrub, after all, is more than a symbol or an occasion for disagreement—it is a living being, with actions that have a bearing on the ecological pattern of which it is a part. Rubber boots methods might require the empirical and analytical tools of conventional ecological science alongside those of ethnography and others from the humanities. If Clifford Geertz (1973) taught anthropologists to envision the discipline not as a scientific one in search of universal laws, but as an interpretive one in search of meaning, what does one do with biophysical processes? Given that the methods of ecological science are underwritten by a positivist interest[1] in universal laws—a predictive science—what use are they to anthropologists?

This essay, like other contributions in this volume, asks what it might mean to practice ecology humanistically—without reducing the human to a merely "natural" actor nor reducing the shrub to a mere "symbol" in human cultural production. What might a critical, interpretive ecology (see also Lave, Biermann, and Lane 2018; Caple 2017) look like? Conversely, what might it mean to practice environmental humanities ecologically, in such a way that it becomes legible to natural scientists? In an effort to better understand the problems posed by these questions, I scrutinize some empirical work I have done to interpret shrub patterns in the Lesotho highlands. This is an experimental essay, exploring the possibilities and pitfalls of an interdisciplinary science that takes positivist ecology and critical anthropology equally seriously. I argue that drawing together semiosis, materiality, and history is key to this critical ecology project and that, in dialogue with Mathews (2016; see also chapter 1), the concept of "form" can serve as a boundary object (Star and Griesemer 1989) to facilitate such a project.

The essay works across conventions not only of empiricism and analysis but also of genre. The jargon I employ borrows from ecology and anthropology, and the structure and argumentation address both audiences. I employ the putative transparency of natural science reportage as well as the narrative reflexivity and political concerns of the humanities. In toggling between positivism and reflexivity, my aim is to test the possibility that ecological science can be both a site of cultural production worthy of critique and an empirical project worthy of use. After providing some political and ecological context for my discussion, I show how two groups of research subjects differently interpreted the pattern of shrub dominance in the Lesotho highlands, both doing so through reference to different aspects of the shrub's bodily form. I then describe the methods I employed to address questions posed by those contradictory accounts. Having initially sought to determine which account was more accurate, in line with a positivist ecological science that strives toward universal truth, I came to realize that each account explained distinct aspects of ecological process. That is, my inquiry into the ecological substance of these two claims revealed that each illuminated different moments in the history of the landscape. While the question of which account was right is an important one, this research experience showed me how humanistic attention to the materiality and semiosis of shrub form can illuminate the conditions under which specific landscape histories are true and the ways that history shapes ecological process.

After all, ecological formations, like social formations, are of their time— they are materially, historically, semiotically specific. Excavating my informants' narratives and ecological science approaches to shrub practices, I try to render sensible the ecological laws that structure plant intimacies and the world histories that make those laws matter.

THE SETTING

Lesotho is a mountainous country completely surrounded by South Africa. The Drakensberg and Maloti Ranges run directly through the country and cover some 85 percent of its territory. Even Lesotho's "lowlands," where most of Lesotho's population lives, are quite high, with no point below one thousand meters above sea level (masl). Geologically, the lowlands are part of a sandstone formation, while the highlands are derived from basalt, giving the two regions distinct flora and geomorphology. In addition, the highlands are much wetter and colder, with some areas featuring as much as twelve hundred millimeters of rainfall annually and persistent snow on ridgelines during wintertime. The highlands (2,000–3,184 masl) are naturally grassland, though settlers and conservationists have long accused Basotho (i.e., people from Lesotho) of having wantonly destroyed all their forests (Showers 2006).

Lesotho maintained its sovereignty by deftly negotiating the colonial period—both settler colonialism by the white Europeans who would establish the white ethnostate of apartheid South Africa and the administrative colonialism of British indirect rule. Afterward, the neocolonialism of structural adjustment and development (Ferguson 1994) rendered the country largely powerless to the regional political economy. Lesotho has become a periphery to the South African core (Wallerstein 2004; Hoag, forthcoming). The country currently resembles one of apartheid's "Bantustans," the ethnic homelands that were established as part of South Africa's labor reserve system (Ferguson 2006), with generations of Basotho leaving home to work in the mining, construction, and domestic help industries in South Africa. Back home, many people keep livestock to make ends meet.

With the turn to democracy in South Africa, work abroad dried up for citizens of Lesotho, as the South African government favored its own workers over foreign ones. Livestock production, especially of small stock (sheep and goats), which generate cash from wool or mohair and can be sold over the border at a moment's notice, shifted from a supplement to a primary income (Hoag, forthcoming). Hence shrubs encroaching into these important grasslands is a problem.

What Is a Dwarf Shrub?

It would be helpful here to clarify natural science jargon used in this essay as well as the natural science consensus about the phenomenon of shrub encroachment. The term *dwarf shrub,* sometimes called a *subshrub,* is a functional type. A "functional type" is a nontaxonomic category that groups species according to their form rather than their evolutionary lineage. A dwarf shrub in this sense refers to a plant with a woody stem, no taller than two meters in height at maturity (dwarf shrubs are distinct from shrubs primarily by size). "Tree" is another functional type, meaning woody vascular plants taller than around six meters at maturity and that emerge from a single primary stem (the trunk). "Herbs" or "herbaceous plants" refer to nonwoody vascular plants, including grasses. Dwarf shrubs occur in most parts of the globe, from arid rangelands in Africa to tundra in the Arctic, often in nutrient-poor soils. They are particularly well represented in southern Africa, where they are found in dramatic diversity. In addition to the dwarf shrub communities of the Western Cape known as *fynbos,* several other shrub-dominant

FIGURE 2.1. The dwarf shrub *Chrysocoma ciliata.* Photograph by the author.

communities are found, including *renosterveld,* karroid shrubland, and Afro-montane shrubland (Mucina and Rutherford 2006). Lesotho's dwarf shrubs inhabit what plant biogeographers call the Drakensberg Alpine Centre (Carbutt 2019), located in mountains that form the boundary between Lesotho and South Africa, a site of endemism and also speciation for plants that occur as far north as East Africa.

Like other woody plants in savannas, shrublands, and grasslands where fire occurs with some regularity, dwarf shrubs sometimes feature horizontal woody subterranean branches with meristems (i.e., stem cells) that enable them to regrow even after losing all their photosynthetic tissues, an adaptation that also allows them to survive freezing, dry winters in Lesotho. Lesotho's dwarf shrubs are mostly from the Asteraceae plant family. Though scientists know little about how these shrubs are pollinated in Lesotho, small flies are probably key, if other alpine systems are an indication. In Lesotho, seed dispersal is probably mostly abiotic (i.e., wind and water) for shrubs as well as grasses and sedges at high altitudes in the Drakensberg (Steve Johnson, pers. comm., 2015) and also probably occurs by graminivorous

FIGURE 2.2. *C. ciliata* establishing on the roadside. Photograph by the author.

birds, such as siskins and canaries (Clinton Carbutt, pers. comm., 2014). The Asteraceae family is known for its high diversity, for its cosmopolitan distribution, and for producing large numbers of seeds on a single plant. Aster "flowers" are actually a composite inflorescence, often featuring dozens of individual florets. Most feature a pappus with fine hairs, a structure attached to seeds that aids in dispersal (as with dandelions). In the early summer in the Lesotho highlands, seeds of *Chrysocoma ciliata*—one of the most common shrubs—can be seen flying around rangelands in abundance. Some Asteraceae shrubs are early successional species that grow rather quickly in disturbed sites. For example, I spotted *C. ciliata* growing on a berm for a new road at almost 0.75 meters tall just two years after construction had begun, and it is not uncommon to see them colonizing roadside space (see Figure 2.2).

What Causes Shrub Encroachment?

Shrub encroachment is a new and old problem in Lesotho. Concern first arose in the early twentieth century, when British conservationists from the colonial administration claimed that the some 14 percent of the highlands had been invaded by shrubs (Driver 1999).[2] Most people with whom I spoke, however, suggested that they had become a problem in the past twenty years (also see Mokuku 2004). Conservation bureaucrats and livestock owners in the Lesotho highlands during my field research were both concerned by the problem of shrub encroachment. Both groups engaged the shrub as an indicator species of rangeland degradation and decline. In theorizing landscape change, however, they focused on different aspects of the plant's body and drew different conclusions about the causes of shrub encroachment. I turn to these differences now.

Conservation Bureaucrats: The Livestock Problem

"Have you noticed any changes in the mountain rangelands over the years, or do they look more or less the same?" It was a question I asked many people during field research. The man with whom I was speaking was a former civil servant named Bokang who had worked for many years on rangeland issues.[3] We were sitting in a breezy outdoor café in the capital, Maseru.

He said, "Yes, definitely. They are badly degraded today—much worse than when I first began working. When you go to the mountains, you see *sehala-hala* everywhere—this 'bitter Karoo bush.' And you know what the Karoo desert is like. The shrubs are a sign of desertification." That response was one that I would hear regularly from conservation bureaucrats in government and foreign nongovernmental organizations. He carried on telling me about the history of grazing associations in Lesotho and their importance to rangeland improvement. To him, management had been the critical problem burdening Lesotho for decades: that livestock exceeded the carrying capacity of pastures and that local chiefs were unable to control grazing adequately.

The symptom of this problem was the shrub *C. ciliata* (known in the Sesotho language as *sehala-hala* and in English as bitter Karoo bush). The shrub was not only a symptom—a sign—but also a cause. Specifically, Bokang and other conservation bureaucrats referred to two related features of the plant as enabling its spread: its palatability and its competition for light resources. They explained that, because the plants tasted badly to livestock—

FIGURE 2.3. Conservation bureaucrats on the roadside. Photograph by the author.

even goats more or less avoided them—they were given a competitive advantage for light resources when the grasses around them were eaten. The toughness and sour taste of the shrubs' leaves and their height relative to grasses around them called forward an ecological story about their spread that made their increasing dominance and the effects of livestock almost self-evident. It was a landscape history that Bokang told, through reference to shrub practices, situated in relation to particular political and historical tropes (e.g., Karoo desert). And he felt the problem was solvable through better management, the focus of his department.

HERDERS AND LIVESTOCK OWNERS: THE RAIN PROBLEM

By contrast, herders and livestock owners told landscape histories through reference to shrubs' evergreen leaves and woody stems. A walk I took with a man named Ntsupa through one shrub-encroached valley was illustrative of this. Ntsupa and I stood on the ridgeline looking out over the valley, a so-called cattle-post area *(motebong)* where people keep livestock (mostly

FIGURE 2.4. Landscape vista with Ntsupa. Photograph by the author.

sheep and goats) during the winter months. I had asked him to hike with me through the area to tell me about the landscape's history—whether it had changed, how it was managed, and so on. In the lower reaches of the valley, dense stands of shrubs created a homogeneous mass of vegetation. But they thinned out as they reached the ridge, where we stood. Ntsupa explained to me that the entire area used to be primarily grassland just twenty or so years ago. Shrubs had always been present, he said, but they were confined to small patches along the river or in the small valley coves.

We came upon a herder whom Ntsupa knew. He seemed as old as Ntsupa, perhaps older, with a herder's blanket and a wide-brimmed, woven-grass hat. He and Ntsupa were clearly old friends, both leaning on their walking sticks as they made some joking small talk. The man asked us what we were doing, and I explained that I was interested in learning about changes in the landscape. We stood knee-high in shrubs—mostly *sehala-hala* on this north-facing slope. He agreed with Ntsupa about the recent encroachment. What is the reason for their spread? I asked. *Pula e lesiko,* they said almost in unison—there have not been rains. The rains used to come much earlier, they explained—in October, at the beginning of spring, rather than in December or January, well into summer. When these rains do finally arrive, these days they fall as destructive thunderstorms rather than steady rains that fall over many hours and percolate into the soil. Nor has there been as much snow during winter, Ntsupa added. In the past, snow would fall and melt during the daytime, giving moisture to the grass over the course of the winter. They explained that, with drought, the shrubs have an advantage: their hardy roots hold water when the grasses have nothing. Being evergreen, too, they can photosynthesize and use that stored moisture to grow throughout the year. He hit a shrub with his stick as though to emphasize the stiffness of its stem. Plus, the man added, all their leaves can die off and their roots can still regenerate, giving them additional resistance to drought.

FORM AND FABULATION

My informants' landscape historiography focused on plant practices and plant response to disturbance or shifting environmental conditions—they theorized plants as agents. Because of historical commitments within the social sciences to bracket nonhuman relations from "the social" (Tsing 2013), the question of plant action is typically left to ecologists. Anthropologists are authorized instead to interrogate plants' symbolic function—say, as "indicator species" used by humans to understand or intervene in the world. To be sure, shrubs are seen as indicators of rangeland decline. But more than symbols, they are living things. They stretch out into sun space, they poison neighboring plants with allelochemicals, and they soak up soil moisture. This is not so much "new materialism" as it is simply "materialism," or the recognition of the fact that plants are alive. Can anthropologists restrict themselves to investigating only the human symbolic work in landscape historiography, *even when our informants are raising plant practices for us*? What would be the politics of such a narrow multispecies anthropology? What might it mean to investigate shrub ecology humanistically?

Anthropology and ecology seem natural partners. Both blossomed in the early twentieth century, at a time when holism and the interconnectedness of institutions or ecological processes came to concern scientists and the fields sought each other out for tractable metaphors to understand the phenomena they investigated (Worster 1994; Otto and Bubandt 2010). Race, superorganism, community, and culture all bear these traces. Despite their mutual interest in understanding connection, relation, and sociality, the fields are very dissimilar in their approaches to data, theory, and representation; mainstream anthropology is an interpretive practice, and mainstream ecology is a positivist one.

Building on the work of others, I suggest that the concept of "form" can draw history and power into thinking about ecological process and provide a language for articulating how the specific materialities of shrubs inform their contact with environments in historic, embodied, and symbolically informed ways (Kohn 2013; Tsing 2013, 2015; Mathews 2018; Hustak and Myers 2012). Anthropologists use form as a means of thinking about the semiotic: indication, resemblance, typology, and figuration, for example, dwarf shrub as a portent of ominous national futures. Ecologists use form as a means of thinking about functional traits, that is, how an organism's physiological, morphological, or phenological properties indicate strategies for survival in ecological space. From Raunkiaer's (1934) "life-forms" to Westoby's (1998) L-H-S scheme, ecologists have sought to develop a language for understanding how plants differ in their formal abilities to photosynthesize, transpire, and propagate, with implications for how they might thrive under some conditions and not others.

"Form" contains within it both these significations, deferring conflicts between interpretivism and positivism. As with other boundary objects (Star and Griesemer 1989), it allows for collaboration without harmonious agreement between partners. Accordingly, it can operate as a ground for doing ecology humanistically: to interrogate the narratives of our informants to learn how they are shaped by historical and political circumstance and, in doing so, use these narratives as vantage points for seeing different material processes. It is in the movements across significations of plant form that the work of landscape historiography is done: the work of speculation and fabulation (Haraway 2016), the work of ecological theorizing. In this, I follow Mathews (2016; also chapter 1), who shows how the forms of trees point to a history of interaction with humans, soils, parasites, and more. I focus more narrowly on how two aspects of shrub form—shrub leaves and shrub roots—are deployed in analyses and narratives of landscape change. They feature not simply as traces to be read but as mechanisms of change—explanatory tools. These aspects, it turns out, would explain different moments in the landscape's history.

In the next two sections, I describe the methods I used to sort through these conflicting accounts, working to understand my informants' ecological fabulation—their movement from plant form to landscape historiography—while also making such a movement myself.

An Anthropology of Conflicting Accounts

When I went to Lesotho for my preliminary field research, I was a diligent anthropologist. I observed and documented land management practices, including the relevant legal frameworks and their interpretation or application; I hung out with herders and livestock owners; I learned who has access to which parts of and products of rangelands; I hiked through rangelands and made observations about them; I asked people like Bokang and Ntsupa how rangelands had changed over the years. I hoped to approach Lesotho's rangeland history as a more-than-human one, but to do so with some accountability to ecological science methods and audiences, having heard the many calls for an interdisciplinarity that could explode conventional boundaries between the humanities and natural sciences to understand and remedy environmental change in the Anthropocene (Chakrabarty 2009). Having been presented with these conflicting accounts of environmental change, I was not content only with presenting the cultural or political motivations for one account or another. I looked deeply into the ecological science literature on the drivers of change in rangelands in the hope of sorting through claims about Lesotho's rangeland degradation.

No existing literature could answer the question for me. In fact, research in ecological science provides support for both perspectives on the cause of shrub encroachment. As described by conservation bureaucrats, the aboveground competitive advantage of shrubs relative to grasses in grazed pastures is a documented phenomenon (Lett and Knapp 2003). Ecologists, though, have also documented shrubs' ability to access comparatively more of the available soil moisture than grasses—particularly in the lower subsoil—through their extensive root systems (Breshears and Barnes 1999). From the 1980s onward, authors cast doubt on the impact of livestock grazing on land conditions in semiarid and arid ecosystems, suggesting that in such systems, the high coefficient of variability in interannual rainfall means that climate is the primary determinant (Archer et al. 2017). Lesotho's highlands are a unique ecosystem with few analogs, however, having a high coefficient of variation but a mesic (rather than semiarid) rainfall regime (Hoag and Svenning, n.d.). Generally speaking, understanding of shrub encroachment is still limited (Ward 2005), but multiple factors are almost always determinant of vegetation structure and composition (Archer et al. 2017).

In the year before heading to the field as a doctoral student in anthropology, I enrolled in courses in soil science and rangeland ecology, and when an opportunity presented itself, I enrolled in a PhD program in biological sciences with the interdisciplinary research program Aarhus University Research on the Anthropocene (AURA). This opened up my project to many new possibilities but put me in an awkward position. I became responsible to both the interpretive social sciences and the positivist ecological sciences. What conventions of genre and empiricism would govern me? How would I draw ecological methods into anthropology without sacrificing anthropology's commitments to critique, multivocality, and subaltern perspectives?

Efforts to synthesize social and natural sciences into a single intellectual program are not new. Ecological anthropology sought to explicitly integrate ecological principles (e.g., Harris 1971) but employed a functionalism that rendered history, change, and conflict invisible (Asad 1973). The "ethno-" sciences of ethnobotany, ethnobiology, and others have similarly attempted such a synthesis, adopting the genre conventions of scientific quantification and representation while being attentive to the cultural differences in people's use and naming of the natural world. These approaches tend to appeal to the positivist sensibilities of natural scientists; yet, they leave little room for humanist interpretivism, while sometimes ignoring the lesson from science and technology studies that mainstream science, too, is an "ethnoscience."

Political ecology introduced a sharper critical sensibility to this synthesis, showing that natural environments are contested—both their ownership and their definition. Yet, it has become more concerned with establishing itself as a voice outside ecology, critiquing ecology's tropes and biases rather than trying to incorporate its methods of observation and analysis (see Walker 2005). Multispecies ethnography has begun to build an alternative space, decentering humans and approaching nonhumans as subjects rather than objects of interpretive anthropological research. However, important questions are left to be answered regarding the position of natural science and its ways of seeing or explaining, some of which this edited volume aims to address. Can ecological science be read critically as a site of cultural production, while also being used as an empirical project (see Swanson 2017)? One way it could, I suggest, is to work from "form," holding multiple theorizations of landscape change in tension to scrutinize their historical situatedness and their vantage on biophysical process.

THE CONTINGENCY OF ECOLOGY

Humanists know to be skeptical of "capital S" science "vision from everywhere and nowhere," as Donna Haraway (1988, 584) has put it, having demonstrated how knowledge is situated—always culturally, historically, politically specific. African landscape change—and the role of Africans in driving it—has been misunderstood by scientists repeatedly, sometimes because of racist or antipoor biases and sometimes with serious political implications for land access or management (Leach and Mearns 1996). But recognizing that science is situated cannot be the end of the story, and I decided that I needed to faithfully employ ecological science methods. After all, ecologists have paid painstaking attention to the ways that nonhumans act within social fields through relations of predation, mutualism, parasitism, and competition. Surely their methods could be useful to multispecies anthropologists.

To examine the relative importance of different possible determinants of shrub density in Lesotho's pastures, I set up a study in the area where I walked with my informants (see Hoag and Svenning, n.d.). In dialogue with my collaborator and advisor (Jens-Christian Svenning), I delimited a study area of approximately twelve square kilometers in ArcMAP and generated a set of random points. At each of them, I measured shrub density—the percentage of shrub cover in a five square meter area—and I also measured some soil and topographic properties, such as soil depth, soil moisture, and soil nutrient loads; slope; aspect; elevation; and hill position. I then mapped the locations of heavily grazed areas, like the cattle posts where animals sleep and water points, such as springs and rivers, where they congregate. Combining that information with topography, which affects how livestock move through the landscape, I created a map of grazing intensity so that I could describe the grazing intensity at each of my sample points. I then used statistical regression to determine which variables explained whether a sampled point had lots of shrubs or just a few. The analysis showed that the factors that best explained shrub density were grazing intensity, soil moisture, and soil nutrient load. Where livestock grazing and soil nutrients are high and soil moisture is low, you will find more shrubs. So, in a sense, both conservation bureaucrats and livestock owners appear to be correct: livestock grazing is important, and so are soil resources.

What are we to make of this? First, it is worth noting that the temptation to see natural sciences as the final word is mistaken. Despite the quantitative character of the results, ecological science results are often murky. This is partly because ecologies are so complex. It is not uncommon for ecology results to explain less than 50 percent of the variation in data, meaning that good ecology can leave half of the data unexplained—and this assumes that the study measured the right variables, at the spatial and temporal scales appropriate to the research question.

Second, these results do not show historical process. Instead, they explain the dispersion of data produced by the study—a study carried out at a given

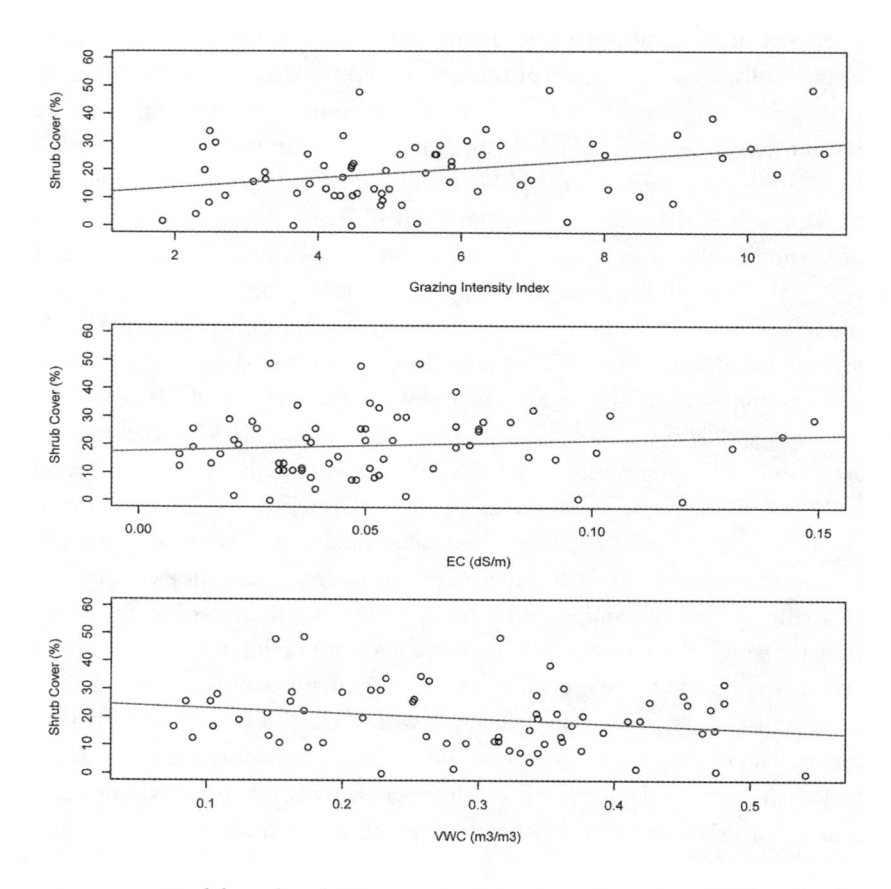

FIGURE 2.5. Model results, plotting grazing intensity, soil nutrients (EC), and soil moisture (VWC). From Hoag and Svenning (n.d.).

moment in time for a given set of variables. While they can be used to describe interactions between species, they do not tell us why the species are there in the first place or which historical events may have given some an advantage over others. For example, the structure of a forest might be determined more by a major drought twenty years ago—or by state-supported settler colonialism featuring fire prevention—than by the competition for soil moisture between different plants in the forest. Indeed, since the rise of quantitative ecology in the mid-twentieth century, history has largely been excluded from view in ecological science in favor of this kind of theorizing

(Kingsland 1995). I do not mean that ecologists do not believe in history but rather that their notion of history is particular and brackets certain processes that humanists would include. In ecology, the dominant mode of inquiry is to delimit variables, sample for these variables, and use the relative statistical importance of each to infer past ecological process. This contrasts sharply with anthropological sensibilities, where contingency and political struggle are key to explaining why the world today looks as it does. Settler colonialism, for example, is not a "proper" ecological variable until it can be quantified.[4]

The knowledge produced by this study was situated, and recognizing its situatedness can make it more useful to humanists. It was situated in reference not only to my identity as a white, cisgender man from the United States but also to its organizing rationale. To do the study as an ecologist, I needed to figure out what to count and at what spatial scale to do the counting. Because of a commitment within anthropology to recognize a multiplicity of ontologies (e.g., Mol 2002), anthropologists are able to make observations about a natural world that is full of emergent objects rather than preformed things. My studied decision regarding what to count was a form of natural history—structured and unstructured exploration of landscape patterns or of organisms in their natural environment—a practice that is critical to ecological science but often bracketed as merely preliminary research (Mathews 2016; Tewksbury et al. 2014). Mathews (2016; see also chapter 1) shows that humanists can mobilize natural history and insights about the situatedness of science to do ecological storytelling critically and interpretively using ecological science insights about nonhuman processes. Beyond simply comparing different perspectives, anthropologists can use distinct landscape readings to probe the findings of ecological science—how things emerge as seeable or countable from specific positions and how differences between the ethicopolitical positions of knowledge makers call into question any effort to render those knowledges symmetrical (Lyons 2014).

SITUATING ECOLOGY

The humanities help us recognize the relational quality of objects in the field—to interrogate the narratives of our informants, whether they are scientists or so-called everyday people—to learn how they are shaped by historical and political circumstance. This allows us to get inside those narratives and treat them as vantage points for seeing the world (Tsing 2017). In an effort to sketch the outlines of such a situated story, I'll situate the narratives of my informants. Then, I'll square them with some additional data speaking to claims of overgrazing (livestock statistics) and changes in precipitation (rainfall statistics).

THE NARRATIVES

I was taken to many shrub-encroached pastures by conservation bureaucrats, and they gave me consistent answers regarding the causes of landscape change and the solutions to fix it. The language was strikingly homogeneous, and it was basically just as Bokang described it—with shrub unpalatability calling forward a story of grazing intensity and mismanagement. Reading descriptions of shrub encroachment from the colonial period—in Lesotho as well as South Africa—I was struck by just how much they echo the language of these conservation bureaucrats today.

For example, the conservation bureaucrats with whom I spoke were mostly decent people who wanted to do their jobs well. But they knew little about the history of the district where they worked, given that they grew up mostly in the lowlands. Nor did they necessarily have formal training in rangeland ecology. In the massive area over which their office presided, they typically visited just a few villages per week, and little of their work was devoted to learning from villagers about the histories of their specific rangelands. Instead, they mostly helped resolve conflicts over grazing land or promoted grazing associations. When I went along with them on trips—even when they visited a rangeland—we didn't stray far from the truck.

In sum, their narratives were characterized by a roadside view of rangelands, an interest in "management," an emphasis on the personal responsibilities of herders, and an ecological language that endures from the colonial period.

Rural livestock owners have more of a historical sensibility about the landscapes in which they live, of course, but these are not necessarily deep-time engagements. Basotho are somewhat recent arrivals to the highlands, having been present for just a few generations, with most settlement of the highlands not taking place until the early twentieth century. They regularly told me that shrubs only became a problem in the previous twenty or so years, when rains diminished and became erratic. This is the same period during which mining labor declined dramatically, along with the entrepreneurial dreams that those earlier economic arrangements entailed (Hoag, forthcoming). While working in the mines, families would build up herds that could pay out annually after men returned home due to retirement or injury. Herders' lives were deeply caught up in this political economic shift away from labor migration. They often described their livestock histories to me with reference to work in South Africa—through stories of how their herds were begun when they were working in the mines; telling how they sought (in the case of younger herders) to one day leave the cattle post to find one of those rare jobs in South Africa; wanting to teach me Fanakalo, the pidgin language spoken in the mines; or even simply wearing hard-hats and other mining paraphernalia. Since the decline of labor migration, what was once strictly a retirement strategy is now a primary occupation for many.

As with the political economic shift away from mining, the environmental shift they described—the droughts and the delays in the onset of the rains that made shrub roots significant—are out of their control. This was a *structural* explanation of climate change in the context of political economic change rather than an agentive one focused on management decisions like livestock numbers. Shrubs are indicators of diminished forage in a specific, historical context.

LIVESTOCK AND RAINFALL

With respect to what we know about livestock numbers, grazing expanded dramatically in the highlands during the late nineteenth and early twentieth centuries (see Figure 2.6). This was for a few reasons, including land pressures in Lesotho's western lowlands as white Afrikaner farmers began expropriating land with British help, British promotion of highlands wool production for colonial monetary gain, and commoner Basotho turning to wool production in the highlands to circumvent the authority of an aristocratic class of chiefs who governed the use of lower-lying pastures (Hoag, forthcoming).

When sheep arrived in the highlands at that time, they found a territory whose pasturage was not accustomed to intensive grazing. Scholarship in rangeland ecology (Melville 1997; Milchunas and Lauenroth 1993) is clear that the effects of a sharp increase in grazing in historically ungrazed pastures are much more significant than the effects in pastures with a long history of intensive grazing. This is because the vegetation found in ungrazed pastures isn't evolutionarily preadapted to enduring the impacts of intensive livestock grazing. In essence, such systems do not have a high tolerance threshold for grazing and are more susceptible to the invasion of grazing-tolerant species like shrubs.

While the increase in stock during the early twentieth century is clear, livestock statistics mostly show periods of boom and bust. We do see a substantial increase in stock numbers in the late 1980s and early 1990s, though the plant community at that time was probably less responsive to an increase in grazing than it was in the early part of the twentieth century. As early as 1971, a longtime botanist of Lesotho wrote, "Surveying the country as a whole, it is true to say today that every acre of land in Lesotho is subject to human use in some form or other" (Jacot-Guillarmod 1971, 29).

With respect to precipitation, rainfall statistics show that the period 1979–96 experienced the highest incidence in drought for more than two hundred dred years (Chakela 1999), with more recent data suggesting that agricultural

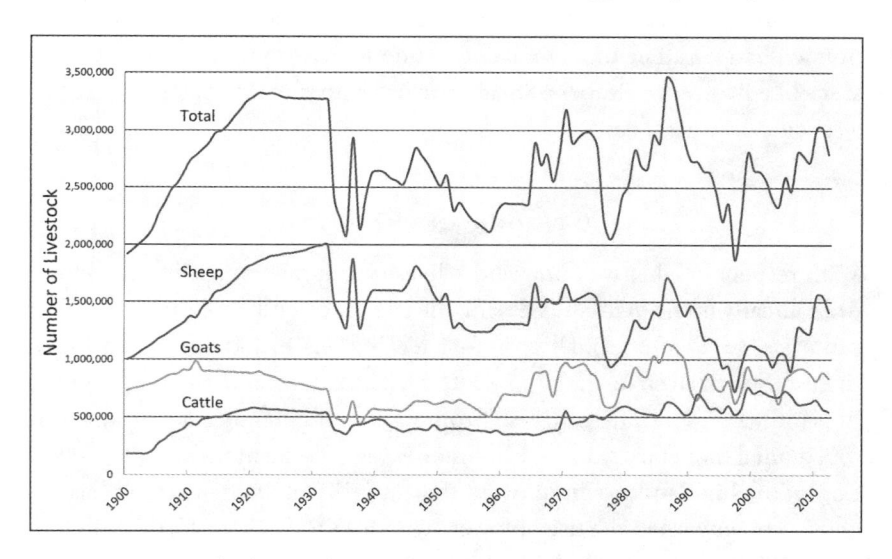

FIGURE 2.6. Lesotho livestock statistics, 1900–2013. From Lesotho Bureau of Statistics.

droughts are worsening (Mokuku 2004). Those with whom I spoke pointed especially toward delays in the onset of spring rainfall. Figure 2.7, which shows early spring rainfall (October–November) as a percentage of total annual rainfall, indicates a trend toward greater rainfall, but with greater dispersion in the data, meaning more variability. Rainfall data show a decline in early spring rainfall during the previous decade, and the opinion among rural people in Lesotho is unanimous that climate change is making rains erratic. By the middle of winter, Lesotho's pastures are nearly empty of forage by my observation and that of nearly everyone with whom I spoke, and it is not uncommon for cattle posts to become water stressed and even abandoned.

Discussion

The dwarf shrub is not only an empirical object but also serves as a conceptual tool for theorizing a transformation from livable to unlivable worlds

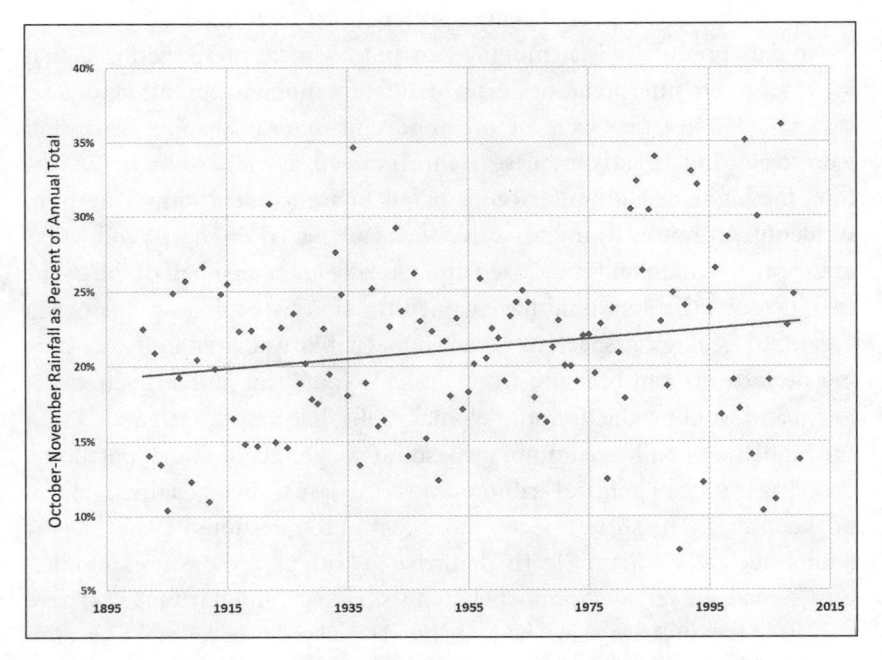

FIGURE 2.7. October–November rainfall as percentage of annual total, 1900–2010. From Lesotho Meteorological Society.

in Lesotho's post–labor reserve era. Both narratives entailed a declensionist landscape history, complete with a description of change and identification of a causal mechanism in shrub form. Conservation bureaucrats emphasize that livestock grazing gives shrubs a competitive advantage over herbs for light resources, placing blame for rangeland degradation on rural people for their allegedly irresponsible management practices. Herders and livestock owners tell a different story: one primarily of climate contingencies beyond their control. The droughts that are more common with recent climate changes, they explained, expand shrub populations, when woody species store water longer in their thick stems and roots.

Both narratives are probably correct in certain ways. Shrubs' extensive root systems likely help them compete with other plants for soil resources, and livestock grazing probably aids them in outcompeting grasses for sunlight. But these discrete plant community dynamics only make sense in light of the historical processes that configure them. Each narrative emerges out of a specific, historical context and operates at distinct temporal and spatial scales. Neither universally explains shrub population numbers.

No data point here is a smoking gun, but my own interpretation is that my informants' interpretations explain different moments in this landscape history—that livestock were a more important factor in shaping vegetation patterns during an early increase in shrub spread, when Basotho began settling the highlands and herbaceous plants unaccustomed to grazing were suddenly confronted by intensive livestock grazing. When sheep production in the pristine highlands increased dramatically in the first half of the twentieth century, shrubs would have gained traction by exploiting this newly organized ecological space in which unpalatability was vital. It was during periods of land pressure from settler colonialism and British efforts to expand wool production for colonial profit that Basotho expanded into the highlands—and as commoner Basotho sought economic opportunity, breaking free from political economic forces that marginalized them (Hoag, forthcoming). The shrub species most central to questions of land degradation, such as *C. ciliata,* clearly do thrive in both grazed pastures and disturbed roadside verges. Prominent botanists (Hilliard and Burtt 1987, 14) have suggested that invasive plants had inhabited small valley coves in the Lesotho highlands before being led out into the rangeland at large along roads and livestock paths. The dwarf shrubs moved along disturbance pathways, along the roadsides of capitalist production (Kalwij, Robertson, and van Rensburg 2015), flourishing where soil conditions were favorable and livestock grazing was high. The effect of grazing intensity has likely diminished today, however, if the plant community has shifted toward one that is more tolerant of livestock. Increasing drought and delayed summer rains in the past few decades likely explain this more recent expansion of shrubs into previously grass-dominant zones.

Two distinct movements of ecological fabulation—from shrub form to landscape historiography—emerge out of historical moments when specific functional traits come to matter in ecological process. Livestock-oriented accounts of shrub encroachment by conservation bureaucrats emphasizing agency and management appear homologous with the first increase in shrub spread during a spike in livestock population numbers, and rainfall-oriented accounts by livestock owners emphasizing structural forces appear homologous with the more recent increase during a period of intensifying drought.

Conclusion

Recognizing the failures of environmental science to account for power and culture in ecological models and in conservation strategies, many ecologists are eager to learn from humanists. This is an important intellectual challenge to understanding environmental change in the Anthropocene, the planetary-scale crisis that will be felt by all humans but caused by few. This chapter has shown how humanists might need ecological science too. Programs for drawing ecology into anthropology have historically tended toward either quantitative scientism or anthropocentrism, however. Rubber boots methods point in a different direction, leveraging curiosity, collaboration, and critical description to make observations about multispecies worlds. In the process, humanists should be at least partially accountable to ecologists as they work to authorize this emergent knowledge practice that incorporates biophysical processes, history, and semiosis. Ecological science does not offer silver-bullet answers to questions about ecological process, but ecologists have devised methods of observation that should not be ignored—tools for describing and understanding plant social relations, for example, including those that are invisible to the human eye and conventional ethnographic methods.

My chapter does not resolve the question of how to authorize this kind of interdisciplinary work in which natural science empirical strategies operate alongside critical, interpretive ones. In line with other contributions from this edited volume, however, it does pose questions that will need to be answered, such as how humanists should go about producing and analyzing biophysical data. Rather than simply deconstructing scientific knowledge and its biases, I argue that scientific knowledge can be used—and indeed that it might be unintelligible without the qualitative observations and interpretive techniques of anthropology. First, I approached conservation bureaucrats, herders, and livestock owners as serious ecological theorists. Second, I mobilized insights about the situatedness of knowledge to scrutinize the relationship between their accounts, on one hand, and positivist ecological science, on the other. I've done so by working from the material and semiotic bases of "form" and its use in interpreting ecological patterns. As a material and semiotic phenomenon, form operates as a boundary object for positivist and interpretive approaches, removing the obligation to flatten all differences between "matter" and "meaning."

Disciplinary empirical and analytical differences are knotted together with genre differences. Where ecology's genre sensibilities removed from view the historical trajectories within which ecological "laws" are caught, I found that my ecological science results were of little value without the interpretive and historical tools of anthropology. My results suggested that an ecological mechanism immanent to the shrub's bodily form enabled its spread during one historical period, whereas another mechanism enabled it later. Dwarf shrubs tracked the history of settler and administrative colonialism as well as the country's transition into an era in which labor migration was no longer possible and which was marked by continued livestock density and climate change. Livestock grazing aids shrub spread, though probably less so today than in the early decades of the twentieth century. Additionally, the competitive advantage presented by shrubs' extensive root systems to compete for soil resources seems more important in recent decades, when drought and erratic rains appear more common.

Lesotho's mountain rangelands are not bucolic nature spaces. They are postcolonial, postindustrial landscapes, and these multispecies communities express something of that colonial and industrial past—emanations of a regional political economy that situates Lesotho on the periphery and South Africa at the core.

NOTES

1. This interest has led some ecologists to call for the abandonment of community ecology in favor of other spatial scales (see Ricklefs 2008).

2. It is worth noting that the shrubs in question in Lesotho's highland landscapes are *encroaching,* not invading from a faraway place. These species have been present in Lesotho for at least a century, and probably longer, but appear to be expanding.

3. All names are pseudonyms.

4. For example, the niche-neutral theory debate described by Hubbell (2008; see Wennekes, Rosindell, and Etienne 2012) envisions "history" respectively either as mechanistic determinism or randomness. For anthropologists, history, by contrast, refers to contingency, political struggle, and interpretation. Emerging work on "priority effects" (Vannette and Fukami 2014) is closer in this sense. It is also worth noting that concerns about baselines are long-standing within ecology (e.g., Pauly 1995).

REFERENCES

Archer, Steven, Erik Andersen, Katharine Predick, Susanne Schwinning, Robert Steidl, and Steven Woods. 2017. "Woody Plant Encroachment: Causes and Consequences." In *Rangeland Systems: Processes, Management and Challenges,* edited by David D. Briske, 25–84. Cham, Switzerland: Springer Open.

Asad, Talal, ed. 1973. *Anthropology and the Colonial Encounter.* Atlantic Highlands, N.J.: Humanities Press.

Breshears, David, and Fairley Barnes. 1999. "Interrelationships between Plant Functional Types and Soil Moisture Heterogeneity for Semiarid Landscapes within the Grassland/Forest Continuum: A Unified Conceptual Model." *Landscape Ecology* 14, no. 5: 465–78.

Caple, Zachary. 2017. "Holocene in Fragments: A Critical Landscape Ecology of Phosphorus in Florida." PhD diss., University of California, Santa Cruz.

Carbutt, Clinton. 2019. "The Drakensberg Mountain Centre: A Necessary Revision of Southern Africa's High-Elevation Centre of Plant Endemism." *South African Journal of Botany* 124 (August): 508–29.

Chakela, Qalabane, ed. 1999. *State of the Environment in Lesotho—1997.* Maseru: National Environment Secretariat, Ministry of Environment, Gender, and Youth Affairs, Government of Lesotho.

Chakrabarty, Dipesh. 2009. "The Climate of History: Four Theses." *Critical Inquiry* 35, no. 2: 197–222.

Driver, Thackwray. 1999. "Anti-erosion Policies in the Mountain Areas of Lesotho: The South African Connection." *Environment and History* 5, no. 1: 1–25.

Ferguson, James. 1994. *The Anti-politics Machine: "Development," Depoliticization, and Bureaucratic Power in Lesotho.* Minneapolis: University of Minnesota Press.

Ferguson, James. 2006. *Global Shadows: Africa in the Neoliberal World Order.* Durham, N.C.: Duke University Press.

Geertz, Clifford. 1973. *The Interpretation of Cultures*. New York: Basic Books.

Haraway, Donna. 1988. "Situated Knowledges: The Science Question in Feminism and the Privilege of Partial Perspective." *Feminist Studies* 14, no. 3: 575–99.

Haraway, Donna. 2016. *Staying with the Trouble: Making Kin in the Chthulucene*. Durham, N.C.: Duke University Press.

Harris, Marvin. 1971. *Culture, Man, and Nature*. New York: Thomas Y. Crowell.

Hilliard, Olive Mary, and Brian Laurence Burtt. 1987. "The Botany of the Southern Natal Drakensberg." *Annals of Kirstenbosch Botanic Gardens* 15.

Hoag, Colin. Forthcoming. *The Fluvial Imagination: On Lesotho's Water-Export Economy*. Berkeley: University of California Press.

Hoag, Colin, and Jens-Christian Svenning. n.d. "Livestock Grazing Intensity and Soil Resources Determine Shrub Densities in Lesotho's High-Altitude Rangelands." Unpublished manuscript.

Hubbell, Stephen. 2008. *Unified Neutral Theory of Biodiversity and Biogeography*. Princeton, N.J.: Princeton University Press.

Hustak, Carla, and Natasha Myers. 2012. "Involutionary Momentum: Affective Ecologies and the Sciences of Plant/Insect Encounters." *Differences* 23, no. 3: 74–118.

Jacot-Guillarmod, Amy. 1971. *Flora of Lesotho*. Lehre, Germany: Cramer.

Kalwij, Jesse, Mark Robertson, and Berndt van Rensburg. 2015. "Annual Monitoring Reveals Rapid Upward Movement of Exotic Plants in a Montane Ecosystem." *Biological Invasions* 17, no. 12: 3517–29.

Kingsland, Sharon E. 1995. *Modeling Nature: Episodes in the History of Population Ecology*. Chicago: University of Chicago Press.

Kohn, Eduardo. 2013. *How Forests Think: Toward an Anthropology beyond the Human*. Berkeley: University of California Press.

Lave, Rebecca, Christine Biermann, and Stuart Lane, eds. 2018. *The Palgrave Handbook of Critical Physical Geography*. Cham, Switzerland: Palgrave Macmillan.

Leach, Melissa, and Robin Mearns, eds. 1996. *The Lie of the Land: Challenging Received Wisdom on the African Environment*. Portsmouth, N.H.: Heinemann.

Lett, Michelle, and Alan Knapp. 2003. "Consequences of Shrub Expansion in Mesic Grassland: Resource Alterations and Graminoid Responses." *Journal of Vegetation Science* 14, no. 4: 487–96.

Lyons, Kristina. 2014. "Soil Science, Development, and the 'Elusive Nature' of Colombia's Amazonian Plains: Soil Science, Development, and the Colombian Amazon." *Journal of Latin American and Caribbean Anthropology* 19, no. 2: 212–36.

Mathews, Andrew. 2016. "Ghostly Forms and Forest Histories." In *Arts of Living on a Damaged Planet: Ghosts and Monsters of the Anthropocene*, edited by Anna Lowenhaupt Tsing, Heather Anne Swanson, Elaine Gan, and Nils Bubandt, G145–56. Minneapolis: University of Minnesota Press.

Mathews, Andrew. 2018. "Landscapes and Throughscapes in Italian Forest Worlds: Thinking Dramatically about the Anthropocene." *Cultural Anthropology* 33, no. 3: 386–414.

Melville, Elinor. 1997. *A Plague of Sheep: Environmental Consequences of the Conquest of Mexico*. Cambridge: Cambridge University Press.

Milchunas, D. G., and W. K. Lauenroth. 1993. "Quantitative Effects of Grazing on Vegetation and Soils over a Global Range of Environments." *Ecological Monographs* 63, no. 4: 327–66.

Mokuku, Chaba. 2004. *Lesotho: Second State of the Environment Report, 2002.* Maseru, Lesotho: Ministry of Tourism, Environment, and Culture, Government of Lesotho.

Mol, Annemarie. 2002. *The Body Multiple: Ontology in Medical Practice.* Durham, N.C.: Duke University Press.

Mucina, Ladislav, and Michael Rutherford, eds. 2006. *The Vegetation of South Africa, Lesotho and Swaziland.* Pretoria: South African National Biodiversity Institute.

Otto, Ton, and Nils Bubandt, eds. 2010. *Experiments in Holism: Theory and Practice in Contemporary Anthropology.* New York: Wiley.

Pauly, Daniel. 1995. "Anecdotes and the Shifting Baseline Syndrome of Fisheries." *Trends in Ecology and Evolution* 10, no. 10: 430.

Raunkiaer, Christen. 1934. *The Life Forms of Plants and Statistical Plant Geography; Being the Collected Papers of C. Raunkiaer.* Oxford: Clarendon Press.

Ricklefs, Robert. 2008. "Disintegration of the Ecological Community." *American Naturalist* 172, no. 6: 741–50.

Showers, Kate. 2006. "From Forestry to Soil Conservation: British Tree Management in Lesotho's Grassland Ecosystem." *Conservation and Society* 4, no. 1: 1–35.

Star, Susan Leigh, and James Griesemer. 1989. "Institutional Ecology, 'Translations' and Boundary Objects: Amateurs and Professionals in Berkeley's Museum of Vertebrate Zoology, 1907–39." *Social Studies of Science* 19, no. 3: 387–420.

Swanson, Heather. 2017. "Methods for Multispecies Anthropology: Analysis of Salmon Otoliths and Scales." *Social Analysis* 61, no. 2: 81–99.

Tewksbury, Joshua, John Anderson, Jonathan Bakker, Timothy J. Billo, Peter W. Dunwiddie, Martha J. Groom, Stephanie E. Hampton et al. 2014. "Natural History's Place in Science and Society." *BioScience* 64, no. 4: 300–310.

Tsing, Anna Lowenhaupt. 2013. "More-Than-Human Sociality: A Call for Critical Description." In *Anthropology and Nature,* edited by Kirsten Hastrup, 27–42. New York: Routledge.

Tsing, Anna Lowenhaupt. 2015. *The Mushroom at the End of the World: On the Possibility of Life in Capitalist Ruins.* Princeton, N.J.: Princeton University Press.

Tsing, Anna T. 2017. "The Buck, the Bull, and the Dream of the Stag: Some Unexpected Weeds of the Anthropocene." *Suomen Antropologi* 42, no. 1: 3–21.

Vannette, Rachel L., and Tadashi Fukami. 2014. "Historical Contingency in Species Interactions: Towards Niche-Based Predictions." *Ecology Letters* 17, no. 1: 115–24.

Walker, Peter. 2005. "Political Ecology: Where Is the Ecology?" *Progress in Human Geography* 29, no. 1: 73–82.

Wallerstein, Immanuel. 2004. *World-Systems Analysis: An Introduction.* Durham, N.C.: Duke University Press.

Ward, David. 2005. "Do We Understand the Causes of Bush Encroachment in African Savannas?" *African Journal of Range and Forage Science* 22, no. 2: 101–5.

Wennekes, Paul L., James Rosindell, and Rampal S. Etienne. 2012. "The Neutral–Niche Debate: A Philosophical Perspective." *Acta Biotheoretica* 60, no. 3: 257.

Westoby, Mark. 1998. "A Leaf-Height-Seed (LHS) Plant Ecology Strategy Scheme." *Plant and Soil* 199, no. 2: 213–27.

Worster, Donald. 1994. *Nature's Economy: A History of Ecological Ideas.* Cambridge: Cambridge University Press.

CHAPTER 3

Tracking as Method

*Perspectival Sensibilities in a More-Than-Human
Desert of Tracks*

PIERRE DU PLESSIS

Tracking is a practice and a method that offers a window into the stories of multispecies landscapes. By attending to tracks—actual footprints on the ground—one can gain perspective, albeit it partially, about the positionalities and doings of more-than-human actors. Skilled trackers "step into the shoes" of multispecies others, so to speak, to imagine and speculate about what these actors in shared landscapes might be doing. This became evident to me early in my research in the Kalahari Desert in Botswana during a walk with !Nate, Njoxlau, and Karoha, three San trackers who would become my primary interlocutors. !Nate located an aardvark *(Orycteropus afer)* den and unfolded a story about the creature's activities, dispositions, and relations to other nonhumans in the landscape.

When he spotted the den, !Nate's embodied response indicated that he had been drawn toward something that had happened. He slowed, his two hands outstretched in front of him, palms up, moving slowly from side to side. He walked toward a hole dug into the sand, taking careful steps, and gently reenacted the aardvark's movement. He spoke softly as he moved toward the den. "He came here. He was looking for ants. He went there, far. A porcupine passed. When he's coming back, he is looking. Coming to his house, he doesn't want anyone at his house." At each sign, he shimmied his body to imitate the creature's movements, before kneeling down near the entrance.

!Nate then began talking to the aardvark, asking if it was home. But there was no response. Noticing the tracks of a hare next to the den and the aardvark's scuffed trail, !Nate began laughing. "He don't want the hare stealing his food so he has gone now to make another house!" Karoha and Njoxlau laughed, nodding their heads.

"Ey! These hares!" Karoha exclaimed as !Nate dusted himself off before we continued.

The Kalahari is a desert, but its many tracks remind you that you are never alone. At first blush, the desert's landscapes are exceptional mainly in their expanse and seeming monotony. Today, one can move through parts of the Kalahari for days and not see an animal other than semiwild domesticated cattle and goats and the ever-present little steenbok antelope (*Raphicerus campestris*). Seeing wildlife is a special occasion in these parts, and even more so today as a series of encroachments have come to threaten these environments. But not seeing wildlife does not mean it is not there. In fact, once one learns to notice it, these landscapes are full of multispecies drama.[1] The signs are all around. One only need look to the ground, and the tracks of animals will appear sooner than later, if other more subtle spoor—the cumulative term for tracks and signs[2]—are not noticed first.

The skilled practice of tracking is a method of attending to these tracks, signs, and spoor that draw you into populated more-than-human worlds and the mundane, though dramatic, stories of Kalahari landscapes. Indeed, tracking is a method for critical landscape description, a form of noticing that is both highly empirical and imaginative because it alerts us to the empirical features of landscapes, plants, and animals and entails a curious historical and speculative imagination that is attentive to how landscapes tell stories of multispecies interaction under changing economic and political conditions. Tracking, I argue, is an Indigenous form of multispecies noticing that, like natural science methods, may provide an important tool for understanding and becoming sensitive to Anthropocene landscapes. Its attention to more-than-human worlds allows for an empathetic (partial or double) perspectivism that also offers potential insights into collaborating across difference, whether different species or even different knowledge traditions and disciplinary perspectives. This is key to tracking as a rubber boots method. It is particularly important as a rubber boots multispecies method, because it draws attention to the value of diverse knowledge practices and methods—

not just, but also, those defined by institutionalized Science—in making contributions to landscape studies in the Anthropocene. Tracking, in this sense, much like the rhythms of gumboots dances that resist the violences of histories being erased described in the Introduction to this volume, is a decolonial rubber boots method that tells stories about the tracks of those that often go unseen or unnoticed, recognizing that even the tracks of conquest do not stand alone despite their hegemonic ambitions. As Njoxlau told me, "tracking is good because it tells you that you are here and that you are not alone."

A Desert of Tracks

This chapter draws on more than ten years of research with !Nate, Karoha, and Njoxlau, each a renowned tracker from the western region of Botswana's Kalahari Desert. I first started walking Kalahari landscapes with them in 2009, while they were conducting spoor surveys for a wildlife conservation project that aimed to formally establish a protected conservation corridor. Working with scientists, these trackers collected data in this project, and others that followed, to provide evidence about animal population densities that would support the case for protecting the corridor from further encroachments (cf. Keeping 2014; Keeping et al. 2018, 2019). Tracking surveys, financed and organized by wildlife conservation organizations and researchers, are one of the ways these trackers utilize their knowledge to bring attention to the importance of wildlife and their movements in the Kalahari. In doing so, they actively work against the transformation of these spaces into inhospitable landscapes for more-than-human life. This is important work, because it brings attention to multispecies presences in Kalahari landscapes. In a contemporary context in which hunting has been banned, the surveys also provide a way for people to continue using and developing their tracking skills. Furthermore, as employment opportunities are generally few and far between, participating in surveys has become a vital source of cash income for some San trackers.

During this time, however, I learned that tracking has much to offer as a method for learning about landscapes and landscape change that the spoor counts fail to capture (Du Plessis 2010). To further explore what is missing in these survey counts, I returned to the Kalahari for more than twelve months between 2015 and 2016 to study tracking as an "art of noticing" (Tsing 2015) landscapes with !Nate, Karoha, and Njoxlau. !Nate passed away in 2016, but his lessons—much at his urging—have inspired me to continue exploring tracking as a way of knowing Kalahari landscapes with Karoha and Njoxlau. Indeed, before he died, !Nate insisted that I should "keep on tracking" with Njoxlau and Karoha to show people that the desert is not empty and lost to the growing encroachments of cattle and other extractive industries. I remember !Nate saying that if we do not keep tracking, "people will say there is nothing here." He was gesturing toward his understanding of tracking as a tool to resist a variety of erasures: of wildlife, of landscapes, but also of San communities and their knowledges.

In part because of its vast and subtle landscapes, the Kalahari is sometimes framed as underutilized by growth-oriented politicians and entrepreneurs. Today, a variety of anthropogenic encroachments, largely promoted by the development and growth industries of Botswana's postcolonial state, threaten Kalahari landscapes and ecologies. Wildlife populations have declined significantly since at least the 1950s, when networks of fences and roads began to proliferate through the desert. Cattle have moved in and disrupted wildlife migratory routes, eating up much of the graze, while mineral and natural gas prospectors seek to tap into the natural resources that lie deep beneath the surface. And yet, the Kalahari remains a desert of tracks. Indeed, despite these encroachments, a more than seven-hundred-kilometer wildlife corridor in the western Kalahari represents one of the longest contiguous, uninterrupted wildlife dispersal areas in the world, even as it is steadily narrowing. Still, the Kalahari Desert is not the figure of desertification that has come to feature so prominently in Anthropocene imaginaries. It ranges between intermittent zones of sandy savannah grasslands, shrublands, woodlands, and gradients between. It is full of multispecies life, even as it is squeezed in by cattle—those "shock troops" of empire, as Deborah Bird Rose (2004, 85) and Milton (1997, 200) describe them—extractive capitalist industries, and infrastructures, which my interlocutors track too. Learning how to track these changes is, today, more important than ever.

The practice of tracking and my subsequent elaboration of it as a form of critical landscape description attend to these transformations and their multispecies histories through an ongoing attention to tracks and traces in the present. Coming to understand tracking as such has involved a great deal of collaboration, patience, slowing down, and bumbling. My collaborators, !Nate, Njoxlau, and Karoha, are my teachers, and they continue to track today, even though they are forbidden to hunt by state conservation laws. But as we will see, and as they often remind me, tracking is so much more than just a process of identification, counting, and cataloging. For them, it is also much more than work. Tracking is a method for noticing and relating to nonhuman landscape companions, of those one lives with, in these shared environs. As such, it is critically attentive and adaptive to environmental, but also social, change. In this context, tracking is an art of noticing that can be relatable to science but is also "more-than-scientific."

Tracking is sometimes represented as the skill of simply identifying and following the footprints of animals. And while it is true that footprints, or tracks, are the foundation of tracking and provide lots of important information about the identities and activities of animals, this is not all that tracking is. Scholar of tracking Louis Liebenberg (1990, 3) writes, "The art of tracking involves each and every sign of animal presence that can be found in nature, including ground spoor, vegetation spoor, scent, feeding signs, urine, feces, saliva, pellets, territorial signs, paths and shelters, vocal and other auditory signs, visual signs, incidental signs, circumstantial signs and skeletal signs." In other words, tracking attends to all variety of signs that emerge in landscapes through the movement and activities of multispecies actors.

Tracking is empirical, but it is also empathetic. This is a critical aspect of tracking as a rubber boots method. Through a grounded engagement with tracks and trackways, trackers partially project themselves into the metaphorical shoes of more-than-human landscape companions. In the opening vignette, for example, !Nate pieced together a story of a grumpy aardvark that had abandoned its home due to the irritating behavior of a hare. This assessment came about through an attention to a variety of spoor—the aardvark tracks, when it moved and at what speed, what it was looking for, the den, the lack of response from within the den, and how these signs were situated in relation to the tracks of the hare—that together allowed for an interpretation of a social situation. By being drawn into this story, !Nate partially adopted the perspective of the aardvark, with whom he could empathize as another being whose house was disturbed in a shared landscape. He analyzed the material, empirical signs, but it was his empathetic sensibility that enabled his imaginative interpretation of the social relations between the aardvark and the hare.

To interpret the tracks, !Nate brought us into the coinhabited landscapes of the Kalahari. In mimicking the aardvark's movements, he enacted a degree of empathy that made its behavior socially relatable. Tracking, in this sense, involves becoming empathetic toward the perspectives of more-than-human actors to interpret and understand what they are doing and what may have motivated them to do so. It is a method that involves multi-species empathy—a perspectival sensibility that, along with the practical aspects of tracking, is something that ethnographers can learn from as a rubber boots method in the Anthropocene when considering how to explore more-than-human worlds and perspectives.

Nonhumans have methods, too, and tracking helps to understand the methods of those lifeways by attending to their material tracks and traces in landscapes. Importantly, however, while tracking can attend to individuals, by situating them in relation to the sociality of more-than-human doings, it is also oriented toward landscapes. Landscapes are not just the grounds across which things move; they are vehicles for both establishing and understanding the specificity of more-than-human relations. Tracking in this sense is an "art of noticing" and a kind of collaborative, multi-species "participant observation." It is an immersive practice that draws trackers into landscapes that serve as the relations between more-than-human presences. The landscapes *are the relations.* More precisely, tracking is an empathetic method attuned to how more-than-human doings shape and are shaped by landscapes that enrolls trackers by virtue of their own movements and noticing in shared environments. It is thus also a situated practice, adaptive to change, through which empathetic, partially connected, and even collaborative relationships between humans and nonhumans are formed. And because the cast of characters is not determined preemptively and landscapes are constantly changing, even if in subtle ways, tracking requires shifting between perspectives in the unfolding everyday dramas of landscapes.

At a time when more-than-human ecologies are increasingly difficult to trace as they become silenced, paved over, encroached upon, and eradicated, tracking is a practice that can teach us to notice and follow multispecies lifeways even when their presence is not immediately evident. If tracks can remind us that one is never alone in the Kalahari, perhaps tracking is also an important method for reminding people how to pay attention to increasingly threatened and invisibilized more-than-human worlds. It offers lessons about how to attend to more-than-human environs and socialities at a time of planetary environmental crisis, when people need reminding that, like a desert of tracks, humans are never alone in the Anthropocene. Tracking can teach us how to attend to and care for those more-than-human lifeways at a time when our own tracks of environmental exploitation and human exceptionalism have sedimented into the planet's geology in violent and unlivable ways. Doing so, however, requires careful consideration about politics of knowledges and what it means to collaborate across different modes of knowing and being in the Anthropocene that will be central to framing tracking as a rubber boots method.

Tracking as More-Than-Indigenous and More-Than-Science

Introducing the importance of developing rubber boots methods in the Anthropocene, the editors of this volume ask, how might we learn to track the histories of more-than-human socialities? This question has driven much of my research in the Kalahari Desert and on tracking as a method for writing multispecies landscape histories. However, before delving further into the tracks of the Kalahari, let me first address the figure of the rubber boot. In this volume, rubber boots conjure up associations with grounded—or, rather, on-the-ground—fieldwork approaches, such as natural history and field biology, and the literal immersion of our feet into our fields of multispecies, more-than-human study. The implication is that such approaches are necessary in the Anthropocene, affording better opportunities for scholars from the humanities to bridge the disciplinary divides that separate the natural—traditionally the domain of natural scientists—from the cultural. Rather than collapsing disciplinarity, I suggest, this requires a willingness to adopt a kind of disciplinary *and* epistemological perspectivism—a perspectival sensibility—that emerges in practice, through our methods, of exploring ecologies together with their social worlds.

While rubber boots methods give us the impetus to refocus our attention toward the traditional domains of the natural sciences, it is important to be reminded that our methods need not be limited to those of the institutionalized Western knowledges. The disciplinary divisions are part of the same tradition that has marginalized non-Western knowledges. Science and technology studies, anthropology, and postcolonial scholars have long challenged the dimensions of power involved in knowledge production that naturalizes the nature–culture divide. The Enlightenment knowledge traditions that have reified Nature–Culture have privileged their own authority to lay truth claims, while often marginalizing other ways of knowing to the realm of cosmology and belief (de la Cadena 2010; Blaser 2009). Challenging this tendency, scholars have encouraged more nuanced considerations of the authority granted to the sciences, all the while reminding us that these are not the only knowledge traditions to take seriously as modes of knowing the world (Watson-Verran and Turnbull 1995; Turnbull 1997; Verran 2001, 2013).

The question of how to track the histories of more-than-human socialities references a practice of learning about and following nonhumans—that literally attends to the ground—that predates the post-Enlightenment schism in the sciences by millennia: tracking. Tracking has primarily been studied by social scientists and humanists as a cultural form within hunter-gatherer societies, rather than as a broader mode of knowing the natural world. With few exceptions, the knowledge has been "othered" and studied as a category of exotic "Indigenous" or "local" knowledge. One important scholarly exception here is the work of Louis Liebenberg (1990), who has argued that the art of tracking is the origin of human scientific thinking. Liebenberg, also working in the Kalahari, describes tracking as an empirical practice of interpreting animal signs, or spoor, which at the most skilled levels are used to develop hypotheses and theories—what Liebenberg calls "speculative tracking"—about what specific animals have been or may be doing and where they are located. It is empirical, interpretive, and speculative. As Liebenberg describes it, tracking could, too, be endorsed as an "art of noticing" (Tsing 2015) and a rubber boots method, but, I suggest, with an important twist. We should approach it carefully, without either entirely subsuming it into the realm of science or exoticizing it as confined to a bounded sphere of "Indigenous knowledge."[3]

Tracking—in the Kalahari—is also part of a "cosmological package" (Mavhunga 2014). It is not the purely scientific secular method Liebenberg describes, even though the endeavor to recuperate the scientific value of tracking has been an important political project of drawing attention to the depth of Indigenous knowledge practices marginalized as other-than-science. This chapter draws from and is inspired by Liebenberg's work but uses it as a point of departure to explore contemporary practices of tracking as a method for knowing, relating to, and tracing the histories of more-than-human socialities as taught to me in the Kalahari. In doing so, I emphasize that what I have learned is situated in terms of both my experience with trackers in changing Kalahari landscapes and as a multispecies ethnographer concerned with the challenges facing the planet and its environments. In learning to track, I was also tracking the trackers—not in the surveilling sense, but by following and mimicking their tracking, I learned to extend my ethnographic empathic sensibilities beyond the human, partially, to navigate between interconnected human and nonhuman landscapes in the Kalahari.

Tracking is a skill that is developed and refined over a lifetime, but it is one that anyone can begin to learn. Learning requires more than just a field guide, because experience is the best teacher.[4] My own efforts to learn to track have been slow and difficult due to my comparatively limited experience and my own situated attachments that make aspects of my collaborators' and teachers' situated knowledges (Haraway 1988) difficult to approach. This is part of the process of learning, and allowing space for "epistemic disconcertment" (Verran 2013) is an important and necessary aspect of collaboration. And, because cosmologies are not universal, deploying tracking in different contexts might, too, require traffic between cosmologies, wherein science and even multispecies ethnography may be part of a set of relations.

If we are to track and trace the histories of more-than-human socialities in the Kalahari, it will serve us well to avoid reproducing natural–cultural divisions that dismiss cosmologies to the realm of belief, for cosmologies, too, have material histories that sediment in landscapes. Space must be allowed for an empathetic traffic between ways of knowing without reducing the diversity of knowledges to one or presuming that different ways of knowing can become fully knowable and experienced. On several occasions, for example, my interlocutors would sense the presence of lions through a tingling sensation and sweating under their arms long before seeing a lion or its tracks directly (Du Plessis 2018). They attribute this embodied sensing to connections developed over time from dwelling in shared landscapes. But they also describe it as a remnant of a primordial time when animals and humans were not entirely distinguishable and were therefore relatable persons—relations that resonate into the present. This has sometimes been described as a "nonscientific" aspect of tracking and surely would not count as evidence on a tracking survey. Dismissing these dimensions, however, would be a mistake. These histories of more-than-human relations are important to the practice of tracking and are marked in the landscape (cf. Guenther 2020). Learning about these histories, even as I could not experience such relations, was crucial to developing my understanding of tracking as a situated practice and mode of relating to more-than-human socialities in the Kalahari.

If tracking attends to landscapes, it is important to remember that landscapes are not simply a jumble of things—they are also cultural artifacts. Cosmologies are necessary alongside "community" knowledge to learn about these landscapes. As a rubber boots method, I argue, the tracking knowledge of my San teachers cannot be legitimized solely in terms of its science merit. Instead, engaging with tracking as a rubber boots method requires a kind of epistemological, or ontological, perspectivism. This means creating space that allows for the partial convergence of and traffic between diverse ways of knowing and worlding. This, in itself, is something that tracking can teach us about.

TRACKING IS CRITICAL DESCRIPTION

It was by walking together with !Nate, Karoha, and Njoxlau that I came to understand tracking as a mode of critical description that extended beyond identifying individual tracks. They began by familiarizing me with the shapes of particular tracks during our walks, but the real lessons emerged when they described the movement and temporality of tracks, which they taught me to situate in relation to the broader landscape. A description of a day of tracking helps to demonstrate this.

On a cool morning in late March, we wandered out of camp just after sunrise. !Nate, Karoha, Njoxlau, and I had set up a small camp deep within a Wildlife Management Area (WMA). Situated between two game reserves, these WMAs comprise parts of one of the longest uninterrupted wildlife dispersal ecosystems in the world. But this corridor is closing quickly as cattle and their ranchers are moving in. Parts of these areas are in the process of

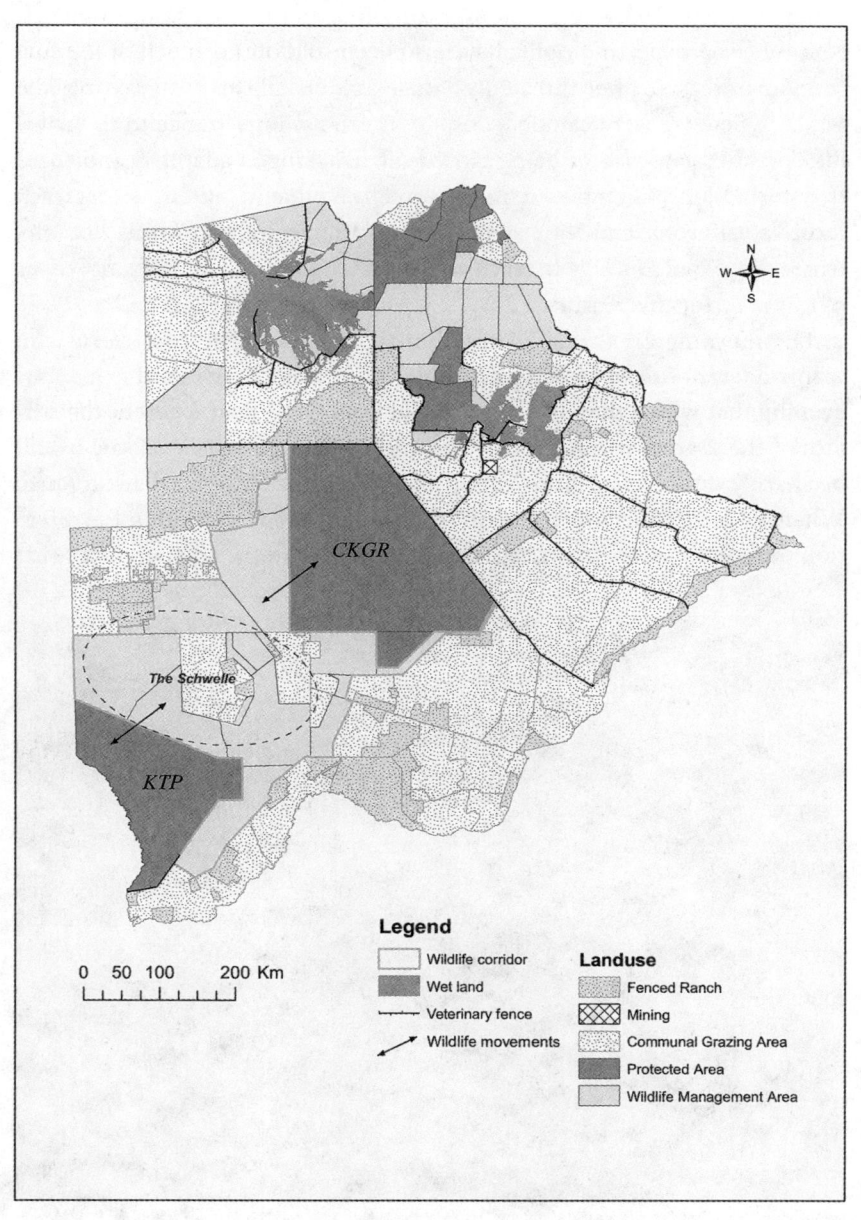

FIGURE 3.1. The Kalahari Wildlife Corridor in Botswana. Map by Marie-Charlotte Gielen; reprinted with permission.

being re-gazetted into either cattle-grazing areas or privatized cattle ranches. Natural gas prospecting rights have also been sold off for much of the surrounding areas, further threatening the corridor. All this is to say that the WMA where we were camped, despite its crucial importance to the wildlife corridor, is at risk of being closed off. Tracking is adaptive, and these transformations and movements are something that my interlocutors track too. !Nate, Karoha, and Njoxlau had chosen this remote site for us to camp, where they would be able to teach me about the flora, fauna, and landscapes of this area, for this reason.

The morning air was still crisp, but the desert sun would overwhelm within a few hours, so we set out quickly. There was a light rain the previous evening that wiped the sand clean of old tracks and held together the surface of the sand as a damp crust. This meant that the tracks we saw would be sharply defined and from within the hours since the rain had stopped. When we passed our first set of tracks, Karoha waved his hand in the direction the animals were traveling, shaking it rapidly in the air to indicate that

FIGURE 3.2. Fresh hartebeest track. Photograph by the author.

the tracks were fresh, from earlier this morning. With the same hand, he issued another hand signal identifying one set of tracks as belonging to a hartebeest antelope and another to a pair of gemsboks.

The identification, movement, and freshness signals were performed together to offer both a temporal and directional assessment of the animals' passing. The hand signs are used as a nonverbal method for relaying information quickly and silently to avoid being detected by nearby creatures. They both relate empirical information and enact an empathetic sensibility, exemplifying the trackers' reflexive attention to their own positionality in relation to other nonhumans that share these landscapes. How we moved not only guided our perspectives of the landscape at any given moment but might affect the movements of others. The hartebeest's cloven hooves (Figure 3.2) were well defined in the tracks, indicating that they were just a few hours old. They were spaced at a regular walking gait, suggesting that the animals were unstartled when they passed. The pair of gemsbok tracks were even more fresh and moving at a faster pace. Together the two sets of tracks told the trackers of the recent history of this multispecies landscape. Something in the last few hours had happened to encourage the two different sets of pace: another unseen animal, human disturbance, or a change in the weather perhaps?

Karoha pointed to the edges of the hartebeest tracks: they were pockmarked from the prior evening's rain. But the inner impressions of the tracks were smooth, showing that the hartebeests had passed after the rain, but still before the gemsbok. The ridges on the outer edges of the tracks were still sharp, and the dried sand that fell into the impression revealed that it had been at least several hours since they'd passed. These were all material signs in and of the landscapes: the displaced sand, water marks, and bits of debris were signs of the animals' movement and the temporality of their movements. Karoha assessed them together in a matter of seconds, never breaking pace, relaying that the hartebeest had probably passed around sunrise, perhaps on its way to lick minerals at a nearby pan.

Had recent rains not helped to date and bring the tracks to life in a multispecies history, there are other clues my tracking teachers taught me to look for. Fresh tracks are usually darker in color than older tracks because the sun tends to lighten the top layers of sand as time passes. Thus, in dry spells of weather, tracks can be dated along the gradient of sand color, which expert trackers can estimate precisely from a few minutes to several weeks. For relatively fresh tracks, knowing the times of day when animals tend to move more regularly is also important—usually at sunrise and in the late afternoon, when the moon rises above the horizon.

We carried on, stepping between tufts of grass, bushes, and the occasional tree, which guided our movements as we found the paths of least resistance. Without stopping, !Nate casually pulled a few berries off to taste. "These ones are old," he said as we continued walking. We would have lingered were the berries fresh. My tracking teachers are alert to all forms of life beyond the species that might have their immediate interest. Attending to flora as well as fauna, they piece together their multispecies stories.

We passed a few small cloven hoof prints of steenbok before moving around a thicket and arriving a set of very fresh tracks with elongated toes. The sand had been recently kicked up by the long toes, but the tracks were scuffed. It was not immediately clear whether they belonged to porcupine or honey badger. !Nate, Karoha, and Njoxlau discussed the tracks, before following the trail a few meters to gain a better context. They quickly agreed that this was a porcupine because, though the tracks were unclear, faint lines dragged alongside the outer edges of each track, where a porcupine's quills would have touched the sand.

Uncertainty in tracking is approached by a heightened sense of curiosity and resolved through collaboration. When my tracking teachers are uncertain about a track, they look for the trail with which they can situate the spoor. They often do this collaboratively, gently debating each other's interpretations and critical descriptions until they come to an agreement. It involves a willingness to adapt one's own initial assumptions; as Karoha once told me, "If you think too much, your mind will tell you what you want it to be.... That's why we look for more spoor. I listen to Njoxlau or !Nate, and they listen to me." Being too committed to your assumptions, as Karoha alludes to, can blind you to other possibilities. Curiosity and collaboration in tracking involve a degree of humility, a willingness to change one's mind through continued noticing and a sensitivity toward different perspectives, including those of nonhumans, but also those of whom you are collaborating and tracking with. Collaboration, curiosity, embodied noticing, and this sensitivity toward diverse perspectives are all essential to tracking as a method of critical landscape description.

PERSPECTIVAL SENSIBILITIES

!Nate, Karoha, Njoxlau, and I rounded a thicket shortly after leaving the porcupine tracks. As we did, the three trackers immediately froze and bent into a crouch. This time, Karoha shook his hand even more rapidly, signaling very fresh gemsbok tracks. Freshly bent grass and especially clear track prints told him so. Looking around carefully, they took slow, silent steps forward. I was much clumsier and may have cracked a small twig as I carefully attempted to follow my teachers' movements. Njoxlau indicated that he was going to the right, behind a silver terminalia bush. We gathered behind the bush, crouched. Njoxlau then quietly said to me, "You see?" and pointed into the distance. I had almost missed the two tan gemsbok, camouflaged in the bush fewer than fifty meters in front us. They were at high alert, looking in our direction. The wind was blowing toward us, so it was unlikely that they had picked up our scent in the air, but they may have heard us, or rather, me.

After a few minutes, we slowly stood up, and the two antelope ran off. !Nate, Njoxlau, and Karoha agreed that these were the same two gemsbok whose tracks we had seen earlier in the morning. Karoha explained that when we saw the first set of tracks, he thought we might catch up to the gemsbok. We did not follow the gemsbok trail directly but instead walked in a direction that might intersect with their current location. Though unspoken, Karoha was performing speculative tracking, imagining where they might be based on the signs of their movement. He knew that there was a valley in the direction they were traveling, and, judging the age of the tracks, he guessed that they had gone to that valley to lick for minerals in the morning before retreating into the thicker bush for shade as the sun began to rise. Throughout our walk, while engaging with a multitude of different signs and nonhuman activities, stopping to debate the porcupine tracks, Karoha adopted, partially, the perspective of the gemsboks, not in a transformative way, but relationally, attentive to a landscape of potential encounters. Attuned to the multiplicity of temporal and spatial coordinations, this kind of curiosity, collaboration, and responsiveness to encounter came to typify my experiences tracking in the Kalahari. No track limited our attention to a singular entity. Instead, we tracked unfolding landscape stories that subsequently led us into the worldings of different more-than-human actors.

!Nate, Karoha, and Njoxlau approach their environments as active, lively, animated, and relational, rather than as an "open book" in which they "read everything that is written in the sand" (Liebenberg 1990, v). Instead, they "read between the lines." As Liebenberg writes, "trackers themselves cannot read everything *in* the sand. Rather, they must be able to read *into* the sand" (v). This involves an interpretive and imaginative process of coming to understand and know nonhuman others through the traces they make in the world more than establishing a definitive account of marks on a palimpsest. "Reading into the sand" is a kind of empathetic perspective taking. Nature is not simply an open book to be read, Liebenberg argues, and such understandings have led to crucial misunderstandings of tracking. Those understandings are part of the larger ontological problematic that treats the environment and nonhuman worlds as passive and, under the moniker of "nature," the background upon which inscriptions are made, from which knowledge can be extracted or interpreted like a text.

For Liebenberg, "reading between the lines" is part of the process of scientific knowledge making. He argues that the most advanced form of tracking—what he calls speculative tracking—is scientific in that it utilizes empirical material to propose hypotheses about animal behaviors, timelines, and locations. It involves an imaginative capacity to project oneself into the position of the creature being tracked—to put oneself in its shoes—to anticipate its movements and behavior:

> To track an animal, the tracker must ask himself what he would do if he were the animal. In the process of projecting himself into the animal, he actually *feels* like the animal. The tracker therefore develops a sympathetic relationship with the animal, which he then kills. (ix)

Though this bond might be formed most intensely when hunting, forming partial empathetic relations with other landscape companions is a crucial part of even the most mundane aspects of tracking as an art of noticing and relating to. As Bubandt and Willerslev (2015) point out, empathy is both prelinguistic and a key component of sociality across difference.

An aspect of this for San trackers, which Liebenberg (1990, 93) attributes to the realm of belief or "nonscientific aspects of hunting," involves understanding that all animals were once people, a history that makes them and their subject positions more relatable. As my tracking teachers described to me, nonhuman actors are socially relatable because "we are together, we live in the same places." While stark distinctions are made between people, plants, and animals, they are relatable because of shared histories of living together. As Matthias Guenther (2020, 107) has recently argued, among San, the empirical and the empathetic work to counterbalance each other. These histories of co-species habitation are not limited to the Kalahari—though the cosmological dimensions are specifically situated.

Some practical examples help here. Tracking in the Kalahari includes a high level of risk for one's safety, because humans are not the only predators, and other animals also track. If trackers are not careful, they, too, can become prey. This, combined with a simple curiosity about the life and habits of other nonhumans, is part of a broader approach to being aware of and relating to more-than-human socialities. As !Nate's reenactment of the grumpy aardvark's disposition demonstrated, it is not just when tracking on the hunt that trackers project themselves into the position of an individual animal. Rather, trackers constantly consider, are aware of, and negotiate between the potential perspectives of a variety of nonhuman presences—animals, plants, and fungi—that they may encounter as they move about landscapes. Thus, while walking, trackers allow themselves to continuously adopt and shift between the perspectives, however partially, of nonhuman actors that they may encounter, especially those that pose a threat to their safety. This perspectival sensibility, though, is not restricted to sensational and dangerous encounters. It is a key aspect of being available to mundane rhythms and movements that inevitably arise in crossing paths, much as Karoha demonstrated in leading us to eventually encounter the set of gemsboks.

These two points—that trackers project themselves into the animals they are tracking and that they are also always alert to the fact that they might be tracked themselves—are extremely important to the experiential and empirical realities of tracking in the Kalahari. They also resonate with important theoretical work in anthropology about animist and more-than-human perspectivism. Two of these contributions deserve mention, as they help to demonstrate the significance of adopting the perspectives of nonhuman others and better explore, anthropologically, the question of how multispecies ethnographers might track more-than-human socialities and come to understand them without the benefit of language.

Eduardo Viveiros de Castro (1998, 2004a, 2004b) introduced the concept of "multinaturalism" in developing a renewed approach to understanding perspectivism in his studies of Amerindian animism in the Amazon.
According to Amerindian perspectivism, he argues, the universe is peopled
by subjective agencies, human and nonhuman. Perspectives are inscribed
on different bodies but share a generic soul in that "whatever possesses a
soul is capable of having a point of view, and every being to whom a point
of view is attributed, is a subject; or better, wherever there is a point of view,
there is a 'subject position'" (Viveiros de Castro 2004a, 466). That is, different species all have perspectives, just as humans do; however, those perspectives and what they see are different, depending on their bodies. This disrupts
ontological continuity by understanding humans and nonhumans as sharing
similar perspectives, but from different points of view, such that "individuals of the same species see each other (and each other only) as humans see
themselves, that is, as beings endowed with human figure and habits, seeing
their bodily and behavioral aspects in the form of human culture" (Viveiros
de Castro 2004b, 6).

While this argument has generated significant theoretical and conceptual insights about distributed subjectivities, notions of personhood, and
ontological plurality beyond the great Nature–Culture divide, there is a very
practical dimension to Amerindian perspectivism that scholars often move
past too quickly and that resonates with my argument about tracking. That
is, the Amazon is a dense rainforest that limits one's range of visibility and
therefore makes it difficult to see an approaching predator. Thus, when moving through the rainforest, one must always be ready and able to anticipate
an encounter with a nonhuman other before seeing it. One way to do this is
to assume the perspective of a potential predator. Adopting, or at the very
least acknowledging and attempting to anticipate, the subject position of a
nonhuman other in a very practical way—much as trackers in the Kalahari
do—can be a matter of survival. Even as the ontological considerations have
generated important and complex theoretical insights, this perspectival sensibility is a very practical means of acknowledging and anticipating potential encounters with other actors in shared environs.

Rane Willerslev (2004) finds similarities to Viveiros de Castro's perspectivism among Siberian of Yukaghir hunters, who transform themselves into
elk to be more successful when hunting them. Mimicking elk and adopting

their perspective make it easier to anticipate their behavior but also make one less likely to be detected. But, while perspectivism helps to explain conceptions of the body and identity, he argues that it is perhaps too abstracted from the real-world experiences of his informants (629). Willerslev attempts to make Viveiros de Castro's perspectivism more practical through the concept of mimesis, "which registers both sameness and difference, of being Self and being Other" (630). Yukaghir hunters adopt the perspectives of their prey through mimesis, developing a form of "mimetic empathy" (630), but do not transform into elk in any absolute sense. Willerslev suggests that there is always the possibility of transforming into the animal in the processes of imitating it, but this is difficult and risky business. The aim of the hunter is not to fully adopt the perspective of the animal, but to adopt it partially, and just enough to achieve what Willerslev calls a "double perspective." This allows the hunter to assume the perspective of their prey while still maintaining their own. It involves not giving into a single point of view and allowing for an engagement with many.

In much the same way, this partiality is a significant aspect of tracking as a rubber boots method. Tracking in the Kalahari involves partially adopting the perspective of that which is tracked. Being sensitive toward other perspectives, or empathic—even if for deceptive purposes—is a more general quality of social relationality, much as described by Bubandt and Willerslev (2015). Difference becomes relatable in the "entanglement of alterity with the empathic imagination" (Bubandt and Willerslev 2015, 13). This is a central component of tracking that is both embraced and foregrounded in the practice. As such, tracking approaches a similar doubling of perspective by both bringing one closer to and maintaining distance from a nonhuman other in the recognition of difference. It involves a kind of "empathetic perspectivism": a willingness to consider and become empathetic toward the perspectives of nonhumans through a close attention to the tracks and traces they make in the world, precisely because they are different, as part of the process through which shared landscapes are socially constituted through more-than-just-human relations. Tracking, then, marks a partial break with phenomenology: from reading minds to reading landscapes, in which practices tell (partially) of behavior or doings that enact landscapes. It is a practice of interpreting and relating to the material semiotics of landscapes through the relational traces of more-than-human actors and the histories that their movements both make and reveal. Stepping into and then out of the perspectives—or boots—of those they track is how trackers read multispecies histories into the sand.

What trackers track are not just animals but historical landscapes and the multiplicity of movements that constitute those landscape histories and with which landscapes emerge. They pay careful attention to the conditions through which assemblages of relations may gather to make life possible for a given entity or set of relations. That is, tracking is also a practice of gaining perspective, in the more traditional sense. A tracker does not have to adopt the perspective of a Kalahari truffle as if presuming cognitive intentionality, for instance, to know that it likes to live with certain species of grass and bush and that it emerges after late summer rains that are followed by cool air (Du Plessis 2022). Instead, they can engage with and interpret the material-semiotic social relations through which they emerge. It is in this way that tracking gains perspective about multispecies histories as a form of critical landscape description without necessarily presuming cognitive intentionality. Tracking involves a perspectival sensibility attentive to the diversity of landscape relations.

Conclusions

While footprints were my empirical entry point into a desert of tracks, I quickly learned that tracking involves much more than noticing just tracks. It is a method that itself attends imaginatively to the lives and histories of more-than-humans. As such, it explicitly explores and establishes lively relations in the process of becoming empathetic toward the diversity of perspectives—and methods—of landscapes actors. The practice is not limited to observing tracks as if they were out there in the world, separate from the person tracking. Tracking is embodied, and the method involves immersing oneself, much as !Nate did with the aardvark place, into assemblages of material signs that make multispecies socialities relatable. While tracking presumes difference and change—and this is crucial—it pulls you into the common grounds of shared landscapes, where familiarity of habit emerges and becomes relatable.

One of the tasks, or challenges, for multispecies ethnography in the Anthropocene has been to find ways of immersing oneself into more-than-human worlds, to find ways of describing their doings and telling their stories. Tracking as a rubber boots method is one way of learning to story the lively worlds of more-than-humans, while also being attentive to the colonial and extractive histories with which they are entangled. After all, tracking as a method is also itself a product of history: of San subjugation to colonial power, and today the growth and development of the cattle industry, extractivism, hunting bans, and even wildlife conservation research, evidenced in the transformation of San landscapes into shrinking WMAs. As !Nate said shortly before he passed away, if we do not keep tracking, "people will say there is nothing here." And, while much of this involves drawing from and retooling different disciplinary methods to study co-species socialities, we might also want to design methods for the Anthropocene that can deal with the more-than-human in terms that do not presume a universal *anthropos,* that involves methods and knowledges that are transdisciplinary, nondisciplinary, and vernacular but that are also more-than-Western.

Tracking is one such method that is a powerful tool for noticing and understanding more-than-human doings. And it is an approach that is adaptive to change in ways that encourage a perspectival sensibility that enriches critical description in the Anthropocene. But it is not the only one. Andrew Mathews (chapter 1) presents another, arguing that combining the oral histories of farmers, the insights from local ecologists, and his own landscape observations is a form of critical description. Here I suggest that Indigenous San methods of tracking that read multispecies lives into the desert sand as part of a practice of noticing are both empirical and imaginative and provide a different kind of critical description. They also remind us that we have as much to gain from collaborating with Indigenous interlocutors as we do from collaborating with natural scientists. Tracking the desert or snorkeling coral reefs with local informants, as Nils Bubandt (chapter 6) shows, also bring us into a more-than-secular Anthropocene.

Tracking teaches us how we may become more empathetic toward more-than-human others—their perspectives and their histories of anthropogenic decimation—in our arts of noticing. Tracking has become necessary as a wildlife survey tool, for instance, such that the method is itself a result of anthropogenic change evidenced by declines in wildlife populations in the encroached landscapes of the Kalahari. This empathetic perspectivism also offers clues for how we might better collaborate across disciplines, knowledge traditions, geographies, and histories. In other words, from tracking, we get a method, but we also get a lesson about how more-than-human ethnographers might engage in forms of transdisciplinary and epistemological perspectivisms in their collaborations across the natural sciences and humanities, between researcher and "informant," and across diverse knowledge traditions—embracing a willingness for our assumptions to be unsettled. This presumes, not a total perspectival shift, but rather a willingness to partially adopt a "double perspective" or, at least, to be empathetic and sensitive toward the situated knowledges and perspectives of different approaches to understanding landscapes and their contingent histories. This is not simply a romantic call for more empathy—after all, the most intense forms of mimetic empathy can culminate in killing and can also be politically deceptive (Bubandt and Willerslev 2015). But as tracking shows, learning to understand and relate to different positionalities, human and nonhuman, through their tracks and traces makes it possible to consider their histories and their world-making practices.

Tracks tell the stories of their more-than-human engagements and the socialities with which their lives are entangled. Learning from them, we can see that tracking offers a number of tools for engaging with more-than-humans as landscape actors and draws attention to landscapes themselves as the emergent relations between diverse ecological actors. Tracking as a rubber boots method is a practice of critical landscape description that draws these aspects together to attend to landscapes, multispecies histories, and the diversity of perspectives through which they emerge. Tracking reminds you that you are never alone in a desert of tracks. In the Anthropocene, tracking as a rubber boots method teaches us how to pay attention to, and critically describe, more-than-human worlds, across difference, that we humans live in the company of.

NOTES

1. The use of "drama" here draws from Mathews's (2018, 409) suggestion that in thinking dramatically about landscapes, "we can cultivate a dramatic form of attention that sustains multiple, competing stories of social and ecological change, of encounters between people, plants, soils, and diseases, while remaining alert to the limits of each account, to the excess that it fails to capture."
2. Referred to as spoor in Afrikaans as well as in international hunting language.

3. We should also be wary of reproducing the series of bifurcated denunciations that led to the impass of the "indigenous knowledge–science wars" (Green 2012).

4. But, for an excellent field guide to begin learning about tracking, see Liebenberg (1990).

References

Blaser, Mario. 2009. "The Threat of the Yrmo: The Political Ontology of a Sustainable Hunting Program." *American Anthropologist* 111, no. 1: 10–20.

Bubandt, Nils, and Rane Willerslev. 2015. "The Dark Side of Empathy: Mimesis, Deception, and the Magic of Alterity." *Comparative Studies in Society and History* 57, no. 1: 5.

de la Cadena, Marisol. 2010. "Indigenous Cosmopolitics in the Andes: Conceptual Reflections beyond 'Politics.'" *Cultural Anthropology* 25, no. 2: 334–70.

Du Plessis, Pierre. 2010. "Tracking Knowledge: Science, Tracking and Technology." Thesis, University of Cape Town.

Du Plessis, Pierre. 2018. "Tingling Armpits and the Man Who Hugs Lions: Dangerous Ghosts of Sameness and Different Differences." In *A Non-secular Anthropocene: Spirits, Specters and Other Nonhumans in a Time of Environmental Change*, edited by Nils Bubandt, 97–106. AURA More-Than-Human Working Paper 3. Aarhus, Denmark: Aarhus University Research on the Anthropocene (AURA). http://anthropocene.au.dk/working-papers-series/.

Du Plessis, Pierre. 2022. "Tracking Meat of the Sand: Noticing Multispecies Landscapes in the Kalahari." *Environmental Humanities* 14, no. 1: 49–70.

Green, Lesley. 2012. "Beyond South Africa's 'Indigenous Knowledge–Science' Wars." *South African Journal of Science* 108, no. 7–8: 44–54.

Guenther, Mathias. 2020. "The Hunting-Field and Its Doings: Subjectivities of Territory of the |Xam Bushmen of the Northern Cape." *Southern African Humanities* 33: 99–117.

Haraway, Donna. 1988. "Situated Knowledges: The Science Question in Feminism and the Privilege of Partial Perspective." *Feminist Studies* 14, no. 3: 575–99.

Keeping, Derek. 2014. "Rapid Assessment of Wildlife Abundance: Estimating Animal Density with Track Counts Using Body Mass-Day Range Scaling Rules." *Animal Conservation* 17, no. 5: 486–97.

Keeping, Derek, Julia H. Burger, Amo O. Keitsile, Marie-Charlotte Gielen, Edwin Mudongo, Martha Wallgren, Christina Skarpe, and A. Lee Foote. 2018. "Can Trackers Count Free-Ranging Wildlife as Effectively and Efficiently as Conventional Aerial Survey and Distance Sampling? Implications for Citizen Science in the Kalahari, Botswana." *Biological Conservation* 223: 156–69.

Keeping, Derek, Njoxlau Kashe, Horekwe (Karoha) Langwane, Panana Sebati, Nicholas Molese, Marie-Charlotte Gielen, Amo Keitsile-Barungwi, Quashe (/Uase) Xhukwe, and !Nate (Shortie) Brahman. 2019. "Botswana's Wildlife Losing Ground as Kalahari Wildlife Management Areas (WMAs) Are Dezoned for Livestock Expansion." BioRxiv. https://www.biorxiv.org/content/10.1101/576496v1.

Liebenberg, Louis. 1990. *A Field Guide to the Animal Tracks of Southern Africa.* 1st ed. Cape Town, South Africa: D. Philip.

Mathews, Andrew S. 2018. "Landscapes and Throughscapes in Italian Forest Worlds: Thinking Dramatically about the Anthropocene." *Cultural Anthropology* 33, no. 3: 386–414.

Mavhunga, Clapperton Chakanetsa. 2014. *Transient Workspaces: Technologies of Everyday Innovation in Zimbabwe.* 1st ed. Cambridge, Mass.: MIT Press.

Milton, Shaun. 1997. "The Transvaal Beef Frontier: Environment, Markets and the Ideology of Development, 1902–1942." In *Ecology and Empire: Environmental History of Settler Societies,* edited by Tom Griffiths and Libby Robin, 199–212. Melbourne, Australia: Melbourne University Press.

Rose, Deborah Bird. 2004. *Reports from a Wild Country: Ethics for Decolonisation.* Sydney, Australia: UNSW Press.

Tsing, Anna Lowenhaupt. 2015. *The Mushroom at the End of the World: On the Possibility of Life in Capitalist Ruins.* Princeton, N.J.: Princeton University Press.

Turnbull, David. 1997. "Reframing Science and Other Local Knowledge Traditions." *Futures* 29, no. 6: 551–62.

Verran, Helen. 2001. *Science and an African Logic.* Chicago: University of Chicago Press.

Verran, Helen. 2013. "Engagements between Disparate Knowledge Traditions: Toward Doing Difference Generatively and in Good Faith." In *Contested Ecologies: Dialogues in the South on Nature and Knowledge,* edited by Lesley Green, 141–61. Cape Town, South Africa: HSRC Press.

Viveiros de Castro, Eduardo. 1998. "Cosmological Deixis and Amerindian Perspectivism." *Journal of the Royal Anthropological Institute* 4, no. 3: 469–88.

Viveiros de Castro, Eduardo. 2004a. "Exchanging Perspectives: The Transformation of Objects into Subjects in Amerindian Ontologies." *Common Knowledge* 10, no. 3: 463–84.

Viveiros de Castro, Eduardo. 2004b. "Perspectival Anthropology and the Method of Controlled Equivocation." *Tipití* 2, no. 1.

Watson-Verran, Helen, and David Turnbull. 1995. "Science and Other Indigenous Knowledge Systems." In *Handbook of Science and Technology Studies,* 114–39. Thousand Oaks, Calif.: SAGE.

Willerslev, Rane. 2004. "Not Animal, Not Not-Animal: Hunting, Imitation and Empathetic Knowledge among the Siberian Yukaghirs." *Royal Anthropological Institute* 10, no. 3: 629–52.

CHAPTER 4

Plants of Internal Colonization

*Critical Descriptions of Agrarian Change through
Plant Agencies in South India*

DANIEL MÜNSTER

Walking the gardens of natural farmers in the highlands of Kerala offers a glimpse into an alternate history. What if small-scale farmers had embraced agroecology rather than the temptations of agrochemistry and monoculture? As I followed Mr. John, a retired teacher, through his five acres of tropical forest garden, the scene changed with every slow and barefooted step that we took. Seemingly proud and eager to share his knowledge, he helped me notice that edible, medicinal, or simply beautiful plants were growing everywhere around us, above and below. John was an admirer of the environmentalist and biologist Rachel Carson (1962) and followed the teachings of the Japanese farmer and alternative agronomist Masanobu Fukuoka (2009). His one-acre garden produced an abundance of diverse crops: pepper, areca nuts, coffee, ginger, grass for his cow, cardamom, coconut palm, jackfruit, and a variety of tubers and vegetables, and plenty of small edible plants grew in different spots of his surroundings. I had met Mr. John, widely known in the region by the honorific John Master, to talk to him about the series of farmer suicides that had afflicted Wayanad district in recent years. He was among the first interlocutors I met who explained farmer suicides not primarily in economic terms. Instead, he drew my attention to the role that plants played in individual and collective stories of suicide, death, and ruin. As he gave me the tour through his plot, he singled out leguminous plants and explained their nitrogen-fixing capacities. John also pointed to edible plants that have the capacity to grow and produce food in abundance, and he explained the importance of old trees for providing shade and moisture to the plot. To his regret, however, these plants were no longer cultivated in Wayanad, a region now famous for chemical-based, intensive, cash crop cultivation, such as ginger. By cultivating "old" plants, John highlighted the possibility of an alternative to the cost- and chemical-intensive farming practices of mainstream agriculturalists.

As we walked and talked, John urged me to notice the contrast of his garden plot to the fields of some of his neighbors, whom he called "chemical farmers." He made me smell the moist soil, notice earthworm castings, admire a healthy vine of pepper, and see the botanic diversity all around as he challenged me whether I had noticed the effects that generations of different cash crops have had on the region's landscape. Had I seen banana and areca nut take over wet rice valleys? Was I aware that most dryland fields in Wayanad once featured gigantic trees and that settlers like himself had cleared them? Had I realized that frogs, fish, and crabs had disappeared from most fields in Wayanad?

John explained that he maintained his homestead garden also as a kind of "model farm," a way to inspire hope and demonstrate to his neighbors and friends that mixed horticultural fields can yield a diverse but rich harvest throughout the year without requiring a lot of money and work. He was well aware that the security of his government pension had once allowed him to take the risk and experiment with Fukuoka's "do-nothing farming," a style of farming that minimized human interventions, such as tilling, pruning, or fertilization, and instead nurtured symbiotic relationships between plants and soils. Mr. John also acknowledged that many other settler farmers, many of whom were Syrian Christians like himself, felt that they had no other choice but to enter the treadmills of monoculture, credit, and chemicals that produced Wayanad's landscape of agroecological crisis in the highlands of Kerala. Visiting John's fields, I began to see that an alternative trajectory of settler agriculture would have been possible, and I began to ask whether the contemporary agrarian crisis was the outcome of specific "styles of farming" and the cultivation of particular crops in the postcolonial history of the district.

FOLLOWING FRONTIER PLANTS: MULTISITED FIELDWORK IN A LANDSCAPE OF CRISIS

In 2008, the year I stayed on John's farm, Wayanad was infamous as a hot spot of farmer suicides. During my fieldwork, part of my research involved tracking down families in which suicides had occurred and trying to learn from family members and neighbors about the possible causes of suicide, which I had initially assumed to find in the political economy of small-scale farming in globalizing India (Münster 2012, 2015a; see also Vasavi 2012). This phase of the research led my friend and field assistant Joby and me to farm visits across the district. We had obtained a list of names of 435 "farmers' suicides" that had been officially recognized by the state of Kerala between 1999 and 2007. Visiting the "victim families," as Joby and other nongovernmental organization workers often called them, and conducting semistructured interviews with family, neighbors, and friends, we hoped that recording individual life trajectories would help us make sense of the suicides. My initial research was aimed at understanding the connection between suicide, biopolitics, and precarious livelihoods (Münster 2012, 2015a), but soon my interlocutors reoriented my curiosities to the materialities of farming and to the multispecies history of plants growing in this landscape. As Heather Swanson (2020a, 16) puts it, "in good fieldwork, one must open oneself to the curiosities of others and allow them to alter one's own." In my case, this involved modes of attuning to historicity and the agency of plants in the history of settler migration to Wayanad since 1945.

Following plants and how they were remembered by a variety of people meant that my fieldwork became largely multisited, in the sense that it did not involve deep immersion in a face-to-face community through stationary fieldwork as I had practiced previously in an ethnographic study of a multicaste village in Tamil Nadu (Münster 2007). This time, fieldwork recurred over the span of eight years and covered a district of more than two thousand square kilometers. Settlement patterns in Wayanad were largely dispersed, with individual houses spread out in the landscape, each surrounded by a garden or fields. A typical fieldwork encounter was thus a prearranged visit to a house, with more or less structured conversations about settler histories and successive styles of planting. During these visits, my interlocutors directed my interest to their struggles in appropriating the land on which we were walking and to the role that plants had played in the process of migration and settlement, and later in the boom and bust of cash crop economies. The settler areas are distributed across the district, making the rubber on the tires of my modes of transportation as methodologically important as the rubber below my feet.

It took me a while to recognize the landscape of Wayanad as a frontier in ruins. As I was traveling hundreds of kilometers with my companions by car, on the backs of motorcycles, on local buses, and by foot, hiking to houses in remote parts of the hilly district, my companions taught me to read the cultivated landscape as formed by the political history of migration and the multispecies history of key crops. First, I learned to see the difference between wetlands and drylands—in Malayalam, called *vayal* and *kāra,* respectively. This distinction, which is widespread across Asia (Geertz 1972; Kirch 1994), describes both the landscape and its appropriate plants. Whereas rice was the plant of the *vayal* wetlands, the *kāra* drylands were traditionally used for the mixed cultivation of garden fruits in homestead gardens *(tottam),* such as the one preserved by Mr. John. Crisscrossing the district collecting stories of how people and plants made places, I learned to see this historical landscape beneath a current reality in which dryland crops were increasingly being cultivated in wetlands for the purpose of fast profits. Focusing on plants helped me to "better 'read' the landscape" (Tsing, Mathews, and Bubandt 2019, S188).

I learned to see how settler plots were different from those of the landowning cultivating castes of Chetties, clustering around paddy wetlands that used to be communally owned. Colonies of Paniya, the Adivasi laborers with hardly any land of their own, were located near the fields of their former Chetty landlords or on marginal lands. The settler houses, in contrast, were more likely to be built on dryland and feature a variety of styles of cash crop farming. I also learned to notice agrarian ruination. When farmers showed me around their fields, I learned from them to notice the material signs of ruination in the fields themselves; former wet rice fields that were now planted with cash crops like ginger or banana and encroached on from the margins by tree plantations of areca nut *(Areca catechu)* or rubber *(Hevea brasiliensis).* The formerly cool and moist wetlands *(vayal)* now looked dry, hot, and exhausted. I learned to recognize disease-affected former pepper plantations by long rows of *murikku* trees *(Erythrina variegata),* which used to be planted to support pepper monocultures. I learned to recognize Jeeps with laborers and cultivators leaving for the neighboring state of Karnataka to plant ginger on leased land (Münster 2015b) because the soil in their own villages was contaminated with pathogens.

At some point during my intermittent fieldwork, the focus on suicide cases made way for oral histories of migration and for the study of the historical agency of plants in the process of internal colonization. I continued with farm

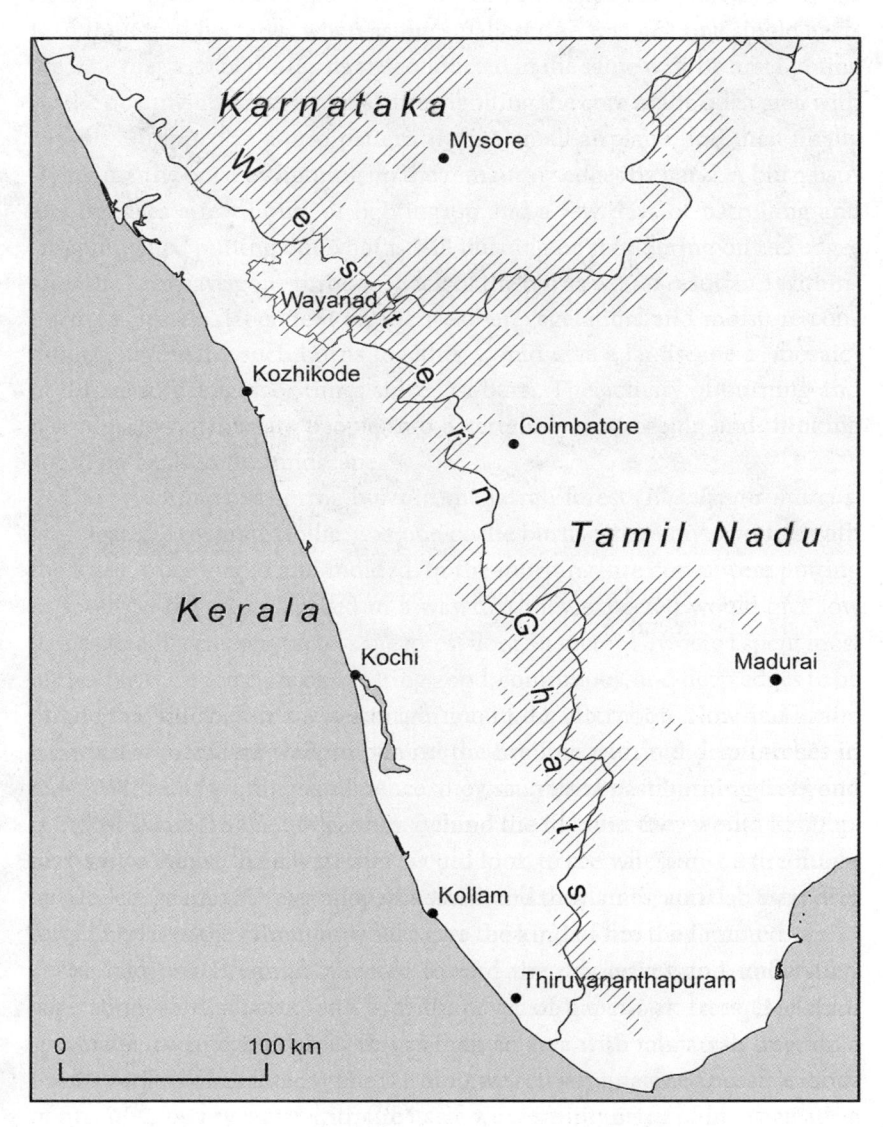

FIGURE 4.1. Location of Wayanad district in South India. Cartography by B. Kraus.

FIGURE 4.2. A Syrian Christian church in the eastern part of Wayanad, 2011.
Photograph by the author.

visits but increasingly looked for people who had witnessed different phases
in the settlement of Wayanad since 1945 and for interlocutors with special
knowledge about individual plants.

The settler frontier was based on the promise of fertile forest soil, struggles for land, and the allure of cheap fertilizers, which began arriving in
Kerala in the wake of the so-called Green Revolution (Cullather 2013) in the
1960s. Windfall harvests of pepper and other boom crops based on chemical fertilizers and monocropping made so many farmers rich in the 1980s
that the region was called Kerala's "mini Gulf." The agricultural frontier came
to an end in a ruinous frenzy of engaging in increasingly risky monocul-
tural and chemical ways of farming until the troubles began in the 1990s. I
call this conjuncture a frontier in ruins, following environmental historian
Don Worster's work on the American frontier.

In *Dust Bowl*, Worster (2004, 100–106) calls the 1930s agroecological disaster of the Dust Bowl in the American Great Plains a "frontier in ruins" of the

settlers' own making: "The Dust Bowl ... was the inevitable outcome of a culture that deliberately, self-consciously, set itself the task of dominating and exploiting the land for all it was worth" (4). My work takes inspiration from Worster's focus on how agricultural settler frontiers may end in human-made disaster, fueled by the logics of capital, colonization, and cash crop production. Following two sets of case studies, suicide cases embedded in life histories and plant cases situated in histories of migration, was facilitated by (self-)critical conversations within the settler community itself. I experienced a frontier society caught in a moment of self-reflection. Conversations about the present ecological failure of agriculture and its connection to human tragedy and suicide frequently turned to crop histories gone wrong.

Attuning agricultural anthropology to the "multispecies crisis" of the Anthropocene requires methods that go beyond the study of people's relations to land, labor, and markets by cultivating a curiosity for the plants, animals, and microbes involved in making agrarian landscapes (see the Introduction to this volume). Moving beyond human settler relations to land, I suggest, allows anthropology to more fully recapture the tragic historicity of those settler relations. Walking the fields of settler farmers and engaging them in conversations about historical successions of crops, I was aiming to reconstruct the anthropological history of how cassava and lemongrass were complicit in making the frontier landscape; how pepper, ginger, banana, and agrochemicals coproduced an ecology of suicide; and how native cows and their microbes became the species of symbiotic recuperation of soils. In agriculture, I argue, stories about the natural–cultural composition of the world and about the material basis of cultivation abound. My human interlocutors, first- and second-generation farmer settlers, were acutely aware of the distributed multispecies agency in the making of the frontiers. Boots and roots made the frontier landscape in concert.

My multispecies methodology is inspired by scholarship in environmental history that focuses on frontiers as projects of more-than-human world making. Particular plants have agency in them as frontier cultivars. Rubber (Dove 2011; Peluso 2012), cocoa (Ruf 2001; Li 2002), palm oil (McCarthy and Cramb 2009; Brad et al. 2015), and cardamom (Fitzpatrick 2011) afford their own natural–cultural properties to the ways land grabs and deforestation have been organized at historical frontiers. Forests and mountains become central figures in resource frontiers or forest frontiers, as are the migrants and companies who grab the resource-rich forests (Tsing 2003, 2004; Peluso and Vandergeest 2011). I thus associate with a strand of multispecies studies that is less preoccupied with novel forms of "immersion" in the lives of non-human others by honing the senses toward multispecies "attentiveness" (van Dooren, Kirksey, and Münster 2016). Emergent plant ethnography (Hartigan 2019) has involved playful attempts at interviewing (Hartigan 2017, 259), listening to, and sensing plants (Boke 2019; Myers 2015). I am interested in how plants make historical landscapes, inscribe themselves in the projects of territorialization (Besky and Padwe), and contribute to changing land relations in frontier conjunctures (Li 2014). Instead of tuning in to plants, I turned to my interlocutors' memories of plants to "read histories in material forms" (Swanson 2020b). I turned to farmers as scientists and historians of a frontier in ruins. My approach is aligned with recent anthropological calls for attunement to multispecies landscapes: "The multidimensional crises of our times call for an anthropology ... that takes landscapes as its starting point and that attunes itself to the structural synchronicities between ecology, capital, and the human and more-than-human histories through which uneven landscapes are made and remade" (Tsing, Mathews, and Bubandt 2019, S186).

For the remainder of this chapter, I sketch the more-than-human history of Wayanad's postcolonial frontier landscape. I hope to show that a historical-anthropological understanding of dramatic changes to the landscapes of the Anthropocene benefits from a double methodological focus on both human and nonhuman agents of change as coproducing a frontier in ruins as well as the potential for repair. I focus on the role of plants, chemicals, and animals in the internal colonization of this agrarian hinterland since the end of the Second World War.

INTERNAL COLONIZATION OF MALABAR (CIRCA 1945–1975)

Agrarian frontiers are sites of more-than-human encounters. Calling Wayanad a settler frontier invites uneasy questions about nativity and belonging—human and nonhuman. Newcomers, settlers, and pioneers bring their own cultivars, chemicals, and aesthetics to encounter people, plants, and wildlife embedded in landscapes and ecologies. "Native" landowning Hindu castes and Adivasis of Wayanad (who had at some point in history also migrated there) used a variety of terms for these cultivating newcomers. *Kuṭiyēṟṟakkar* is the common term and means both "migrants" and "settlers," two English terms also frequently used. In official publications, the settlers are often referred to as "colonisers from Travancore" (e.g., Devassy 1966, 181). Despite the large numbers of Hindus among the settlers, settlers are also frequently called *kristyāni,* which means "Christians." The agrarian frontier I describe here is a postcolonial frontier, from the perspective of mostly second-generation Christian settlers. The Syrian Christians of Kerala comprise several different denominations that follow the Syrian liturgy due to their historical ecclesiastic orientation to the Patriarchate of Antioch and claim to be able to trace their origin to conversions made by the Apostle Thomas when he, according to legend, evangelized in India in the first century A.D. The (then) princely state of Travancore, which merged with Malabar and Cochin to form Kerala state in 1956, is the historical heartland of Kerala's autochthonous Christians. A combination of need and opportunity initiated a land rush by Christians from Travancore to Malabar (including Wayanad) from approximately 1945 to 1975. Poverty, food shortages, and political tensions with an increasingly Hindu majoritarian state in Travancore (Devika and Varghese 2011) "pushed" this relatively well-educated and entrepreneurial community to seek the opportunity for cheap land that opened up after the end of British colonial rule. In Malabar, which was under "direct" rule, the colonialists had given land titles to large feudal land claims (called *janmam*). These extensive lands, often forested, attracted the peasant settlers, who were given land by the state, bought cheap plots from absentee landlords and temples, or encroached on land with future promises of land reforms in mind.

When the migration of agricultural settlers coincides with the expansion of state control (with the formation of Kerala) or development plans to expand agriculture, it may be justified to call the migration to the frontier zone a form of "internal colonialism" with far-reaching consequences for entire landscapes. As Scott (2009, 12) puts it, internal colonialism

> involved a botanical colonization in which the landscape was transformed—by deforestation, drainage, irrigation, and levees—to accommodate crops, settlement patterns, and systems of administration familiar to the state and to the colonists.

In contrast to Scott's Southeast Asian cases, however, Wayanad's internal colonization had less of a planned, state-led character. Rather, it was made up of a multitude of individual efforts in grabbing land, mostly through legal means but also through encroachments, but always with the help of pioneer crops. The remainder of this chapter describes the contribution of more-than-human species to making the frontier. Settlers brought tapioca, coconut palm, areca nuts, and improved varieties of pepper alongside new chemical species, such as dichloro-diphenyl-trichloroethane, better known as DDT, an insecticide that opened the landscape by killing mosquitoes during the large-scale malaria eradication campaigns. Later, during the era of the Green Revolution, new chemical inputs arrived, such as mineral fertilizers and pesticides, along with more improved varieties of high-value cultivars, such as pepper. I then turn to pepper and ginger as neoliberal crops that materially shaped the spate of suicides that had become emblematic of Wayanad's agricultural crisis since the 1990s.

Frontier Crops: Cassava and Lemongrass Aid Internal Colonization

In Wayanad, two plants, cassava and lemongrass, can be regarded as frontier plants during the land rush from 1945 to 1975. Their qualities made them an excellent means of taking land and accumulating money in the initial stages of the settler frontier, when land tenure was insecure and the needs of subsistence slowly faded into the desire for accumulation. To put it into a simple formula, the frontier had one plant to feed the settlers and another plant to eat the forest. Both crops were "agents of deforestation," as Ruf (2001) puts it. The two cultigens also share an essential property: they both begin to yield after a few months. Cassava and lemongrass lived to their full potential in a situation of insecure land tenures and forceful appropriation of forest land by the settlers.

Manihot esculenta is known in Wayanad as tapioca and *kappa* and to agronomists by the name cassava. The plant, native to tropical South America (where it is known as manioc and yucca), was introduced as a food crop to Goa and the coast of Kerala by Portuguese settlers in the sixteenth century (Lebot 2009, 8). In contemporary India, the starchy root is not very popular as a vegetable, since rice and wheat dominate as sources of carbohydrates and carriers of flavor. People eat cassava only in the states of Andhra Pradesh and Kerala, where it is perhaps most accepted as a staple food crop and cash crop and only since the 1850s (Crosby 2003, 193). In the rest of India, as in most of the world food system, cassava has become an industrial crop and raw material for the extraction of tapioca starch and its derivatives (e.g., sago) for further industrial use in processed foods, in textiles, or in the manufacture of bioplastics (Edison 2001).

The special relationship of Kerala and tapioca was consolidated in the late nineteenth century in the princely state of Travancore. After the Great Famine of 1876–78, which devastated all of South India (see Davis 2001, 25–59), the government of Travancore promoted tapioca as a new diet staple for the poor by exempting it from taxation (Joseph 1986, 98). The first wave of migrants, those who migrated during the late colonial period, knew tapioca as a famine crop. In Travancore, they had learned to eat and cultivate the tuber, which arguably protected Travancore from major famines up until the Second World War, when India's food production was diverted to feed British Indian troops in their war against the Japanese. Some settler families with whom I spoke remembered that it was hunger and food shortages in the 1940s that drove them to relocate to the available lands of Wayanad, with the intention of cultivating tapioca for subsistence.

Regardless of whether the primary motive of settler migration was subsistence cropping or cash cropping, alongside dryland paddy, the first plant they tended to cultivate was cassava *(kappa)*. Elders in the community remembered how it could be intercropped and made part of a forest garden with yams, pepper, ginger, lemongrass, trees, and coffee as companion plants, or it could be monocropped, harvested, and sold for a high price during famine years. During the 1940s, tapioca was also a high-value cash crop and opened up the possibility of personal accumulation. Cassava was instrumental both in feeding the first wave of migrants and in raising the cash for their further expansion on leased, purchased, or encroached lands. Cassava was also the ideal plant for insecure land tenures. Prior to the land tribunals of the 1970s, most settlers were cultivating land with very insecure lease agreements and land that often was contested as belonging to a temple or the Forest Department. The South American plant, which was fast and easy growing, without any natural enemies, thrived on encroachments onto recently cleared forestland.

Cassava is widely used in tropical shifting cultivation systems across the world. James Scott (2009, 24) includes it under crops suited for what he calls "escape agriculture" or "forms of cultivation designed to thwart state appropriation." These include technologies (such as swidden agriculture) and crop choices (such as tubers) practiced historically by communities evading or escaping appropriation by state authority:

> They preferred to plant root crops (for example, manioc/cassava, yams, and sweet potatoes), which were unobtrusive and could be left in the ground to be harvested at leisure. Depending on how secure the site was, they might plant more permanent crops, such as bananas, plantains, dry rice, maize, groundnuts, squash, and vegetables, but such crops could more easily be seized or destroyed. (Scott 2009, 190)

As the case of Wayanad shows, however, the same crops that have a high "escape value" (Scott 2009, 199) are also useful for the appropriation of new lands at the frontier—and for many of the same reasons. Cassava was the ideal frontier crop because it was very forgiving in its cultivation: it could be planted at any time of the year—whenever a new stretch of forest had been cleared by cutting or burning—and would bear tubers six months after planting without much attention or care needed; it could handle a variety of soils but thrived on former forest land; it could be vegetatively propagated, making it easy to multiply by using cuttings from existing plants; and it was less attractive to the wild animals of the forest. In his praise of cassava, James Scott writes, "Perhaps the most striking advantage of cassava, however, is its undisputed status as the crop requiring the least labor for the greatest return" (206). Agronomists claim that cassava has about the highest return of calories per hectare, which makes it also an ideal subsistence crop for intercropping with other cultivars (Leihner 2002). Cassava's harvest is also quite flexible. Although the roots degrade quickly after harvest, once mature, they can be kept in the soil for up to two years as storage roots for perennial plants, and the settlers can harvest them as needed. Once harvested, the plant is also easily preserved by boiling and drying. Dried cassava, locally known as *kothu kappa,* was also a nonfood merchandise in high demand in the textile industry as a source of starch (Mathew 2011).

Given these qualities, cassava quickly spread in the fresh clearings of the forested dry lands *(kāra bhūmi),* occupied by the settlers from Travancore. It made the settlers independent of paddy production in the wetlands *(vayal),* which continued to be largely controlled by the indigenous Hindu castes. According to some accounts, incoming migrants planted tapioca before they even built their first houses of bamboo mats *(tital)* and mud. But tapioca and other tubers attracted the Indian wild boar *(Sus scrofa cristatus)* that lived in the forest. Some settlers soon figured out that planting bitter cassava, locally called *kannan kappa,* had the great advantage that wild boars did not like it. The raw tuber is poisonous and only edible after soaking in water and boiling. After the settlers had reduced the forest cover further and controlled the wild boar population through poaching, they shifted to normal—sweet—tapioca.

Cassava was the essential food crop of the pioneering settlers at the forest frontier. It grew quickly and abundantly in Wayanad and provided plenty of carbohydrates to hungry newcomers. *Kappa* stew became the staple food of rural workers and is to this date the signature dish of Christian identity. It was served twice a day in the fields to workers and cultivators alike. Older settlers remember how the tapioca they had introduced helped settlers and Adivasis survive the food shortages during and after the Second World War. Contemporary Christian settlers consider tapioca a typical delicacy of their community and prefer to eat it cooked as a sticky, starchy stew, prepared with all kinds of meat. As a frontier crop, cassava was a plant of shifting cultivation that helped people stay in one place.

What made cassava the ideal crop for the appropriation of the frontier wildness was its versatility, sturdiness, and ease of cultivation. Like corn, it has a dual identity, "as food and commodity" (Pollan 2006, 26). This dual identity as a nonfood commodity that could also be eaten made it an ideal plant of reckless arrival. The plant does not take sides in the moral-historical question that came up in most of my conversations, of whether settlers came out of need or out of greed. Cassava itself, quite "greedy" for the nutrients of freshly opened forest soils, was happy to offer its starchy tubers to anyone who cared to propagate it. Spreading onto fertile soils that have been built up over millennia, cassava enabled the smallholder frontier by providing quick food and quick cash. It was not the only frontier crop. Other plants, such as lemongrass, also followed the settlers' migration and made their way into Wayanad's ecologies. Lemongrass has inscribed itself into the contemporary landscape by being an agent of deforestation and commercialization.

Lemongrass was the first nonfood cash crop on the land of the settler migrants. In contrast to cassava, lemongrass was exclusively a commercial plant. The olfactory qualities of lemongrass—its characteristic lemony smell—allowed for the addition of value through distillation. With its value concentrated in its aromatic oil, the plant was ideal for capital accumulation in faraway places. Moreover, its cultivation had the beneficial side effect of requiring large quantities of firewood for the extraction of its oils, making lemongrass the ideal crop for settlers who wanted to get rid of the trees, bamboos, and bushes that covered the landscape. In conversations in eastern Wayanad, where land appropriation was most messy and violent, I heard of an additional advantage of lemongrass cultivation in the context of legal and political struggles for land. In the context of imminent land reforms in Kerala that promised occupied land "to the tiller" but also assigned forested land to the Kerala Forest Department, encroachers were eager to show that they were cultivating their plots and that the land was indeed cultivated. It was thus important to rapidly "clear" the landscape of forest shrubs. In the years of legal uncertainty, fields of lemongrass gave the landscape a cultivated look, biophysically marking it as occupied ground.

Several varieties of lemongrass, now grown throughout the tropics, are native to Asia. In Wayanad, cultivators grew the so-called Malabar grass or Cochin grass *(Cymbopogon flexuosus)*. This variety is native to the Western Ghats and known under the trade name East Indian lemongrass. It is a perennial plant that accumulates essential oils in the leaf blade. After two to five months of cultivation, the leaves can be harvested and processed for the first time. Subsequently, the plant will yield two to five times a year for several years, providing a steady income to cultivators.

Lemongrass was a perfect successor for cassava, as it has the ability to grow even on soils that have lost fertility after the cultivation of cassava. The most important feature of lemongrass, however, was that it meshed so well with the settlers' desire to clear the forest and mark it as occupied. A recurrent theme in my conversation with first-generation settlers comprised expressions of negative emotion toward the forest. Older settlers talked about the forest in "those days" as cold, misty, overwhelming, and dangerous. Many spoke of forests in the 1950s and 1960s not only as potential land for settlement but also as a threatening "jungle" hosting dangerous wildlife, malaria, and a harsh climate. Lemongrass served them as a means and justification for the removal of trees. Lemongrass flourishes in full exposure to the tropical sun and can withstand its heat. It does not like the shade provided by forest trees. Thus its yield increased in direct correlation to the disappearance of the forest. Burning forest trees was technically illegal, or at least highly regulated, depending on the type of land. However, as one elderly migrant put it, "[in the 1950s,] the Forest Department was not yet organized." Lemongrass cultivation codeveloped also with the emergent pattern of turning forest not only into food but also into profit. The excess wood and shrubs cleared from Wayanad's dry deciduous forests fueled lemongrass distillation units in need of cheap energy.

As a commercial pioneer crop, lemongrass was ideal because its value could be increased by distilling the essential oils in situ. Its scented oil is known by the international trade term Cochin oil, as the port of Cochin (Ernakulam) is the main export hub. Like cassava, lemongrass has a close association with Kerala. The state holds the monopoly of its production in India; one-fourth of the global crop is distilled there. In 1951, Kerala Agriculture University established the first research station for aromatic and medicinal plants, commonly known as the "lemongrass research station" (Joy et al. 2006). For many frontier settlers, the flip side of lemongrass production was that they were reducing the growing of food in their fields and (retrospectively speaking) were embarking on a path of speculative commercialization that decades later culminated in a suicide crisis. Unlike cassava cultivation, which was simultaneously a subsistence crop and produced for the market, lemongrass was only for the market. This was not the frontier of hungry peasants fighting starvation; it was the frontier of rural entrepreneurs. Lemongrass cultivation needed markets—in this case, the global export markets that had been established by colonialism. With lemongrass, the frontier showed its face as a frontier of accumulation: frontier farming became synonymous with cash cropping.

Transporting high-value oil, as opposed to bulk material and heavy cassava tubers, was also much easier at the remote frontier. Settlers could easily produce lemongrass oil via a process called "steam distillation" (Skaria et al. 2012, 361), in which steam passes through the grass blades, collecting the oil—

with heat gained by burning (smaller) trees and shrubs farmers had cleared from their fields. Cooling the steam in a copper vessel separated the water from the aromatic oil. Lemongrass highlighted differences in class and skill among the settlers. One of my interlocutors pointed out to me that only wealthier settlers could afford such a field distillation unit: "Those who had some money built a distillation plant. Wood was not a problem at the time. Those who had the units had the customers. The leftovers from the distillation process were good manure for the paddy fields."

Lemongrass was the pioneer's cash crop of choice, but its cultivation was not sustainable at all. Cassava and lemongrass profited from what scholars of tropical agriculture call the "forest rent" (Ruf 2001; Li 2002; Hall 2011), which is the additional fertility afforded by recently cleared forest. It is a "cheap" fertility, in Jason Moore's sense, as it can be extracted without the work of soil care. The forest rent seems indispensable for "insecure crop booms" (Hall 2011). Lemongrass produced the first "get-rich-quick stories" (Hall 2011) at the frontier and thereby attracted further migration, and it helped to visually mark a plot of landscape as "under cultivation," which was important for land claims. They were the first smallholder boom crops and initiated the cycles of rural booms and busts that had ruined the frontier by the time of my fieldwork. Colonizing the land with these pioneer crops created the landscape conditions and economic logic for the boom crops directly associated with farmer suicides.

Species of Boom and Bust: Pepper, Ginger, Banana, and Agrochemicals Make Riches and Suicides

In this section I turn to agrarian species that are visibly inscribed into the contemporary landscape. They all featured prominently in the collective endeavors of making sense of what went wrong in Wayanad, conversations into which it tapped when studying suicides. Multispecies history is the history of advancing commercialization of specific plants. The commercialization of agriculture in Wayanad intensified once property titles were granted to most of the settlers in the long process of land reform in the 1970s. It was further supported by the arrival of chemical fertilizers and pesticides. Even after Wayanad closed (or slowed down) as a frontier of land grabs, its agriculture continued to be informed by the logic of crop booms and their tragic busts. However, the dynamic shifted from an insecure boom, characterized by legal, political, and economic uncertainty, to a succession of agricultural booms and busts based on secure land tenure and new crops.

PEPPER: THE VINE OF SETTLERS WHO STRIKE IT RICH

Piper nigrum, black pepper, *kurumuḷaku* in Malayalam, is native to Wayanad and the Malabar Coast. Malabar has a long history of trade with the world (Morrison 2002), and its black pepper is considered to be the best in the world. In global commodity trade, the highest-quality black pepper, the one with the largest drupes, is called "Tellicherry Garbled Extra Bold" after the trading town of Thalassery in Kannur. Despite this long-standing biological and economic connection to Wayanad, contemporary pepper is primarily a plant of the settler frontier. The settlers first introduced the new hybrid cuttings that had been developed at the Pepper Research Station in Panniyur (Kannur district) from 1952 onward. Accordingly, these commercial hybrid varieties of pepper were called Panniyur-1 to Panniyur-5 and were the protagonists of the pepper boom of the 1980s, along with a number of other varieties, such as Karimunda, Kottanadan, Narayakodi, and Sreekara. Today, India is the largest producer of black pepper in the world, and Kerala produces some 90 percent of all Indian pepper.

In the 1970s, many settlers started pepper cultivation. This was a substantial shift in the temporality of cultivation. Whereas lemongrass needed only forty-five days before it could be harvested and processed, pepper vines didn't begin yielding until three years after planting, reaching their maximum after seven years, but they can then yield for up to twenty-five years. This changed the temporal horizon of planning considerably. Pepper is vegetatively reproduced and requires timely planting of the cuttings just before the monsoon begins (Ilyas 1976). Throughout its life-span, pepper then requires the constant care of pruning, hoeing, weeding, and fertilizing by the cultivator/gardener. By the mid-1970s, settlers had the first harvests of pepper. In the 1980s, the plants continued to produce bountifully and now combined with great prices for pepper, so that it became known as black gold. The pepper boom ended only a decade later with a near-total loss of pepper and many livelihoods. During the boom years, pepper fundamentally transformed homestead agriculture.

Initially, most settlers had cultivated pepper in combination with coffee as part of mixed homestead gardening. Pepper was a key species in these flourishing gardens that many settler families had established around their houses. Such home gardens used to be widespread in the midlands of Travancore, where the hills have plenty of dryland *(kāra)* and fewer wet-rice flats *(vayal)*. Individual houses were in the center of these dryland gardens, which had qualities of both field and forest. When settlers reproduced that model, Wayanad's version of agroforestry featured combinations of trees (old forest trees and newly introduced coconut and areca nut palms), bushes (coffee), and, at the lowest level, annual crops of tubers (cassava, yam, elephant foot yams, colocasia), rhizomes, vegetables, and grasses. Pepper grew there symbiotically and flourished, just as it continued to do on Mr. John's farm in 2008. With the high prices of the 1980s, however, many farmers wanted to cultivate nothing but pepper and began remodeling their land in favor of pepper monocultures. In light of the boom in pepper, many settlers with whom I spoke recalled perceiving coffee and other longer-living plants as a burden for their plots, as they wanted to produce only more of the "black gold."

The shift to monocropping was not subtle. An older settler described with horror the removal of coffee bushes. His neighbor had hired a *kumkhi* work elephant and an elephant handler, put large metal chains around the grown coffee bushes, and literally ripped them out of the ground. The next casualties were the existing trees that supported pepper, some of which were remains of the erstwhile forest and others of which were planted for fruit and flowers. As a perennial evergreen plant, the woody (herbaceous) vines of *P. nigrum* require trees to support them for many years. Pepper vines are attached by aerial roots to the supporting tree, but they are not parasitic on the tree, and both the trees and the pepper provide welcome shade and moisture to the rest of the garden. Settlers, however, were unhappy with the sparsely and randomly placed support trees in their fields. Many thus felled or burned the last of the old forest trees. After clearing the dryland of perennials and forest trees, many settlers planted rows of fast-growing *murikku* *(E. variegata)* trees, and some even installed cemented poles on their fields.

One settler described the changes thus: "After 1980 all the agricultural area was converted from food [*bhakṣa*] to money crops [*nāṇya viḷa*]. Someone identified the soil as ideal for pepper; slowly everything became a monocrop.

Even on small patches." In 1986, the price of pepper rose dramatically, generating "a large amount of money-vehicles, houses, *paisa* [money]."

The unexpected and sudden wealth generated by pepper made farming a very attractive occupation and, in contrast to the trend in agricultural India, a very lucrative one as well. Wayanad became prosperous and began to attract large numbers of migrant laborers from Tamil Nadu. But even as agriculture in Wayanad was becoming lucrative, it was also becoming speculative and risky. On dryland *(kārasthalaṃ)*, farmers who had both a homestead garden and dryland crops began monocropping closer to home at the expense of their previously diversified gardens. First it was pepper, then whichever crop was promising the highest returns or the lowest labor costs. Many poor settlers with little land of their own extended cultivation on leased land, often on the basis of loans with horrendous interest rates from cooperative banks, private banks, and moneylenders (see Münster 2012).

The pepper boom owed its success not only to the settlers' enterprising spirit and the invention of pepper monoculture but also to the "cheap" fertility afforded by the forest soils and the newly arriving synthetic fertilizers.

Agrochemicals: NPK and Pesticides Enable Monoculture

In the 1980s, new chemical species of mineral fertilizers and synthetic pesticides entered Wayanad on a large scale, supported by agricultural subsidies and the recommendations of agricultural extension officers, who were trained in the new agronomic paradigm of the so-called Green Revolution. This large-scale modernization program, driven by U.S. interests, replaced traditional agricultures with "scientific" methods on the basis of hybrid seeds, irrigation, and agrochemicals across the Global South. In Wayanad, synthetic nitrogen-phosphorous-potassium (NPK) fertilizers, called *rāsa valaṃ* (chemical fertilizers) in Malayalam, gave an additional boost to the new methods of pepper monocropping developed by settlers in conjunction with local research stations. After the forest rent, farmers now benefited from what might be called the "chemical rent." The chemical rent is that initial boost in productivity and profit provided by the chemical enhancement of soils and plants. The additional yield is the main "temptation of nitrogen" (Müller 2012), which lasts until the long-term side effects of NPK use show in the fields and the true costs of artificial fertilizer are revealed. A farmer explained the fertilizer treadmill like this: "In the beginning, if someone is applying hormones and chemicals it will look nice and give a nice price in the market. Now the fertilizers are more important because the land has lost its fertility."

As is well documented in the literature, excessive nitrogen fertilization depleted the soil by disrupting microbial soil activity, which is responsible for replenishing soil with organic carbon. Ecologists speak of the "silent hunger of the soil" (Environmental Governance Group 2011, 15) that occurs when micronutrients—as opposed to the NPK macronutrients—are mined from the soil, if they are not replenished by microbial activity. Mineral fertilizers also raise the pH level in the soil, making it more acidic, which changes the soil's bacterial milieu. Excess nitrogen fertility increases soilborne diseases and fosters the outbreak of insect pests. Nitrogen and phosphorous also have consequences beyond the field, as they "cascade" (Billen, Garnier, and Lassaletta 2013) from field to field and accumulate in oceans and the atmosphere, disrupting the "biochemical flows" of planetary nitrogen and phosphorous cycles that connect all life (Steffen et al. 2015).

Chemical biocides—the entire spectrum of pesticides, herbicides, and fungicides—arrived a few years after NPK fertilizers but were soon adopted with the same indiscriminate and boom-oriented attitude. Before long, the use of biocides took on catastrophic proportions. In contrast to fertilizers, they are freely traded on the market. Today, when Wayanad's settler farmers look back at the 1980s and 1990s, they describe their own attitude of the time as a mixture of greed and ignorance:

> People applied tons of toxic pesticides [kīṭanāśini] in Pulpally. In our region it was more, because everyone wanted to be part of the boom. The fertilizer traders brought a lot of fake and banned products from other states. The government encouraged us to use chemicals but provided no checks on private sale.

Not everyone agreed with these new styles of farming, which people began calling rāsa krsi—"chemical farming." Especially among the native cultivating castes of Wayanad, I heard many remarks that rāsa krsi was the way of the kuṭiyērṛakkar, the migrants, only, and that the Christians had introduced these "harsh" ways of farming.

Most settlers in eastern Wayanad have stories to tell about how their fortunes shifted dramatically when pepper suddenly disappeared. Beginning in the early 1990s, farmers lost almost all their pepper to *Phytophthora capsici*. Phytophthora, which literally means "destroyer of plants," is a water mold (oomycote) that causes symptoms in pepper locally known as quick wilt or foot rot. The water-loving mold infects soils and plants during the monsoon season and begins to show as black spots on the leaves and by turning the tender runner shoots black. Within a month, the disease spreads to the entire plant, causing it to yellow, wilt, lose all its leaves, and ultimately dry up. In Wayanad, quick wilt lived up to its name and rapidly destroyed entire landscapes of pepper gardens.

One farmer who lost all his pepper in 1995 retrospectively regretted that despite the terminal crisis in pepper, "people did not go back to food cultivation but tried the next cash crop." Rather than being viewed as a cautionary tale, the pepper crisis signaled the beginning of speculative agriculture and the suicide crisis. The logic of boom and bust went on to repeat itself with other crops (vanilla, banana, ginger). In this way, the decline of pepper marks the beginning of the socioecological crisis that has since characterized Wayanad. Instead of returning to mixed cultivation, most settlers were forced to engage in the next hype in the hope of making up for past losses. In the next section, I turn to the two crops my interlocutors associated most directly with farmer suicides: banana and ginger.

Banana and Ginger: Rhizomes of Speculation

Ginger and banana had both been cultivated in South India long before the arrival of colonialism and capitalism. Ginger, or *Zingiber officinale* (from the Tamil *iñciver*), is endemic to the Western Ghats and is, together with pepper, among the earliest recorded spices to be cultivated and exported from southwest India (Ravindran and Babu 2005). Despite this long-established plant relationship, the cultivation of ginger has changed considerably, first with internal colonization, then in response to agricultural crisis. Settlers brought ginger rhizomes along with other new cultivars, such as tapioca, lemongrass, and other pioneer crops, when they migrated to the region. Ginger was valued for being hardy and fast growing. It could be planted in between trees and other shrubby cultivars, such as coffee, and it was suitable for recently cleared forest soils and as an addition to mixed gardens. Later, when the settlers established homestead agriculture, they grew it on the fringes of fields or in small rows alongside other plants, primarily for household consumption.

The genus *Musa,* commonly known as banana, has also been cultivated for millennia in South India. Among the settlers of the 1970s, banana was a crop of the home garden in the "planting continuum," to be planted in single shoots near the houses. The planting continuum refers to a gendered spatial division featuring home gardens closer to the houses and wet-rice paddy fields and cash crop cultivation farther away. In many settler families, women cared more for gardens closer to home, while outside cash crops were an exclusively male domain.

With the crisis of pepper, both ginger and banana moved from the female space of the home garden to the male-dominated sphere of the plantation. In this shift, these plants—its "improved" varieties—were forced into monocultures and speculative economies and were subjected to large amounts of agrochemicals. In monocultures, both ginger and banana are prone to failure. A strong wind, untimely rain, or pathogens, such as the soilborne soft rot afflicting ginger or the pseudostem borer *Odoiporus longicollis* affecting banana, can destroy entire fields within days. Compared to pepper, which takes years of commitment, the temporality of ginger and banana as annual crops makes them quintessentially speculative crops. A single season can make a cultivator rich or go bankrupt. Many farmers who had lost pepper started cultivating ginger and banana based on loans at cutthroat rates. Often cultivated on leased land, typically on converted paddy wetlands, these plants had the sole purpose of making a lot of money in a single season. They became high-stakes crops that played a foundational role in the crisis economy of Wayanad (Münster 2015b). These high stakes, in turn, led many cultivators into using excessive amounts of chemical fertilizers and pesticides. In the process, banana and ginger became infamously toxic plants. During fieldwork, I was often offered bananas with the reassurance that these were grown in the "kitchen garden" and without pesticides, so widespread was the recognition that commercially produced bananas are too toxic to be safe to eat. In particular, the pesticide Furadan (carbofuran) became a common poison in many households, kept for its importance in controlling the pseudostem borer. With Furadan, my research on frontier plants was brought full circle. Furadan was also by far the most common means of committing suicide. And the short-term cultivation of banana and ginger, particularly on the basis of credit and on leased land, was by far the most common immediate reason given to me for farmers' suicides.

Conclusion: Landscape, Land, Soil

A multispecies perspective on anthropological histories of the Anthropocene offers a deeper understanding of agricultural frontier dynamics that have transformed large parts of the tropical Global South into plantations at the expense of forests and indigenous agriculture. A focus on plants and other nonhuman species allows us a fresh perspective on the materiality and temporality of the making and unmaking of an agricultural frontier. The contemporary unmaking of Wayanad's agrarian frontier is marked, among other manifestations, by tragic occurrences of smallholders faced with debt and failing crops committing suicide. Critical descriptions of landscape histories from a multispecies perspective provide an ecological, more-than-human explanation for agricultural crisis. A methodological focus on plants and their properties reveals their agency in shaping the history of migration, welcoming Green Revolution chemicals, structuring market forces, shaping Christian settler culture, and driving smallholders to speculation and ruin. Thinking with plants also helped me understand and "attune" to the way agrarian crisis is inscribed onto the landscape. The material-semiotic qualities of domesticated plants offer an understanding of the "structural synchronicities between ecology, capital, and the human and more-than-human histories" (Tsing, Mathews, and Bubandt 2019, S186) that political-economic analysis alone would not achieve. Remembering migration and individual histories of settling in Wayanad puts the role of plants center stage. My attempt at understanding the suicide crisis of the ethnographic present tapped into ongoing discussions of what had gone wrong. Looking back, my interlocutors correlated specific plants with specific phases in the ongoing drama of frontier making, boom, and crash.

The pioneer plants cassava and lemongrass thrived in the temporalities of insecure land relations and were complicit in feeding encroachers, clearing forest, and creating a landscape for extractivist farming. Agrarian cultivars enabled the grabbing of land, the transformation of forest into dry land, and the commercialization of small-scale agriculture. Internal colonization, much as the larger projects of empire (Anderson 2004), is a multispecies process. Pepper was an unlikely candidate for initiating an agrarian boom and bust. It was a plant of the region, ideally adapted to climate, soils, and plants and flourishing in multispecies gardens. During the 1980s, however, it lured farmers into experiments with monocropping. When cultivated as a monocrop, and being fed only synthetic food, pepper falls prey to slime molds and, in the process, coproduces a scene of agrarian crisis, including the first instances of ruined farmers' suicides. With ginger and banana, forced into monocultures and kept alive with pesticides, we encounter the plants and chemical species behind farmer suicides, whose stories are too often told in reductionist terms as a failure of the state or of globalization. "Ginger is a killer" some in Wayanad say. However, the specularization of agriculture has been produced with and through plants and their people.

As colonialism and capitalism transform multispecies landscapes into territory, property, and resource, specific plants at specific times move center stage. While a majority of settler farmers continue to chase ever-elusive profits from cash crops or the sale of land, a small but vocal group of farmers are committing themselves to exploring symbiotic agriculture based on novel understandings of microbial ecology and traditional techniques (Münster 2021). The multispecies engagements of agonistic farmers who practice natural farming as an ethicopolitical counterpoint to chemical farming force us to acknowledge an alternative world in which lively soil is connected to the rest of life via invisible and largely unknown microbial relations. In anthropological storytelling, especially of the hopeful kind, it matters how we conceive of how the well-being of people and other beings is entangled with landscape, land, or soil.

NOTE

Research for this chapter was funded by the Cluster of Excellence "Asia and Europe in a Global Context" (EXC270-JRG-C15) at Heidelberg University. I would like to thank all participants of the 2019 AURA conference at Aarhus for their comments and encouragement. Special thanks to Nils Bubandt, Rachel Cypher, Astrid Oberborbeck Andersen, and the anonymous reviewer for feedback on earlier drafts.

REFERENCES

Anderson, Virginia DeJohn. 2004. *Creatures of Empire: How Domestic Animals Transformed Early America.* Oxford: Oxford University Press.

Billen, Gilles, Josette Garnier, and Luis Lassaletta. 2013. "The Nitrogen Cascade from Agricultural Soils to the Sea: Modelling Nitrogen Transfers at Regional Watershed and Global Scales." *Philosophical Transactions of the Royal Society: Biological Sciences* 368, no. 1621: 1–13.

Boke, Charis. 2019. "Plant Listening: How North American Herbalists Learn to Pay Attention to Plants." *Anthropology Today* 35, no. 2: 23–27.

Brad, Alina, Anke Schaffartzik, Melanie Pichler, and Christina Plank. 2015. "Contested Territorialization and Biophysical Expansion of Oil Palm Plantations in Indonesia." *Geoforum* 64: 100–111.

Carson, Rachel. 1962. *Silent Spring.* Boston: Houghton Mifflin.

Crosby, Alfred W. 2003. *The Columbian Exchange: Biological and Cultural Consequences of 1492.* Westport, Conn.: Praeger.

Cullather, Nick. 2013. *The Hungry World: America's Cold War Battle against Poverty in Asia.* Cambridge, Mass.: Harvard University Press.

Davis, Mike. 2001. *Late Victorian Holocausts: El Niño Famines and the Making of the Third World.* London: Verso.

Devassy, M. K. 1966. *Village Survey Monographs Cannanore and Kozhikode Districts: Census of India 1961, vol. 7, Kerala, Part VI A.* Trivandrum: Government of India.

Devika, Jayakumari, and V. J. Varghese. 2011. "To Survive or to Flourish? Minority Rights and Syrian Christian Community Assertions in Twentieth-Century Travancore/Kerala." *History and Sociology of South Asia* 5, no. 2: 103–28.

Dove, Michael. 2011. *The Banana Tree at the Gate: A History of Marginal Peoples and Global Markets in Borneo.* Yale Agrarian Studies. New Haven, Conn.: Yale University Press.

Edison, S. 2001. "Present Situation of Future Potential of Cassava in India." In *Cassava's Potential in Asia in the 21st Century: Proceedings of the Sixth Regional Workshop, Held in Ho Chi Minh City, Vietnam, Feb 21–25, 2000,* edited by Reinhardt H. Howeler and Swee L. Tan, 61–70. Rome: CIAT.

Environmental Governance Group. 2011. "Lives, Livelihoods, and Environment of Wayanad: A Preliminary Study." Bangalore.

Fitzpatrick, Ian Carlos. 2011. *Cardamom and Class: A Limbu Village and Its Extensions in East Nepal.* Cinnabaris Series of Oriental Studies 4. Kathmandu: Vajra.

Fukuoka, Masanobu. 2009. *The One-Straw Revolution: An Introduction to Natural Farming.* New York: New York Review of Books.

Geertz, Clifford. 1972. "The Wet and the Dry: Traditional Irrigation in Bali and Morocco." *Human Ecology* 1, no. 1: 23–39.

Hall, Derek. 2011. "Land Grabs, Land Control, and Southeast Asian Crop Booms." *Journal of Peasant Studies* 38, no. 4: 837–57.

Hartigan, John. 2017. *Care of the Species: Races of Corn and the Science of Plant Biodiversity.* Minneapolis: University of Minnesota Press.

Hartigan, John. 2019. "Plants as Ethnographic Subjects." *Anthropology Today* 35, no. 2: 1–2.

Ilyas, Muhammad. 1976. "Spices in India." *Economic Botany* 30: 273–80.

Joseph, K. V. 1986. "Migration and Economic Development of Kerala." PhD thesis, University of Kerala, Trivandrum.

Joy, P. P., Baby P. Skaria, Samuel Mathew, Ancy Joseph, and P. P. Sreevidya. 2006. "Lemongrass." Odakkali: Aromatic and Medicinal Plants Research Station.

Kirch, Patrick Vinton. 1994. *The Wet and the Dry: Irrigation and Agricultural Intensification in Polynesia.* Chicago: University of Chicago Press.

Lebot, Vincent. 2009. *Tropical Root and Tuber Crops: Cassava, Sweet Potato, Yams and Aroids.* Crop Production Science in Horticulture 17. Wallingford, U.K.: CABI.

Leihner, Dietrich. 2002. "Agronomy and Cropping Systems." In *Cassava Biology, Production, and Utilization,* edited by R. J. Hillocks, J. M. Thresh, and Anthony Bellotti, 91–113. Wallingford, U.K.: CABI.

Li, Tania Murray. 2002. "Local Histories, Global Markets: Cocoa and Class in Upland Sulawesi." *Development and Change* 33, no. 3: 415–37.

Li, Tania Murray. 2014. *Land's End: Capitalist Relations on an Indigenous Frontier.* Durham, N.C.: Duke University Press.

Mathew, Joshy. 2011. *Tradition, Migration, and Transformation: Agrarian Migration to Wayanad a Socio-historical Perspective, 1928–2000.* Kannur, India: Institute for Research in Social Sciences and Humanities.

McCarthy, John F., and Robert A. Cramb. 2009. "Policy Narratives, Landholder Engagement, and Oil Palm Expansion on the Malaysian and Indonesian Frontiers." *Geographical Journal* 175, no. 2: 112–23.

Morrison, Kathleen D. 2002. "Pepper in the Hills: Upland–Lowland Exchange and the Intensification of the Spice Trade." In *Forager-Traders in South and Southeast Asia: Long-Term Histories,* edited by Kathleen D. Morrison and Laura L. Junker, 105–30. Cambridge: Cambridge University Press.

Müller, Birgit. 2012. "Farmers, Development, and the Temptation of Nitrogen: Controversies about Sustainable Farming in Nicaragua." *RCC Perspectives* 5: 23–30.

Münster, Daniel. 2007. *Postkoloniale Traditionen: eine Ethnografie über Dorf, Kaste und Ritual in Südindien.* Kultur und soziale Praxis. Bielefeld, Germany: transcript.

Münster, Daniel. 2012. "Farmers' Suicides and the State in India: Conceptual and Ethnographic Notes from Wayanad, Kerala." *Contributions to Indian Sociology* 46, no. 1–2: 181–208.

Münster, Daniel. 2015a. "Farmers' Suicides as Public Death: Politics, Agency and Statistics in a Suicide-Prone District (South India)." *Modern Asian Studies* 49, no. 5: 1580–605.

Münster, Daniel. 2015b. "'Ginger Is a Gamble': Crop Booms, Rural Uncertainty, and the Neoliberalization of Agriculture in South India." *Focaal: Journal of Global and Historical Anthropology* 71: 100–113.

Münster, Daniel. 2021. "The Nectar of Life: Fermentation, Soil Health, and Bionativism in Indian Natural Farming." *Current Anthropology* 62: S24.

Myers, Natasha. 2015. "Conversations on Plant Sensing: Notes from the Field." *Nature-Culture* 3: 35–66.

Peluso, Nancy Lee. 2012. "What's Nature Got to Do with It? A Situated Historical Perspective on Socio-natural Commodities." *Development and Change* 43, no. 1: 79–104.

Peluso, Nancy Lee, and Peter Vandergeest. 2011. "Taking the Jungle out of the Forest: Counter-insurgency and the Making of National Natures." In *Global Political Ecology*, edited by Richard Peet, Paul Robbins, and Michael Watts, 254–85. London: Routledge.

Pollan, Michael. 2006. *The Omnivore's Dilemma: A Natural History of Four Meals.* New York: Penguin Press.

Ravindran, P. N., and K. N. Babu, eds. 2005. *Ginger: The Genus Zingiber.* Medicinal and Aromatic Plants—Industrial Profiles 41. Boca Raton, Fla.: CRC Press.

Ruf, François Olivier. 2001. "Tree Crops as Deforestation and Reforestation Agents: The Case of Cocoa in Côte d'Ivoire and Sulawesi." In *Agricultural Technologies and Tropical Deforestation,* edited by Arild Angelsen and David Kaimowitz, 291–315. Wallingford, U.K.: CABI.

Scott, James C. 2009. *The Art of Not Being Governed: An Anarchist History of Upland Southeast Asia.* New Haven, Conn.: Yale University Press.

Skaria, Baby P., P. P. Joy, Samuel Mathew, and Ancy Joseph. 2012. "Lemongrass." In *Handbook of Herbs and Spices,* 2nd ed., edited by K. V. Peter, 2:227–28. Cambridge: Woodhead.

Steffen, Will, Katherine Richardson, Johan Rockström, Sarah E. Cornell, Ingo Fetzer, Elena M. Bennett, Reinette Biggs et al. 2015. "Planetary Boundaries: Guiding Human Development on a Changing Planet." *Science* 347, no. 6223: 1259855.

Swanson, Heather. 2020a. "Curious Ecologies of Knowledge." In *Curiosity Studies: A New Ecology of Knowledge,* edited by Perry Zurn, Arjun Shankar, Pamela L. Grossman, and John L. Jackson, 15–36. Minneapolis: University of Minnesota Press.

Swanson, Heather. 2020b. "Multispecies Research." In *SAGE Research Methods Foundations,* edited by Paul A. Atkinson, Sara Delamont, Richard A. Williams, Alexandru Cernat, and Joseph Sakshaug. London: SAGE. https://doi.org/10.4135/9781526421036833388

Tsing, Anna Lowenhaupt. 2003. "Natural Resources and Capitalist Frontiers." *Economic and Political Weekly* 38, no. 48: 5100–5106.

Tsing, Anna Lowenhaupt. 2004. *Friction: An Ethnography of Global Connection.* Princeton, N.J.: Princeton University Press.

Tsing, Anna Lowenhaupt, Andrew Mathews, and Nils Bubandt. 2019. "Patchy Anthropocene: Landscape Structure, Multispecies History, and the Retooling of Anthropology: An Introduction to Supplement 20." *Current Anthropology* 60, no. S20: S186–97.

van Dooren, Thom, Eben Kirksey, and Ursula Münster. 2016. "Multispecies Studies: Cultivating Arts of Attentiveness." *Environmental Humanities* 8: 1–26.

Vasavi, A. R. 2012. *Shadow Space: Suicides and the Predicament of Rural India.* Gurgaon, India: Three Essays Collective.

Worster, Donald. 2004. *Dust Bowl: The Southern Plains in the 1930s.* New York: Oxford University Press.

PART II

CURIOSITY

CHAPTER 5

Drip Torch Inquiries

Meta-Questions for Ambiguous Forests

JON RASMUS NYQUIST

The southwest forest region of Western Australia is an area dominated by tall eucalyptus forests that often burn. In this region, wildfire shapes landscapes and affects people. When tree crowns erupt in black smoke and a thundering blaze, when leaves on the forest floor crackle with yellow flames, when gusty winds fan elongated tongues through thickets and scrub, fire not only alters forests but also affects those who live here, their lifeways and their forms of thought. When sparks leap from tree to tree, when embers smolder for days in hollow logs, when fresh green shoots sprout among charcoal remains, the forest also nudges thought, sways people to see it in its ways. This chapter is an attempt to foreground how knowledge and thought are shaped in encounters with more-than-human landscapes. I ask, what does it mean to think in fiery ways in the Anthropocene?

In many ways, the Australian southwest is similar to the West Coast of the United States, the Mediterranean region, parts of South Africa, and the Australian east coast. These are all referred to as Mediterranean climate regions. They have broadly similar seasonal patterns, and they all have biotas that are widely understood to be fire prone and fire adapted. But the southwest forests region stands out in one regard. In the southwest of Western Australia, the majority of fires in forests, woodland, and heath are "prescribed." Forest managers in the state Parks and Wildlife Service have systematically burned large parts of the region's forests since the early 1950s. These practices form a stark contrast to most other similarly fire-prone places where the predominant approach has been to suppress wildfire and target the sources of ignition. And while landscape burning is practiced in many places in the world—by Aboriginal people in northern Australia collaborating with state agencies in burns that are both part of traditional land care and part of efforts to limit large bushfires that release a lot of CO_2 and by Indigenous groups in the United States seeking to bring more cultural burning into landscapes, to mention just a couple examples—as a concerted and institutionalized state effort, landscape burning over a large area is decidedly a rarity.

The aim of the West Australian forest managers is to lower the "fuel loads" (of leaf litter and other vegetation) on the forest floor, and they do so by creating certain spatial patterns across the landscape, "mosaics" of different fuel ages whose ideal effect is that if a fire were to start, it would never travel very far before running into an area that had been recently burned. A wildfire isn't necessarily stopped by a recently burned patch, but fire managers expect that a recent burn will help slow the fire down and make it easier to contain. Over the years, they have developed systems and routines for

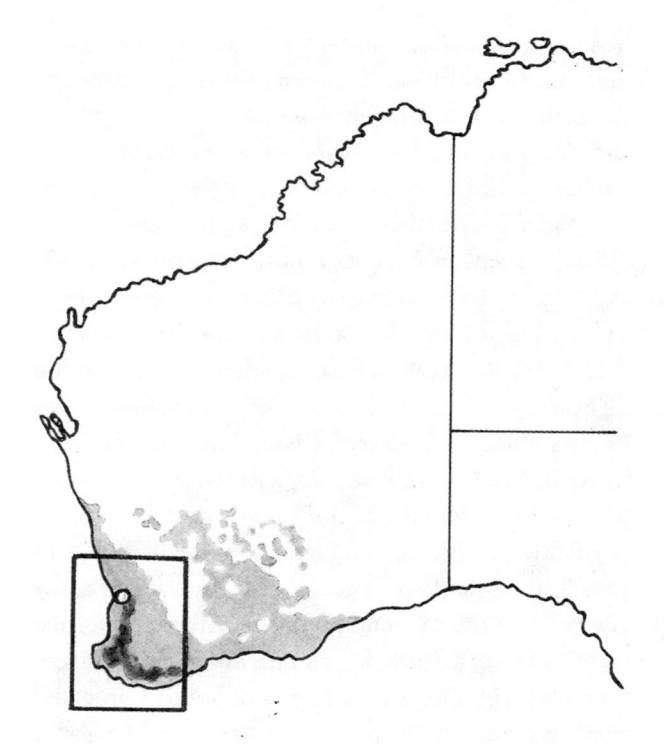

FIGURE 5.1. The southwest of Western Australia is dominated by forests, woodland, and heath and includes approximately 2.3 million hectares of publicly owned forests. Map created by Ragnhild Gjerstad.

burning and burn planning, and fire managers have formed close ties to the landscape, as they work day after day in the forest with fire as their tool, becoming familiar with the landscape as a place that burns.

While the Australian east coast was violently ablaze throughout much of the 2019–20 bushfire season, on the opposite coast, the southwest experienced a season that was not out of the ordinary. The differences between the west coast and the east are many and complex, and I would be hesitant to say anything firm about what role prescribed burning plays. Environmental historian Stephen Pyne wrote in the early 1990s that fire in the Australian southwest is "endemic, not demonic" (Pyne 1991, 49) and "chronic rather than catastrophic" (Pyne 1991, 296)—in contrast, that is, to the more demonic and catastrophic east coast. The southwest has features that has lent it more easily to management by burning—it has a gently undulating topography, it is relatively sparsely populated, and it has a simple land tenure situation where most of the forested areas are publicly owned and vested in a single agency. But things are changing in the southwest. A trend toward larger, more frequent, and more damaging bushfires has so far culminated in the Waroona fire of 2016, which burned an area of more than seventy thousand hectares, including the town of Yarloop. These days, one of the most concerning changes people anticipate is for the already fire-prone southwest to become even more affected by damaging wildfires. More and more people live close to forests and in rural–urban interface areas, producing more ignitions and more places to encounter fire. Forest structure and hydrology have been affected by more than a hundred years of logging and forestry activities, producing forests that have fewer permanent water sources and that are dominated by smaller trees. Annual rainfall has declined by around 15 percent since the 1970s, summers are getting hotter, and warm and dry days more frequently extend further into fall. West Australians who work in the forests live in complicated times, when many factors come together to produce new conditions for fire.

Most of the fire managers and crews that burn the forest and fight fires in the Australian southwest work for the state Parks and Wildlife Service, an agency populated by some with environmental science degrees, some with forestry degrees, some with a background in environmental management from other Australian states, and a strong older guard that has been with the department ever since it was the Forests Department in the 1980s and earlier.[1] Many in the fire crews, who do most of the hands-on work, are locals who own small farms, former employees of recently closed sawmills and coal mines, or young people taking seasonal work.[2] During my fieldwork, I participated in many prescribed burns, often as part of a fire crew. Sometimes I would be the one holding the drip torch, the tool with which we started the burns; sometimes I would be one of those who patrolled and observed to see how it was burning; sometimes I would put fires out; and sometimes I would observe how the forest was growing back.

Over time, as I took part in more burns, I began to get accustomed to the habits of fire in the southwest. I experienced different instances of how the forest responded to our actions and how we could respond to its actions. I noticed how some trees and bushes tend to burn in rapid hot flashes, whereas others burn more slowly, at least at first. Through practice, I learned what kind of vegetation to put a spot in to ensure that it would catch and in what sorts of places it might start slower or even not start at all. I learned to pay attention to slopes (fires burn much quicker uphill), to aspect (in the southern hemisphere, slopes with a northerly aspect will usually be drier), and, perhaps above all, to wind. Fire managers see the wind on forecasts and maps, but they also see the wind as it moves the landscape—in leaves in the tree crowns, in the color and motion of smoke, and in the flames themselves. And I learned that fire has dynamics of its own. I experienced what fire fronts do when they meet—I felt the heat and rapidity of a "junction zone" where fire fronts feed into each other and grow—and I saw how fire can "pulse," climb up in the canopy, then come back down as it loses touch with ground fuels, then build back up again, and so on. I came to see fire as something that is dependent on, feeds off, and synthesizes its surroundings but is not reducible to them. And I learned what happens after fire. I saw numerous examples of the different ways that plants sprout, flower, and grow back. All in all, I learned gradually to see the landscape as a place that has burned and a place that will burn again.

In these practices, the drip torch put me in touch with both fire managers and the forests that burn. A drip torch is a tool to light fires, but for me, it is also, like rubber boots, a figuration. It stands for a way to extend oneself and reach out to landscapes. Tipping the drip torch toward the ground to start a small spot fire is a proposition, an inquiry, and an opening. How the forest responds is never determined, and in such encounters, knowledge is open to be shaped. Drip torch inquiries also stand for an approach that highlights how knowledge can be affected by the world with which we interact. Through drip torch inquiries, curiosity figures as a necessary openness to surprise and to being affected.

It is from my experiences with both fire and fire managers that I articulate a fiery method for the Anthropocene. In this chapter, I take inspiration from the fire managers I met in Western Australia to sketch a fiery mode of attention. The ways fire managers think are, in certain ways, fiery. This is not to say that they necessarily know the forest and fire correctly, accurately, or in the best way that they can be known. Nor is it, on the other hand, to say that they fall short of knowing fire and the forest in accurate ways. I take knowledge to be a qualitative phenomenon—it does not simply come in more and less. Landscapes are not simply known or not known but are known in particular ways. This is an exploration of the *forms* and *styles* of thinking and knowing that, for these Australian fire managers, fire appears to inspire. For me, this means trying to extract from what fire managers say and do those elements of their ways of knowing that are peculiar to an engagement with fire—those elements that are, let us say, fire shaped. Hence my focus lies not primarily in exploring how their practices are influenced by natural science methods, or European-style forestry; how their practices carry a settler-colonial legacy; or how they see the forest in modern ways. Instead, my knowledge and awareness of these things, in combination with my own embodied experiences with fire and the forest, are what allow me, in a partial way of course, to distill those elements of their ways of thinking and acting that are shaped by fire. *My* method, then, is a method of abstracting from a combination of their practices and mine a mode of engagement: one example of a bushfire-based mode of relating to the world.

In this approach to knowledge, I take inspiration from recent writings on human–nonhuman relations that emphasize in different ways how the world plays an active part in our representations of it. I am inspired by Myers's (2015) discussion of how plant scientists, through their interactions with plants, can come to think in "plantified" ways; by Govindrajan's (2018) way of firmly situating people's values, desires, and anxieties involving animals within practical interactions with real animals of flesh and blood; and by Despret's (2004) stories of how lab mice, a jackdaw, and a horse actively affected the scientific knowledge practices in which they were involved. We become a little bit animal by interacting with animals, a little bit plant by interacting with plants, and a little bit fire by working day in and day out with fire. People open themselves, including their ways of thinking, to mutual transformations (cf. Despret 2004) with other creatures, landscapes, and elements.

On a more general level, I draw on the idea that people are "worn into" (Verran 2001) ways of thinking and acting. Routines, apparatuses, and interactions with materials and other beings all affect modes of thought through connections that are short of clear causality and far short of determinism. Thought styles (cf. Fleck 1935) are historically situated, but they also take shape from people's interactions with the world. Far from floating through the material world like purely thinking beings, humans, like an old pair of shoes, take on some of the shape of what they interact with. A forest that burns is something that inspires and nudges people toward certain kinds of thought. Interacting routinely with burning routines and with landscapes that burn is to be worn into patterns of thinking.

What is a fiery kind of thought? What is a fiery way of paying attention? A fiery mode of engagement? In the following, I sketch four elements of a mode of engagement into which people have been worn by burning, through fire routines, and by interacting closely with landscapes that burn. First among these elements is an attention to patterns. Burning practices and experience with the landscape patterns that systematic burning can create make people prone to experience places as meaningful through what these places are adjacent to. Second is a way of seeing scale, more specifically, of seeing large things come together in the very small and local. This, I argue, gives a way of thinking about causality as ambiguous and to be attuned to and to foreground underlying rather than immediate causes. Third is a heightened attention to dramatic emergence and a familiarity with the unexpected. Through experiences with fire, one can become accustomed to events that dramatically exceed expectations and attuned to things that destabilize assumptions. And fourth is a particular way of looking at the past, not as a model for the future, but rather as a set of precedents for ways of being wrong. Above all, drip torch inquires is a method that sees patterns at the same time as it looks for that which can undo them.

Attention to Pattern: Patchworks and Adjacency

In the 2016–17 burning and fire season, when I did most of my fieldwork, Parks and Wildlife burned roughly 250,000 hectares across the southwest region in some fifty different prescribed burns. Such burns come in all shapes and sizes. The largest one in which I participated covered nearly fifteen thousand hectares, whereas the smallest one was just one single hectare. The majority of the burns are conducted in the same way: by first lighting up the downwind edges by hand, then igniting the core of the burn area with capsules dropped in a grid pattern from a small airplane, and then finally "tying up" the burn by lighting up the remaining edges by hand. A burn usually involves a few hours of lighting up and a few days of patrolling and "mopping up," putting out what is still burning or smoldering on the edges after the fires have gone through. Each of the burns would result in a within-burn patchiness dependent on the weather, vegetation, and moisture conditions, and many such burns together would give a landscape a "mosaic" of different fuel ages, or times since last burn. The activity of burning and planning to burn wears people into a certain way of seeing and thinking about patterns in the landscape.

The Leach burn, a spring burn in the jarrah forest *(Eucalyptus marginata)* near the town of Collie, was one of the burns where my own ties with the forest were forged and molded. In the morning, fire crews were putting in spots on the edges, spaced in a way that they expected would give low flames steadily creeping across the forest floor. In sector C, where I spent most of the day, our focus was on creating good, continuous, and deep edges to be on the safe side before the aerial ignition in the afternoon. Now and again, I saw crew members walk in behind the fire line with red drip torches in hand. With unassuming confidence, they sauntered past burning trees and over low flames in the understory. Behind the fire line, they would light up, perhaps to adjust the edges; they would look to see where more fire might be needed; gauge the vegetation, the wind, and the flames; and dab their drip torch in places they thought would give the kind of fire they wanted.

We had our attention directed toward the dry leaves and understory vegetation—a dry creek with a dense cover of paperbark trees (*Melaluca* sp.) made us expect different things than an area with tall jarrah trees on a patchy carpet of leaf litter. While burning, we actively imagined possible kinds of fire. With our eyes and with fire itself, we were inquiring of the vegetation how it might burn. But we were equally focused on what was outside the burn area. We were always on the lookout for any "hop-overs." The forest bordering our burn on the west side had been burned only two years earlier,

FIGURE 5.2. A prescribed burn with low flames burning leaf litter and vegetation near one of the edges of a large burn area. Photograph by the author.

which gave the fire managers a sense of ease, an expectation that if they had a hop-over in that area, it wouldn't burn too intensely or spread very fast. But to the southeast was a patch that hadn't burned in forty-eight years, and this was something they knew to be extra aware of.

When fire managers encounter the forest, they see with their experiences of having been among flames, but they also have patterns in mind. In their fire-oriented routines, fire managers constantly engage with patterns on maps and plans, in practice and in the landscape. Patchworks of various kinds were always one of the first things they would mobilize to make sense of the region. In their offices, at prominent places in important rooms, enormous maps often covered most of the wall. They showed patchworks, districts sliced into blocks, squares, and slivers, often in different colors. One map I usually encountered in the district offices was the "fuel age map," a colorful patchwork where squares, blotches, and other shapes had different colors to indicate how long it had been since they were last burned. For fire managers, it was clear that these maps were a key element in making sense of fire in the southwest. The maps would set off tales about what they had achieved in their burning in the recent years and where they needed to focus in the coming years. Sometimes the maps were catalysts for worry. In Blackwood district, for instance, the maps were far more monochrome than what the fire managers would have liked, dominated by the same beige color that represented all fuel ages of six years or more. Fuel ages from one to five years each had its own color, subtly suggesting that uniformity happens by itself if they leave the forests unburned.

Both the maps and the activity of burning involve an attention to patterns. One of these patterns is what I call the "whole-of-forest mosaic." At its simplest, this could be a situation where the entire forest region is divided into units, each one burned every six to eight years and distributed in time in such a way that each year would require one to burn a more or less equal percentage of the forest. Fire managers see the forest as a mosaic of patches that differ based on their propensity to burn. Another kind of pattern comes from an awareness of sites in their *adjacency* to one another. Often the most crucial places to be aware of for a fire manager are the places adjacent to where one is situated, the places to or from which fire may spread—the two-year-old patch to the west and the forty-eight-year-old area to the southeast. The landscape, then, comes to have an affective pull on fire managers. They come to experience the forest, to feel safe or concerned, through what I call favorable and unfavorable adjacency.

Fire managers encounter adjacency in the field all the time. The Ross burn, a spring burn in which I took part, was next to a fire scar only two years old. We could see it across one of the dirt roads that acted as the boundary for the burn—jarrah trees with the almost furlike appearance given by shoots along their stems, plenty of bracken fern in the understory, and not much leaf litter on the ground. This was a *favorable* adjacency; it filled the fire managers with a sense of calm, a confidence that they could easily extinguish a hop-over if they had to. *Unfavorable* adjacency has an equally strong affective pull on those who burn, making them instead tense and apprehensive, perhaps evoking memories of near misses or even times when things have gone awry. Continuous areas with a lot of litter fuel is, for fire managers, a kind of landscape that anticipates or projects dangerous fires. Many fire managers have a spatial awareness of the district in which they work so that they can stand in one place and be aware of their vicinity according to how long ago different areas were burned. I know this from comments like "we have three-year-old to the north" or "there's some scruffy forty-year-old to the east" or comments about places made dangerous by their surrounding areas. Fire managers can stand in their district and be affectively pulled by an awareness of its variable propensity to burn.

Adjacency and mosaics are patterns of spatial imagination above all related to what is known to be the most dangerous winds. Northerlies tend to be dangerous, as they can be associated with surface troughs in the summer, and the most extreme fire weather often involves northwesterlies in particular. When they plan burns, fire managers seek to "stack" burns from south to north, with blocks that build on each other, taking advantage of burns that are made safer by their proximity to low-fuel areas and patches that together compose low-fuel "corridors" that can slow down fires driven by northerly winds.

Through patchworks and adjacency, any area of the forest becomes relationally defined. No one place in the forest is safe or unsafe in itself but only by its relation to surrounding areas and to the larger patchwork that all areas together make up. This involves a certain kind of spatial imagination, where any patch of forest is a potential conductor of fire moving across the landscape. By burning, the fire managers involve themselves in an attempt to break anticipated wind-driven paths of connection. One feature of a fiery kind of thought, then, is to see landscapes as something made of connections and conduction and to see patterns—patchworks and favorable adjacency—as ways of breaking connections, of inhibiting flows.

THE VERY LARGE AND THE VERY SMALL: CLIMATE, WEATHER, FLAME

Fire makes real what we can usually only see in graphs and other abstractions. It can make long-term trends and long-range connections local and immediate. In 2012, the Indian Ocean Climate Initiative, a long-term research partnership run by the Western Australia state government; the Commonwealth Scientific and Industrial Research Organisation; and the Bureau of Meteorology released the report from stage 3 of the project. If there is such a thing as an authoritative statement on the state's future weather, this might be it. The weather of the southwest, the report explains, is driven to a significant extent by the subtropical jet stream and the cold weather south of Australia. Both drivers are changing in ways that are consistent with what could be expected from increases in greenhouse gases. The report presents a clear causal chain: higher concentrations of greenhouse gases have led to a weakening of the subtropical jet stream and a warming south of thirty degrees, which has led to fewer winter storms and lower rainfall across the southwest. The jet stream, "a belt of strong, upper-level westerly winds" (Bates et al. 2012, 26), is a major force generating storms, and its weakening is associated with a more stable atmosphere. Similarly, the warmer weather in the south has "reduced the equator-to-pole temperature gradient" (27), representing another stabilizing process. Both changes have made it less likely that winter storms, which account for a significant portion of the region's annual rainfall, will form in the southwest. A more stable atmosphere means fewer storms, which means drier vegetation, which means more dangerous conditions for fire.

Furthermore, whereas fewer low-pressure systems were the main cause of rainfall declines from the late 1960s until around 2000, since then, the continued drying trend has been driven by a higher incidence of high-pressure systems that persist further into fall. This is reflected in a 25 percent decline in May rainfall since the year 2000. The atmospheric drivers of Western Australia's weather manifest differently in the summer months, when deep surface troughs—formations of low-pressure areas in the atmosphere—along the west coast and occasional lingering tropical cyclones are important features. The authors of the report speculate that interactions between the El Niño/Southern Oscillation, the Indian Ocean Dipole, and other weather systems may cause more tropical weather to "intrude further south" (35) in the future, possibly bringing more rainfall and more frequent extreme weather events in summer. Decaying tropical cyclones can bring large amounts of summer rain to the southwest, but they can also bring dry lightning and high winds.

In a small oven in the innermost corner of a large shed storing all kinds of firefighting equipment, I saw eucalyptus leaves being dried at a rate much faster than by climate change. Right after we had finished a burn planning meeting in the small south coast town of Walpole, Hayden asked me if I wanted to come have a look at how they do the measurements of the moisture content in the karri *(E. diversicolor)* forest. In among hoses and fire blankets, pumps and generators, rakes and shovels, the incredibly low-tech equipment for measuring moisture in the understory vegetation consisted of an old sooty oven, a clipboard with a data sheet and a pencil, and half a dozen metal cylinders filled with leaves and other vegetation matter from the karri forest understory. Today's samples were from the Ordnance burn, which they were in the middle of. Hayden had been out there the day before to collect the samples, and he explained that the cylinders were supposed to sit in the oven for eighteen hours at 105 degrees Celsius. He took them out one by one and put them on a small scale and wrote the weight on the form. He had entered the moist weight on the form before he put them in the oven, and after he had weighed all the little cylinders, he emptied them and then weighed the empty cylinders as well. Then, later, he would enter all the numbers into a spreadsheet that would give him a percentage for the "profile moisture content." After that, another set of calculations, also incorporating current temperature, humidity, and winds, would give him an expected "rate of spread" for a fire lit under these conditions.

With these routines, Hayden comes to know climate and fire through moist and dried leaves, through forms and calculations, and eventually through concrete effects—how things burn—that hardly ever completely correspond to predictions. With leaves, an oven, a weight, some spreadsheets, and an algorithm, Hayden is asking questions of the forest—What condition are you in? How will you burn? He is aiming to find out something that can inform his actions. In his leaves, Hayden has the potential to see something immediate—how the Ordnance burn might go tomorrow—as well as longer trends and wider areas; he sees a part of the last forty years of an ongoing drying trend and a small effect of weather patterns that link jet streams and pressure systems spanning a large part of the globe, conditions shaped by being situated somewhere between the tropical air of the Indian Ocean and the frigid seas near Antarctica.

The "Southern Australian Seasonal Bushfire Outlook 2017," released by the Bushfire and Natural Hazards Cooperative Research Centre, assessed there to be "potential for above normal bushfire activity" in the southwest region for the 2017–18 season. The year before was also assessed as having "above normal fire potential." For 2016, yet another winter of low rainfall and "an underlying long-term deficit in soil moisture" were the main reasons for the assessment. For 2017, despite an unusually wet summer, the following fall was the driest in five years, and below average rainfall was expected for spring. In an updated hazard note released in November 2017, it was noted that the southwest of Western Australia "has now experienced 12 consecutive cool seasons with below average rainfall."

On Australia Day, January 26, 2017, a tropical low up north extended a surface trough down along the west coast, which brought temperatures in the high thirties, unstable atmospheric conditions, and gusty winds to the southwest. Notoriously dangerous conditions. I could feel the smoke seeping in through the A/C of my car, and as I got closer to the town of Donnybrook, where I was staying that night, I saw the menacing plume from what the radio told me was a bushfire at Gwindinup, a few kilometers south of town. The Watch and Act Warning had by the afternoon become an Emergency Warning, and people close to the fire were urged to take immediate action. In the late afternoon, I saw fire trucks driving around town and a great smoke plume that contained every shade from wispy white to charcoal black. But by evening, the wind had eased, the plume was nearly gone, and in Parks and Wildlife's office in Kirup the next morning, talk of the fire had to compete with how people had spent their Australia Day and with the leftover pavlova someone had brought in to share.

Like most fires, this was a fire that very mundanely drew scales together. It lined up with all the patterns we associate with climate change in the southwest. The soil and vegetation dryness, as conditions under which it burned, can be seen in light of the long-term drying trend, the weakening jet stream and the lower rainfall, the warming south, and the more persisting high-pressure systems in fall. The weather of the day was part of the season with "higher than normal potential for bushfire activity," and the deep surface trough was part of the pattern of tropical weather more often creeping farther south in summer. For an afternoon, the Indian Ocean, the South Pole, and the past forty years had a presence, just as they inhabited Hayden's drying leaves.

To see the very large in the very small is to think with, and to be accustomed to, a more indirect and ambiguous kind of causality. A bushfire isn't simply caused by these things—long-term trends and long-distance connections. And at the same time, these are the most important causes of bushfire. A bushfire actualizes "hyperobjects" (Morton 2013), massively distributed things that outscale us, that are everywhere and nowhere in particular, and that we can only see in pieces and fragments. Fire leads people's attention to causes that are indirect, to that which eases and elicits, inhibits, enables, and constrains—not to the ignition, but to the conditions for it catching alight and taking off. On days like the Gwindinup fire, the ignition is almost incidental. The conditions themselves—of soil, weather, and vegetation—are what embody fire.

Dramatic Emergence and Destabilized Assumptions

There is a video clip that almost everyone who learns to be a firefighter in the Australian southwest has seen. The clip is from a research project in the late 1990s and early 2000s into fire behavior in high-intensity bushfires, and it is used now as a tool to create awareness within the department about the dynamics of forest fire. I saw the clip a few times in preseason training sessions I attended to become certified as a firefighter. It consists of a single grainy shot seemingly in the middle of the forest. At first, low flames are burning calmly, a slow, trickling, oval shape. But we've been told to pay attention to what happens when the wind direction changes. Sure enough, it changes. The fire starts to burn toward the camera; it picks up, then picks up violently. Within seconds, it moves fast; flames become erratic, long tongues, then fill the entire frame, the trees now black outlines in fire. Figure and ground seem to shift, from a fire in the forest to a forest within flames. The time stamp clock in the corner is the only thing that's regular. And then we're told: expect that it can happen even when you least expect it.

A fire can have a volatile relationality, where small differences in the landscape and the weather can be dramatically magnified. A fire can be emergent and exponential; very large fires even affect the weather conditions to which they respond. A fire is linear and predictable only until it's not. With fire, the forest answers in the ways we can expect, only until it doesn't, and we should expect that this shift can happen even when we least expect it. When learning to be firefighters, and later, when working with fire, they are taught to *expect* something they cannot always fully apprehend. They are taught to expect answers from the forest that look different from what they expect. They are sensitized to moments that pry open or elasticize their assumptions.

When using a drip torch, it's not just *as if* the fire is liquid. From a canister through a long metal tube with a wick at the end, ignited kerosene is what drips out onto the forest floor. The kerosene quickly burns off, and soon it is the forest itself that responds to the fire. Most often, we want the liquid flames to start small fires that inspire water metaphors, fires that "trickle" around on the forest floor. Such were the first spots of fire we set at the Driver burn, where crews and managers gathered at mid-morning in early November. The day started with a test fire, a match that we threw down onto the leaf litter by the side of one of the dirt roads that acted as a boundary for the burn. Then we left it alone for an hour or so. Later, I followed Henry, the operations officer for the burn, as he walked in through tall, scrubby under-growth across the ground that had been covered by the test fire. We stepped over the low flame at one end, and Henry walked across to get a rough mea-sure of how much ground it had covered since it was lit, each step arousing a puff of gray charcoal dust around his leather boots. About ten meters per hour, he reckoned, was the rate of spread.

Even with such a sedate morning test fire, it was clear that this would be one of the last burns of the spring season. A surface trough—a low-pressure region in the atmosphere associated with unstable conditions—was on the forecast in a few days, and after that, the soil and the vegetation would almost certainly be so dry that it would exceed what Parks and Wildlife regarded to be safe conditions for burning in spring. Today's burn would be diligently kept an eye on. Every hour, Neil, another fire manager working on the burn, took weather observations and reported over the VHF radio to Henry and to the office. He reported on temperature, dew point, relative humidity, and wind speed, including in gusts, wind direction, and cloud cover. We also took measurements of the moisture content in the leaf litter a few times and saw it drop quite a bit in the late morning hours—such a big drop, in fact, that it caused some concern. Between crew leaders and the different officers working on the burn, there was a continuous interchange of obser-vations about fire behavior. Sometimes they talked about "rates of spread" —in the early afternoon, they reckoned the edges were burning at a rate of about forty to fifty meters per hour—and sometimes they exchanged seem-ingly vague descriptors like "it's willing" or "it's goin' a bit hot," descriptors that got their meaning from shared experiences, shared assumptions of what they could expect from a day like this and about what kind of burn was desirable.

The fire managers engaged in a cooperative questioning of the fire and forest, where provisional answers were emerging from the coordinated activ-ity of field officers, crews, measurements, and standards. The fire manag-ers engaged with the fire through procedures as well as through bodies that

FIGURE 5.3. Flames reach to the crown of a large eucalyptus tree in a prescribed burn that for a moment burned more intensely than the fire managers expected. Photograph by the author.

are used to the same range of heat and intensity, bodies worn into similar patterns. To see a fire as "willing" takes a habituated body, perhaps more so than a well-tried model.

After a few hours, the fire behavior picked up, we got crown scorch from a spot that suddenly flared up, flames reached up to canopies twenty to thirty meters above, and the smoke made my unaccustomed eyes watery. The fire managers exhibited what I read as a particular kind of surprise. Subtle signs conveyed that something was different than they had expected— a tone of voice or a lingering look at the flames that lasted longer than usual— accompanied by evaluations and suggestions they exchanged between each other. They seemed surprised that it flared up, but not surprised by the fire's capacity to take them by surprise. Henry decided to do adjustments, to go with a wider spacing between the spots to be on the safe side. At one point, I heard him say to another colleague that "it's fairly wooly out here, we're coming towards the end [of the season], that's for sure."

When the fire managers take note of large drops in moisture or flames that are especially "willing," they turn their focus toward signs from the forest that exceed or contradict their expectations of fire on that particular day. Throughout the day, they ask the forest, "how will you burn?"—with measurements and with observations of flames and vegetation. But as the fire starts to behave differently, they begin to let their assumptions and expectations stretch. Unexpected flare-ups, sudden drops in moisture, a fire that burns faster than they thought it would, are small events partly outside of patterns, slight pressures from the forest and the fire that nudge and sway habituated forms of thought. Such moments destabilize assumptions and gradually make fire managers more accustomed to having their assumptions destabilized. These moments wear people into patterns of thought.

At first, the forest answers in the ways they expect, but at some point, it begins to answer in a different way—erratically, with wavering indecision, with sudden changes of mood. And pragmatically, these are the most crucial moments. With fire, I learned, we must search most of all for the kinds of answers we are not sure if we are able to grasp, for signs whose appearance we are not entirely sure of. At such points, what an answer from the forest looks like is an open question. These are little pockets where assumptions are relaxed—about what's happening, what we know, and what we're able to know.

This weakening of assumptions often happens through what I think of as almost-patterns. These almost-patterns are situations when fire managers are especially aware, especially sensitized to the unexpected. These are not situations in which they know that something *will* happen or situations where they know *what* might happen. They are situations in which they have a heightened awareness that something unexpected could occur. When these circumstances start to develop, fire managers can find themselves in the seemingly paradoxical situation of expecting to be able to expect less.

One of these almost-patterns is related to soil moisture. On several occasions, I heard about a new challenge that fire managers have with burning in conditions with unusually dry soil. Jim, a fire manager I met in Frankland district near the south coast, told me that in the last few years, they had been seeing some prescribed burns that acted more like wildfires, burns that had been much more intense than expected. At a burn planning meeting, Adrian, from Wellington district, told us about a burn that had done exactly that. On the day of the burn, he told us, it had been very moist—so wet that they could barely get an ignition. The surface moisture content had been very high, but the soil had been quite dry. Then, later in the day, things had drastically changed. Then, "basically everything had caught alight." Erin, in Blackwood district, was yet another one who told me about this challenge. In a small room with a stationary computer, we perused graphs concerning changes in soil dryness. An interesting thing, Erin said, was that in the last few years, even though the graphs made it look like the soil had been fully saturated over winter, the line on the graph was subtly bobbing up and down through the season, meaning that the soil wouldn't have been completely saturated for very long at a time. They were starting to see indications that this had an effect on their burns. Because the ground wasn't saturated for very long, logs and other coarse, woody debris on the forest floor might not have had the chance to get thoroughly saturated in the course of the winter, and then, come spring, they would be drier than expected, and more of these logs would be prone to flare back up in summer or be unusually dry when fall came around. Logs, stumps, and fallen branches out there on the forest floor now seemed to fill the landscape with elements that were too dry, that were dry for longer, and that were out of step with leaf litter and vegetation, out of step with fire managers' expectations.

What was new and unusual wasn't that the soil was very dry. What was new was that the soil was very dry at times of the year when it didn't use to be *that* dry. The expectation that Jim's and Adrian's burns would hardly ignite at all would have been based on measurements and calculations of surface and profile moisture, that is, of how much moisture there was in the leaf litter and other vegetation near the ground. The new kind of situation occurs when soil and logs are drier than what fire managers are used to at a time when the fuel moisture is regarded to be suitable for burning—or, as in both Jim's and Adrian's cases, when the soil is drier than it usually is when the fuel is so moist that they hardly expect a burn to ignite at all. Of course, when it comes to fire, they know they can never take their calculations to be definitive predictions of what will happen. Rather, as Erin told me, such calculations are more of an indication that on one day, they might have to be extra vigilant, whereas on another, they can relax. Now, however, if the soil is very dry, fire managers find that they can get highly intense fire behavior even when their calculations are telling them to relax. And in these situations, with moist fuel and dry soil, fire can be dramatically exponential—turning very quickly from soggy leaf litter that barely ignites into bushfire conditions. Erin, Jim, Adrian, and many others expect that these are situations in which they might have to act with less confidence in their expectations. We can see here a particular way to think about expectations: they are treated as elastic, or at least as conditionally elastic. Expectations are prompted to stretch in certain, almost patterned situations.

For these fire managers, the things they can safely rely on noticing—the signs that align with expectations—can be the least important things to notice. The things that fall neatly into a pattern are the stuff that shouldn't ever be relied on too strongly—not because they are wrong but because they cannot tell them the things about a fire that are most urgent to them. Fire managers are aware of the potential for a cumulative knowledge of the past, and for confidence in such knowledge, to be counterproductive in catastrophic ways.

The Past as Potentially Misleading

What makes up both forest history and forest future is an everyday indeterminacy, a mundane volatility that is part of fire itself but also part of the changing times. Fire brings to attention at the same time the ordinary and the exceptional—the everyday volatility of fire along with fires burning the way they do because of climate change. What sets ordinary and exceptional fires apart is *not* volatility, indeterminacy, intensity, and sudden shifts; instead, these are features that the ordinary and the exceptional have in common, features of fire that have long influenced fire managers to systematically hone their attention to that which is outside of expectations.

The forest shapes knowledge, among other things, through moments of doubt and surprise. Here thought styles are nudged, assumptions are swayed. These are moments that elasticize expectations and moments that destabilize the past. These are moments that make the link between the past and the future less certain. I think of them as moments that turn models into tenuous precedents. The past can no longer tell us what will happen in situations such as these but only what has happened in other situations. The drying climate affects the future, but it can also alter the past. Through moments of unease and apprehension, the forest affects how fire managers think about forest history.

Fire managers cannot always assume that signs and observations come in any of the forms that they can make sense of. When burning the southwest forest these days, they are people learning more ways in which they can be wrong, experiencing more ways in which the past can be misleading. Fire managers in the Australian southwest are in an ongoing dialogue with the forest, but not always to cumulatively build up more knowledge about its history. The past, in fact, can lead them astray. In this sense, forest history is not erased or ignored, but it is meaningful in a particular way: to remind them that their expectations can be dangerously misleading. The drying climate and the conditions of the Anthropocene gradually transform the past from what has happened here before into a collection of things they may no longer be able to trust.

A Fiery Style of Thought

Drip torch inquiries involve certain ways of asking the forest questions, ways of paying attention and being worn into certain ways of thinking and acting. It is not just a way to be attentive to fire but a fiery way to be attentive. From the fire-oriented practices of Western Australian fire managers, as well as my own experiences with fire, I have drawn out some elements of a mode of thinking: one version of a fire method for the Anthropocene. One aspect of this method is to see landscapes as places of connection and conduction and patterns and patchworks as ways to inhibit and direct flows. Another is to be used to seeing the very large in the very small and to be accustomed to thinking in ways that foreground indirect and underlying kinds of causality. A third aspect is to think with assumptions that are elastic and to have one's attention directed toward that which may challenge assumptions. And finally, this mode of thought entails a consciously felt destabilization of trust in the past. This is a method that actively searches for the signs and processes it cannot quite grasp.

Finally, this shows a complexity of thought that is worth calling attention to, especially when the kind of thinking about fire most people are exposed to is fire as something to predict, fight, and control; and the kind of thinking about nature more generally is nature as something that we can simply have and gain some kind of unqualified "knowledge" about. Methods that are attentive to how knowledge is shaped in more-than-human encounters can also help make visible a variety of knowledge forms. If the simplified landscapes of the Anthropocene also come with a flattening of thought forms—perhaps we might call it a plantationization of thought—then a critical intervention is to contribute to a picture of a world that contains a wider qualitative variety of forms of knowledge and thought.

NOTES

1. Parks and Wildlife is the largest section of the state Department of Biodiversity, Conservation, and Attractions. It is meaningful to call Parks and Wildlife the descendant of the Western Australian Forests Department, which since 1985 has gone through five organizational restructurings, in part as a response to conflicts surrounding logging and changes in forest policy.

2. Very few Aboriginal people are involved with fire management in this region. Fire crews are mostly, but not exclusively, composed of men.

REFERENCES

Bates, Bryson, Carsten Fredriksen, and Janice Wormworth, eds. 2012. *Western Australia's Weather and Climate: A Synthesis of Indian Ocean Climate Initiative Stage 3 Research.* Perth, Australia: Commonwealth Scientific and Industrial Research Organisation/Bureau of Meteorology.

Despret, Vinciane. 2004. "The Body We Care For: Figures of Anthropo-zoo-genesis." *Body and Society* 10, no. 2–3: 111–34.

Fleck, Ludwik. 1935. *Genesis and Development of a Scientific Fact.* Chicago: University of Chicago Press.

Govindrajan, Radhika. 2018. *Animal Intimacies: Interspecies Relatedness in India's Central Himalayas.* Chicago: University of Chicago Press.

Morton, Timothy. 2013. *Hyperobjects: Philosophy and Ecology after the End of the World.* Minneapolis: University of Minnesota Press.

Myers, Natasha. 2015. "Conversations on Plant Sensing: Notes from the Field." *Nature-Culture* 3: 35–66.

Pyne, Srephen J. 1991. *Burning Bush: A Fire History of Australia.* Seattle: University of Washington Press.

Verran, Helen. 2001. *Science and an African Logic.* Chicago: University of Chicago Press.

CHAPTER 6

Tidalectic Ethnography

Snorkeling the Coral Reefs of the Anthropocene

NILS BUBANDT

If the tide is right, I go snorkeling every morning and most afternoons when I am on fieldwork these days. And why wouldn't I? Since 2014, I have been working in Raja Ampat, an archipelagic district in the Indonesian province of West Papua that is home to the world's most abundant tropical coral reefs (Veron et al. 2011). Snorkeling these reefs is a stunning experience—an explosion of color, movement, sound, and life. More than 550 scleractinian coral species and 1,400 fish species crowd on these reefs, a world record (Mangubhai et al. 2012). The superabundance is so overwhelming that most explosions of life seem to take place on the periphery of your field of vision and hearing: the darting motion of bright reef dwellers as they disappear into a coral boulder, the clicking of an unseen crustacean, the swoosh of a large school of fish as it changes direction, disturbed by something else that you never manage to notice either.

This sensory overload is not accidental. Snorkeling fieldwork is confusing to a sensory apparatus born on land. Nevertheless, I argue, it is as possible as it is necessary. "Possible" because with practice and the aid of the right prosthetics—snorkel, mask, and fins—this underwater world has become almost as ethnographically accessible to me as it is to the people with whom I study, allowing me access to an incredibly diverse multispecies world that is intimately connected to the world of humans on dry land. And "necessary" because the very same parts of the coral reef that one can reach while snorkeling are also the critical zones that stand at the brink of planetary extinction due to a myriad of intersecting anthropogenic factors (Latour and Weibel 2020). The global ecosystemic collapse of tropical coral reefs is seen by many scientists as a harbinger of things to come in the Anthropocene (Birkeland 2015; Hughes et al. 2017). For this reason, it seems to me, coral fieldwork is critically necessary to understand our time. We humans are, as Astrida Neimanis observes, bodies of water on a watery planet that is radically changing. The rising tides of the Anthropocene require of all of us that we learn to swim (Neimanis 2017, 26), because as Jon Pugh and David Chandler (2021) put it, in the Anthropocene, there are only islands. Swimming here is not only a physical act. It is also a conceptual and philosophical reorientation, part of an "oceanic turn" (DeLoughrey 2017) that recently has led a growing number of scholars to explore the sea as a site for experiments with new kinds of seeing, experiencing, philosophizing, and relating (Raycraft 2020; Jue 2020; Picken and Ferguson 2014; Merchant 2011; Helmreich 2009, 2016).

Most of these calls for reorientation have been based on scuba diving or deep-sea exploration as a method. This article explores a more low-tech method, one used extensively also by the Papuan people with whom I work, namely, snorkeling. Snorkeling is a bodily experience located somewhere between swimming and scuba diving. It is a partial immersion into another element that, like other forms of underwater experience, requires a retooling of human attention and arguably also of anthropological understanding. But snorkeling is also different from scuba diving or submarine ethnography (Helmreich 2007). Unlike diving and deep-sea exploration (as well as conventional anthropological understanding), snorkeling is, for instance, not primarily about depth. Rather, it is about surface. It is an epipelagic activity that takes place at the surface of the ocean, because we humans have to breathe air. Snorkeling therefore requires that you abandon the idea of total anthropological immersion and in-depth understanding. Long-term total immersion when snorkeling is ill advised. It means you are drowning. When you snorkel, the surface is your friend, a site from which to understand how that which goes on below water is connected to what goes on above water. Snorkeling fieldwork entails an anthropological attention to connections across surfaces, elements, and worlds.

Let me describe how I snorkel in order to illustrate what I mean. In the water, I spend most time paddling slowly at the surface at the outer edge of the reef, face down and ideally snorkel up. I try to slow down my breathing, attempting to calm it against the irrational fear that something is coming up behind me. The plan is to time my breathing to whatever happens down there. Then, when I see something interesting below, I prepare to dive. I do so by exhaling as much as I can, drawing in breath slowly and then exhaling deeply, forcing as much carbon dioxide from my lungs as I can. I do this five, six, seven times. Then I draw in one last, large breath and dive down into a different world of sound, color, life. At this stage, depth suddenly matters. I have to continuously equalize the pressure in the Eustachian tubes that connect my nose to my ears to prevent an earache and the risk of a ruptured eardrum. Once at the seabed, equalized and properly buoyant in the water (this is the theory, but none of this necessarily happens every time), I try to stay still on or near the bottom, flailing my arms and legs as little as possible. This is difficult. Kinesthesia, your sense of your own body, changes underwater, because water carries your weight differently. Extra lead weights and an attention to breathing help you to become correctly buoyant and allow you to "hover" in the water. It is hard, but it is worth trying. Not only is even a small level of mastery of the different bodily proprioception that being underwater requires one of the great pleasures of ethnography involving snorkeling and diving (Ota 2006) but the underwater world also behaves differently when you stay still. If you move abruptly or chase underwater life, it will rapidly disappear from sight. But if you stay still, it will likely approach you to take a closer look. In fact, it seems to me that many underwater creatures are as curious about me as I am of them. The magic of this mutual curiosity may last for a while, but I am in this underwater world on borrowed time. The carbon dioxide in my blood invites me ever more insistently to resurface. Much sooner than I would like, I have to return to the surface. After a while, when I have settled my breathing, I will do it all over again. When you do fieldwork with a snorkel, attention to your breathing is key (Braverman 2017).

Snorkeling fieldwork is inherently amphibian, a liminal mode of being both in and out of water. Steven Mentz (2020, 19) argues that swimming crawl-style in water requires you to have one eye in the water and one eye above water, feeling both at ease and entirely mismatched in the ocean. Snorkeling carries this double weight of ease and unease too. Much like swimming, snorkeling is about partial immersion. A day spent snorkeling means that your eyes, ears, and other sensory organs have alternately been under and above water innumerable times. This alternation (and the intimacy and alienation it entails) is for me the point. Snorkeling is an amphibian kind of fieldwork that upsets the conventional distinction between terrestrial field- work and underwater fieldwork (Helmreich 2007, 2009). It is an attention to the underwater world that needs to be acutely aware of and constantly in close reach of the world above water. This amphibian double attention is methodological, epistemological, and political. It is fieldwork where sen- sory connections across surfaces of different elements matter as much as depth (Merchant 2011; Raycraft 2020), where one allows the terrestrial bias of human knowledge and ways of being to be challenged by "wet ontolo- gies" of a multispecies underwater world (Steinberg and Peters 2015; Jue 2020), while still attending to the histories and connections through which the wet and the dry interact (Helmreich and Jones 2018; Mentz 2020).

A day spent snorkeling results in a dazzling montage of epipelagic impres- sions, full of wonder, color, life. But this montage is not mine alone. I tend to snorkel with others in Papua: with Papuan sea cucumber collectors, with spear fishermen, with interested village youths, or with Western snorkeling tourists. For this reason, the montage of underwater multispecies impres- sions on any given day is mixed, mediated, and partially understood through brief above-water conversations about what we saw as well as with later, more reflective conversations with others on land learning about their under- water experiences by telling them mine. A personal experiential phenome- nology of being partly underwater would only tell part of the story, because the sensorial and embodied experience of snorkeling does not exist in a phe- nomenal vacuum. It is mediated by a sociality that is dry and wet, human and more-than-human at the same time. Snorkeling is a snorkeling-with, a being in the water that is always a being-with-others. Snorkeling-with as a multispecies fieldwork method in that sense emerges from Peter Sloterdijk's insight that humans are "ontological amphibians" (Sloterdijk and Heinrichs 2010): animals that move between the elemental worlds of air and water. But it also draws on the fact that humans, as Eben Kirksey (2015) reminds us, are far from the only amphibians.

Snorkeling in Raja Ampat is snorkeling-with because I am clearly not the only epipelagic amphibian being in these waters. Turtles, false killer whales, dugongs, and crocodiles do pretty much what I do in these waters, only bet- ter (Tsing and Bubandt 2020). They, too, have to come up for air. And like me, they, in varying ways, have to negotiate underwater and above-water worlds in sensory, experiential ways across the surface of the water. The Papuan people who live in Raja Ampat also enter into this epipelagic world, also with infinitely more experience and skill than me. Men do so to spear fish or collect sea cucumber; women stand waist deep at low tide on the outer edge of the reef to fish with bamboo poles or scour the low-tide pools at night for squid. But subsistence fishing is not the only activity that draws local Papuans into the underwater world of coral reefs. Some Papuan shamans undertake dream travels to the coral world to find the spirits that hold captive the souls of people and thereby cause them to be sick. The shaman on her or his underwater sojourns (shamans can be both female and male) will try to negotiate with the spirit for the release of the souls of the sick as a way of curing them. These underwater shamanic journeys may last hours or days, because, unlike regular humans, Papuan shamans breathe normally when they travel underwater in their trance states. Add to this already perplexing underwater plethora of multispecies and multibeing life-forms that spirits are not the only strange amphibians in these waters. Every year, thousands of Western scuba dive tourists undertake their own dream travels to Raja Ampat to experience its stunning underwater vistas. Like shamans, these scuba divers can also breathe underwater. Only they need the help of hydraulic tools, tools called "self-contained underwater breathing apparatuses" (or scuba) that today are associated mainly with underwater leisure and tourism but that have long roots in military application. Many other Westerners come, like me, to snorkel. In fact, it was a French free diver, who knew Jacques Mayol, the legendary protagonist of the movie *The Big Blue,* in person, who taught me the technique of free diving that I described earlier.

The tidal zones of Raja Ampat teem with life and dreams in many amphibian forms—human as well as nonhuman. All of these forms come with their own histories—colonial, multispecies, technological, and spirit histories—that connect what goes on underwater to the world above water. This chapter follows the interconnected histories of multispecies living and multibeing dreaming below as well as above the tension-filled surface of a changing sea. Like Raycraft (2020), I am interested in the connections between shamanism and underwater fieldwork as a way to explore the ocean as a space of the otherwise, but in this chapter, shamanism is not a literary metaphor for scuba diving. Rather, it is a troubled historical practice for the Papuans I have gotten to know. I use historical shamanic drawings of underwater spirit worlds to complement and complicate multispecies ethnography. These drawings, as I will show, are reminders that biological creatures are not the only beings in the ocean.

TIDALECTIC ETHNOGRAPHY

Tidalectic ethnography is the name I give to my experiments of snorkeling-with as a method through which to study the critical allure of multispecies and multibeing tropical coral reefs. Tidalectics is a concept coined by Barbadian writer Kamau Brathwaite (2000) in an effort to capture an "alter/native" Caribbean historiography and sensibility. The term is Brathwaite's attempt to formulate a decolonial poesis and philosophy that is less a logic than what he calls a "rhythm-image," like a stone skidding across the surface of water (as cited in Gargaillo 2018). Brathwaite's tidalectics is a kind of island philosophy that seeks to break with the categorical opposites and the ideal of synthesis and resolution that drive Hegelian dialectics. While Hegel's dialectics purport to apply to a universal and "continental" kind of history, born out of the idealism of eighteenth-century Europe, Brathwaite's tidal dialectics insists on being an island philosophy, a specific and materialist rhythm shaped by an acute attention to the hauntings of colonialism in the plantation islands of the Caribbean. Tidalectics, as Brathwaite puts it in an interview from 1990, is

> dialectics with my difference. In other words, instead of the notion of one-two-three, Hegelian, I am now more interested in the movement of the water backwards and forwards as a kind of cyclic, I suppose, motion, rather than linear. (as cited in Mackey 1991, 44)

This empirical and materialist attention to the rhythmic shifts and multi-elemental nature of island seascapes and their colonial histories makes tidalectics an ally of rubber boots methods, an ally that turns the landscape attention of rubber boots to the tidal zone. Tidalectics is an archipelagic imagination, shaped by the rhythms and movements of the tidal seascape, by the history of migration, by colonial slavery, by a global and seaborne white capitalist hegemony, and by the continuing memory of that other place from which one descends. More than an imagination, however, tidalectics is also, as Elizabeth DeLoughrey (2018, 98) argues, a method:

tidalectics provides a methodological tool to foreground islands in history and in modernity, rendering a dynamic model of geography for exploring the complex and shifting entanglements between sea and land, between routes and roots.

This island methodology, for Brathwaite, is very much one where the hauntings of colonialism also entail a multispecies attention. Take these lines from the first verses of Brathwaite's poem titled "Coral":

A yellow mote of sand dreams in the polyp's eye;
the coral needs this pain.
Look closely:
the pearl has limestone ridges, hills,
out of it grows the sun
and the fat valleys of Haiti,
deep mourning waters under the mornes

The coral killers crust my wall of bone
make feet for footprints on this first beach . . .

Even when I was a slave here
I could hear the polyp's thunder
crack of the brain's armour
the ducts and factories sucking
the rivers out, engineering
their courses, as if the stone
were a secret leaf, or a fist curled
in embryo slowly uncurling. (Brathwaite 1967, 232)

In the poem, the corals in the Caribbean tidal zone bear witness to the violence of slavery and environmental destruction of colonialism. The tidal rhythm that allows the grain of sand to become a pearl in the coral polyp, in the manner of an oyster, is in the poem interrupted by another tidal event: the coral-killing feet of colonialism and slavery. As I understand Brathwaite's tidalectics—and here I am greatly informed by the work of DeLoughrey, a longtime champion of the analytical power of tidalectics to move social and natural science beyond the "myth of continents" (Lewis 1997; see DeLoughrey 2001)—it seeks to capture forms of temporal movement, multispecies sociality, and spatiality that challenge the logical bifurcations and colonial heritage of Hegelian dialectics, including the divides between nature and culture, sea and land, nonhuman and human. Tidalectics in this sense is a way of reading corals and colonialism within the same register (see also Elias 2019).

Although Brathwaite's tidalectics emerges out of a specifically Caribbean colonial history, its very insistence on historical specificity has recently allowed tidalectics to become the inspiration for realizing archipelagic thinking elsewhere too. In Hessler and Reymann (2018), tidalectics becomes the site of mooring for a transdisciplinary exhibition and anthology that seek to imagine the oceans and islands differently in situated ways. This curatorial and analytical site of mooring welcomes nonhuman as well as human perspectives in a reflexive decolonial meeting across arts and science. DeLoughrey (2001, 2019), for her part, has argued for an "island heterotopology" that explicitly extends and compares Brathwaite's Caribbean tidalectics to the Pacific Ocean. I extend tidalectics as a methodological tool to the Papuan seascape in the same spirit as a way to attend explicitly to the specific histories and hauntings that are soaked into these waters.

A three-fold curiosity drives my take on tidalectic ethnography in West Papua. First is a multispecies curiosity of how both humans and nonhumans inhabit tidal zones. Here I need biological and ethnobiological methods. Second, a tidal curiosity informs this method that challenges the terracentrism of much science and philosophy. It does so by focusing on the political and historical tensions at the surface where air and water meet, on the amphibian linkages between land and sea in the tidal zones of the Anthropocene. Here research both on land and underwater informs my snorkeling-with. And third, snorkeling-with is driven by a more-than-secular curiosity that is empirical as well as speculative. The abundance of multispecies life on coral reefs easily leads one to assume that biological life is the only kind of life. But the corals of West Papua crawl with all kinds of beings and metahumans that are more-than-biological. There are spirits, ancestors, monsters, and the souls of the sick—if you choose to notice them. Here I learn from shamans and their drawings to see corals otherwise (see also chapter 1, on learning from drawings). In West Papua, tidalectic ethnography tells of tidal worlds that are soaked in multispecies abundance, colonial history, and spirit encounters.

VICARIOUS NATURE

"Can I come snorkeling with you, Uncle?" Willem was in his early twenties and as eager to know what I was doing underwater as anyone I have come across in Indonesia.[1] It was 2018, and I was doing fieldwork in Willem's village in Raja Ampat, staying in the house of his mother's brother. Perhaps that was why Willem insisted on referring to me as "uncle" as well, using the Dutch word *ome* that is still a popular vernacular term of reference in eastern Indonesia, harking back to colonial times, when Holland ruled over West Papua. Colonialism in West Papua is not a distant historical past but part of living memory and contemporary life. The remote western half of the island of New Guinea was controlled by the Dutch until 1962, when a United Nations–brokered agreement handed over control of the territory to Indonesia. In an act of self-determination tightly controlled by the Indonesian military in 1969, some 1,092 carefully selected male Papuan representatives were cajoled, threatened, or lured to unanimously reject Papuan independence and remain part of Indonesia (Saltford 2003).

Every afternoon, when I set off from his uncle's house with my fins, mask, snorkel, and underwater camera, Willem would show up on the beach, holding his own well-worn mask—a gift, he told me, from a French diver—to join me on my underwater exploits. He would point to sharks, turtles, odd-looking sponges, gorgonians, and black corals, gauging my response to get a sense of the appropriateness of his underwater directions. When I asked Willem why he wanted to join me, he told me, "I want to learn from you, Uncle." Willem's interest in my interest in the coral seas was a complete surprise to me. I have been doing fieldwork in eastern Indonesia for thirty years and have often gone snorkeling on the island coral reefs of the region. In the early 1990s, I did so to spear fish and add a little protein to an otherwise bland diet of rice, sago, and boiled cassava leaves but also to get an underwater reprieve from village fieldwork on the island of Halmahera. In Halmahera, no one ever showed any deferential interest in joining me. On the contrary, it was always I who had to beg to join friends and family in the village on their fishing trips, and people would only grudgingly accept.

FIGURE 6.1. Spear gun fishing off Halmahera, 1997. Photograph by the author.

In Halmahera, I had been a cumbersome student. In Raja Ampat, it seemed, I had become a potential teacher. It goes without saying that long-term experience, language skills, and local knowledge are critical to good field-work. But this particular shift in status from student to teacher, I believe, had little if anything to do with whatever underwater skills or local knowl-edge I might have picked up over the years. Rather, my shift in status was an effect of the difference between Halmahera in the 1990s and Raja Ampat in the second decade of the twenty-first century. Halmahera and Raja Ampat are located in the same Papuan seascape. Separated by only 150 kilometers of water, the two regions share much of the same political history. Still, the two places are worlds apart. Halmahera is an out-of-the-way region of min-ing, fishing, and other forms of resource extraction. In the last decade or so, Raja Ampat, by contrast, has become a globally recognized hot spot for marine biodiversity and environmental protection as well as an internation-ally hyped destination for dive tourism. Until the coronavirus pandemic put a temporary but abrupt halt to the influx, some twenty thousand, mainly European tourists flocked to Raja Ampat every year. A concerted effort by

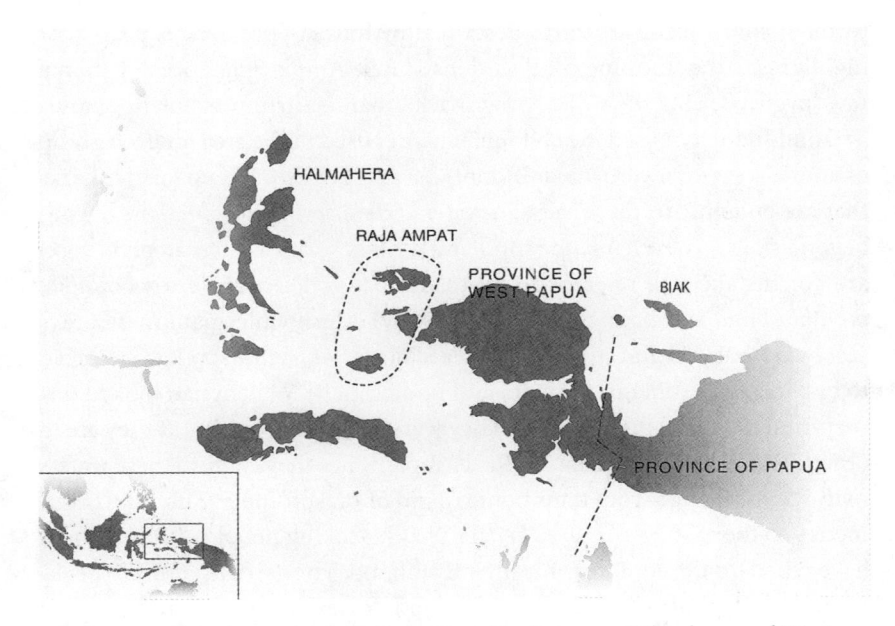

FIGURE 6.2. Raja Ampat and the Papuan waters between Halmahera and West Papua. Map by Louise Hilmar.

nongovernmental organizations and a not entirely sincere Indonesian attempt to compensate for decades of political repression by developing the tourist potential of West Papua have brought a halt to the logging, mining, and illegal fishing that happened in Raja Ampat until the late 1990s, establishing extensive national marine parks and protected forest areas, paid for in part by fees levied on visiting tourists.

It was because of this world of tourism that Willem in 2018 was developing a new and vicarious interest in his own coral sea. And truth be told, I did not feel like a teacher to Willem on our snorkeling trips; I felt more like his object of ethnographic study. He was less interested in what I knew about the ocean than in what interested me in the ocean. Willem was cultivating a vicarious interest in the ocean through me: he wanted to learn what I—as good a stand-in for a white tourist as any—saw and found interesting in the ocean.

"What do white people [Ind. *bule*] look for underwater?" Willem asked after our first trip together. "What do they like to see?" I answered by going through some of the creatures described by tourist operators as particular highlights in the stunning coral gardens of Raja Ampat: giant oceanic manta rays, pygmy sea horses, wobbegong sharks, mantis shrimp, epaulette sharks, and nudibranchs. I used the colloquial names used in the area for each group of animals, a mix of vernacular Indonesian and English. The epaulette sharks that are endemic to these Papuan waters *(Hemiscyllium halmahera),* wobbegong sharks (*Orectolobidae* spp.) and mantis shrimp (*Stomatopoda* spp.) are, for instance, all referred to by vernacular Indonesian terms (*kalabia, ikan hiu batu,* and *udang kutik,* respectively). Meanwhile, giant manta rays *(Mobula birostris)* and nudibranch sea slugs (*Nudipleura* spp.) are referred to by their English names (manta and nudibranch). Willem had looked unsurprised as I listed these species. They were all familiar to him, as they are to other Biak-speaking residents of Raja Ampat. Since they came to these waters many centuries ago from their home island of Biak, some five hundred kilometers to the east (see Figure 6.2), the Biak-speaking people of Raja Ampat have relied on the coral seas for their livelihood. The gardens of Biak speakers in Raja Ampat, who call themselves either Beser or Kafdarún, are small compared to those of their neighbors on Halmahera and other parts of Papua. Beser and Kafdarún people are seafaring people with an intimate knowledge of the sea and its creatures—fishermen whose everyday lives revolve practically and cosmologically around the sea. Before 2007, when 1.7 million hectares of the Raja Ampat water were declared marine protected areas, Willem's male covillagers would regularly hunt manta rays (called *pau* in Beser) and turtles *(wayo)* with harpoons from their outrigger canoes, while his aunts and grandmothers would collect mussels and fish on the tidal areas of the coral reefs. Now both manta rays and turtles are protected species, and hunting them is illegal.

Given his intimate knowledge of the underwater world, my list held no surprises to Willem—that is, until I got to the last one: nudibranchs. "What are nudibranchs?" Willem asked, his interest suddenly piqued. I tried to relay to him my own interest in this fantastic and queer group of creatures. Nudibranchs are oblong, soft-bodied sea slugs that are both carnivorous and hermaphroditic. There are more than three thousand described species. The actual number of species, however, is far higher, and the relatively new and spiraling interest in nudibranchs in the community of amateur scuba divers is turning up new species at a rate that far outpaces the work of taxonomists (Smith 2019, 193). Nudibranchs are alien creatures. Some species endosymbiotically incorporate the chloroplasts and algae of the prey species they consume into their skin as extra sources of photosynthetic power. Other species of the Aeolidioidea superfamily are known to incorporate the tiny toxic harpoons (called cnido-cysts), taken from their jellyfish prey, into their skin and deploy them as defensive weapons (Wägele et al. 2008, 388). There is a Beser word for nudibranchs: *taráir*. While Willem did not know it, other people did. Papuan dive guides range among the most highly skilled in the world, many logging more than ten thousand dives, and both amateur divers and Western scientists rely critically on these Papuan dive guides to find desired or new species. But cultural competence, as Fredrik Barth (1987) points out, is always unevenly distributed. This meant that to Willem, nudibranchs were a novel world. I was essentially describing a creature to him that he did not know from his familiar underwater world. Nudibranchs are inedible. Maybe that was why Willem did not know them: he had had no reason to know them before. Then again, people in Willem's village do not eat pygmy sea horses or mantis shrimp either, and Willem did know them. Whatever the reason for Willem's ignorance of nudibranchs, my mention of them clearly intrigued him: was this a chance to learn something about what white tourists saw in the sea and what brought them to Raja Ampat in the thousands every year? Maybe this creature was part of the secret at the heart of white people's interest in the underwater world? Maybe the nudibranch could be a first step on his journey to becoming an expert underwater tourist guide?

Although he lived in a relatively remote village, Willem was no stranger to the industry of dive tourism. He had worked for a year in a nearby Italian-owned resort, acquiring his PADI Open Water certificate, courtesy of the foreign resort owner. The resorts in Raja Ampat, of which there are now close to fifty, like to advertise their employment of local Papuans as kitchen hands, cleaners, boat drivers, and dive guides. For most young men in Raja Ampat, the job as a dive guide is the best opportunity for a well-paid job that easily dwarfs the wage one might get as a teacher, policeman, or bureaucrat. But in return for paying for the dive certificates, the resorts hold the certificates in escrow and only release them after years of loyal service.

The big dream for people in Raja Ampat—perhaps for men more than women—is to open their own "homestay": a low-budget, Indigenous competitor to the foreign-owned luxury resort that houses Western guests at low nightly rates. This means that many Papuan employees leave their jobs in the resorts as soon as they have saved enough money to begin construction work on their own homestays. Willem also had homestay dreams. But they had been cut short by more immediate problems with the Italian owner, who Willem found to be both unreasonable and hot tempered. "Her name is Maddie, and she is mad," he punned in Indonesian. As a result, Willem's employment in the resort had lasted only a year. Now he was back in the village, with only a beginner's knowledge of scuba diving and without a copy of the official PADI certificate that he needed to pursue his homestay dreams. When I turned up in the village, it was for Willem therefore an opportunity, the first in some time, to continue his sojourn in the underwater world as white people see it. I think that was an important reason for why he wanted to go snorkeling with me.

The next day, Willem and I set out together again. The low tide was at noon, so we waited until it peaked before we set off. This allowed us to reach seascapes of the coral reef that would be out of our depth at high tide. At the climax of each tide, the current is much less fierce. This is an added advantage

FIGURE 6.3. *Doriprismata atromarginata*. Photograph by the author.

for the snorkeler in Raja Ampat, which is renowned for its strong tidal currents. Snorkeling, like all tidalectic ethnography, is all about timing.

We had snorkeled the coral reef in the serenity of the low tide for some time when we got to a muddy underwater sand slope. Immediately upon reaching the slope, Willem stuck his head above water to call me over. "Is this what you mean?" he asked under his breath, lifting the mask from his face to speak. And it was: five meters down, Willem had found a creature that he until that day had never noticed before, despite his long experience with the underwater world as a spear fisherman and saltwater-soaked village youth: a dark-margined nudibranch *(Doriprismatica atromarginata)*. Although this species is common to shallow waters throughout the Indo-Pacific, it was also my first encounter with it, and being no great nudibranch aficionado, I had to look it up afterward to know what I had seen. Willem and I beamed: we had both found something new!

Rubber Flipper Methods

I have brought you to this humble discovery of a common nudibranch to make a tidalectic methodological point: while one might be tempted to draw a hard philosophical and ontological opposition between the land and the sea, between the dry and the wet, underwater worlds are intimately linked to the land. The land–sea connection is well established in ecology and is a key political ecological concern to marine conservation efforts (Halpern et al. 2009; Álvarez-Romero et al. 2011). Tidalectic ethnography, similarly, has a methodological focus on the historical-ecological connections between land and sea. Take Willem's and my shared discovery of a nudibranch. Our discovery was enabled in a transcendence of the boundaries between terrestrial and underwater worlds. Nudibranchs are hermaphroditic marvels of marine science as well as cultural fetishes in a global scuba dive industry. This science–industry encounter deals with the underwater world, but it circulates above water in images, publications, capital transactions, and flight plans. And that circulation in turn impacts how nudibranchs "come to matter" to people like Willem in West Papua. Tidalectic ethnography seeks to trace the ways that land and sea come to matter to each other, across scientific, touristic, or Indigenous registers. Tidalectic ethnography is in that sense about ethnoecology, but also more. It is a retooling of ethnobiology for the changing tidal zones of the Anthropocene (Bubandt and Tsing 2018). Marine science and ethnobiology tend to focus on how the rich treasure of Indigenous knowledge may complement or enrich Western science (Johannes 1981; Narchi and Price 2015). Willem's budding interest in nudibranchs illustrates that what some might call Indigenous knowledge of "things marine" is deeply impacted by scientific and touristic forms of knowledge: Willem explicitly remolded his concerns for and attention to underwater worlds in response to outside interests in these same worlds.

Tidalectic ethnography is interested in rhythms such as these. And rhythms have varying durations. Tidalectic ethnography, therefore, needs to follow how land–sea connections are reshaped by those changing long-term historical links through which underwater worlds are perceived and performed. But it also has to pay attention to the short-term tidal rhythms as locals have always done when they fish and sail. The coral reef is a different world at low tide and high tide, one world during the day and another at night. So tidalectic ethnography entails an attention to the material temporalities of the tide and to the way this affects multispecies life on the coral reef. However, tidalectic ethnography also has to emerge from the water and lift its eyes and ears from the sea to notice the middle-range histories through which underwater sensibilities themselves are changed. Willem was training his senses to vicariously see what white dive tourists were seeing in the rich Papuan waters. Tourists, for their part, were fascinated by the scientific discovery that these waters are the most biodiverse coral reefs on earth and by the allure that many species, still scientifically unnamed and undescribed, inhabit them. This allure was itself refracted by the fragile future of coral reefs in a time of climate change. The regional tourism board of Raja Ampat has sought to capitalize on this "extinction dimension" of dive tourism by naming the region the "Last Paradise on Earth." Indeed, indications that the coral world paradise of Raja Ampat might be irreversibly changing is a common topic of conversation among both tourists and locals: currents seem to be different, the water feels warmer, some corals appear to have begun bleaching, and invasive species have appeared. Paradise, it would seem, has particular allure on the cusp of the Anthropocene (Bubandt 2019). Tracing such sensibilities and changes is also part of a tidalectic ethnography.

At a planetary crossroads, dive resorts in Raja Ampat often promote undescribed species as their special brand of paradise, which they will claim only their own Papuan dive guides are able to find. It was just such a guide that Willem aspired to be. Willem's vicarious underwater interest in nudibranchs was, in other words, historically overdetermined by the specific ways in which marine science and a global tourist industry intersected in West Papua at the time of a coral reef crisis on a planetary scale. A technological hint of this history was the fact that Willem discovered his first nudibranch through the rubber mask gifted to him by a French dive tourist. My own use of a rubber mask, a snorkel, and rubber fins was a similar reflection of this same global assemblage of tourist industry, the exploding allure of coral reefs at the historical moment of their possible extinction, and the colonial legacies of rubber (see the Introduction to this volume). Few Papuans—except the professional scuba dive guides—use rubber flippers underwater. Most people instead prefer to propel themselves underwater by "frog kicking." They do this with an effectiveness that I cannot match, so I need the flippers as a prosthetic to compensate for my inferior underwater skills. I can only reach the same depths as my Papuan underwater friends and collaborators by wearing flippers. One might call this "rubber flipper methods": my flippers are testimony to the colonial–scientific–touristic assembly that is changing Willem's underwater world, but at the same time, they are also a necessary methodological prosthetic to access this world.

Rubber boots methods, as we outlined in the Introduction, insist that the methods of multispecies research must be critically attuned to the terrain of the landscapes and seascapes that they seek to study. Methods are by default a troubled kind of prosthetic in a world of colonial legacies. Taking methods into multispecies worlds, above and below water, merely multiplies the trouble (Haraway 2016). Staying with the trouble of this entails being cognizant of the historical and material conditions that made those terrains as well as the historical processes that put specific kinds of prosthetics into just those terrains. Rubber boots methods are not a one-size-fits-all argument. Wellington rubber boots might be suitable for engaged, grounded research on some boggy grounds. Using Wellingtons in such terrain means attending to their colonial and military legacies. But those same legacies will make rubber boots singularly unsuitable on other terrains (see chapter 14). Rubber flippers, snorkel, and face mask were my "rubber boots" fieldwork tools. They, too, have their own specific histories tied to colonialism and capitalism. Wooden goggles were used in precolonial times in the Persian Gulf and in the Pacific (Marx 1990), but their spread into eastern Indonesia is tied to the influx of Japanese pearl diving operations in the wake of World War II. The first rubber flippers for underwater use were constructed by a French navy commander in the 1920s and brought to Indonesia by tourists and commercial entrepreneurs seeking sea cucumber, rock cod, pearls, and lobster (Wright 1959). Today, scuba diving tourism is the dominant enterprise pushing these prosthetics into Papuan waters.

Being aware of these historical legacies is key. Without this awareness, Willem's interest in nudibranchs is unintelligible. Ironically, these colonially charged prosthetics are what enable my flipper methods in the first place. They allow me to extend participant observation into the multispecies world of the tidal zones of Raja Ampat to explore the intimate, more-than-human relations that shape coral reefs across the border between land and sea. Flipper methods are, to paraphrase Stefan Helmreich (2007), "transductive": they are mediated and made possible by technologies with specific histories charged by colonialism and asymmetric global connections.

However, as it turns out, the Papuan underwater world is more than a transducted colonial-technological world. The Papuan seascape is also an animated spiritscape (see McNiven 2003) through which both white divers and colonial histories are made to make sense—but with a difference. At this point, a tidalectic method that seeks to understand Willem's vicarious interest in his own multispecies, underwater world needs a multisensorial approach that also includes the sixth sense, by which I mean an anthropological attention to the world also when it is revealed as otherwise. To see the coral world otherwise, I need other prosthetics than mask, snorkel, and flippers. This is where I turn to drawings.

ANCESTORS AND CORALS OTHERWISE

Adoption is a common practice in the Biak-speaking areas of West Papua, and like many others, Willem was an adopted child. While his mother lived in the city of Sorong, Willem had been raised by his grandparents in the village. Until his death in 2014, Willem's grandfather, Albert, had been the leader of a millenarian movement. Started in 1999, the movement hoped to bring back the mythical ancestor called Manarmaker, literally "Old-Man-Scabies," a hope that also reverberates through many other millenarian movements in West

Papuan history (Kamma 1972; Rutherford 2003). According to a myth from Biak, Manarmaker had acquired the magical ability to conjure up all things and revive the dead from the Morning Star *(Makmeser),* thereby giving him the power to instate a paradisiacal state called Koreri where "food would arrive by itself" and life was forever. But before he was able to bring this state of paradise about, Manarmaker's in-laws, disgusted by Manarmaker's diseased skin, evicted him from his village in Biak. So Manarmaker sailed west toward Raja Ampat (see Figure 6.2), hotly pursued by his angry in-laws. In Raja Ampat, Manarmaker conjured up a large iron steamboat, boarded it, and sailed to the West. Some say he went to Holland, where he dwells to this day with the Dutch royal family. Manarmaker's presence in Holland was, so people say, what enabled the Dutch to colonize Papua and Indonesia with the superior technology that Manarmaker had bestowed onto them. Before he left Raja Ampat, however, Manarmaker promised his Papuan relatives to return, and it was this glorious return to Raja Ampat that Albert's millenarian movement had promised in 1999, gathering hundreds of followers from dozens of villages throughout the region. The concrete driver behind Albert's movement was his daughter (Willem's aunt), who, at the age of seventeen, had begun to be possessed by Manarmaker. During these possession séances, Manarmaker revealed that he had already shed his diseased black skin and reincarnated himself in a new figure, namely, the Dutch owner of the first dive resort in Raja Ampat. And indeed, as the first white dive tourists began to trickle in and the resort acquired speedboats to whisk them off to the best dive spots, Manarmaker's mythical promise of returning home to Papua with all the riches and technologies of the West seemed in 1999 to finally be realized. Papua would become independent as all nations and peoples would gather there to await the cataclysmic renewal of the world. For Papua, according to a local millenarian interpretation of the Bible, would be the site where the world would be renewed because it was also the Edenic place where the entire world began with Manarmaker.

In this sociocosmological fervor, Albert and his followers began to suspect that the white dive tourists might possibly be angelic beings: biblical figures that were also white incarnations of their own dead Papuan ancestors. This millenarian narrative—where white people are really Black people, where Papua is the real Israel, and where Western technology and power in fact have a Papuan magical origin—is clearly a utopian reversal of a racial, colonial, and Judeo-Christian world order (Worsley [1957] 1970; Lattas 1998; Robbins 2001). It is a world where white and Black, colonial and Indigenous, the West and Papua, are not distinct but each other's condition of possibility. It is a tidalectic notion of transcendence across a tension-filled surface world where water and air meet. It is said in Raja Ampat that the New Jerusalem, of which biblical revelations speak, will rise from its coral reefs. In other words, the millenarian dream of the political return of Manarmaker is also a multispecies dream informed by the magical possibilities of animism. However, hiding among the white tourists, people would tell me, were also emissaries of the Dutch Crown, eager to reclaim Manarmaker's power and the Golden City of New Jerusalem for themselves. Their aim was to prevent political independence and a paradisiacal future for Papua.

This Papuan millenarianism is driven by the logic of animism at every turn. When the Morning Star bestowed its magic to Manarmaker (the same magic that he passes on to the Dutch colonial government), the Morning Star assumed the form of tree-living marsupial, the Biak glider (*Petaurus biacensis*). Manarmaker discovers the Edenic site where the world began (a cave full of youthful ancestors), not in a Dutch Bible, but because an ancestor in the shape of a wild pig shows it to him. In the same way, the Golden City of the Paradise-to-Come will rise from among the coral reefs of Raja Ampat, where the Virgin Mary (who is also the Papuan wife of Manarmaker) guards the real version of Holy Scripture in which the Papuan origin of the world is revealed. Note, however, how the animism of this Papuan millenarianism entails critical references to colonialism at every turn too. The excesses of animist magic are also the excesses of Dutch colonialism in Papua.

Let us revisit Willem's interest in nudibranchs in light of this millenarian narrative. It is, after all, a narrative in which Willem is as soaked as he is in the saltwater of these coral reefs. Trying to figure out what white divers are attracted to in this light is a concern that is part natural history and part divination: it is both about specific animals and about the general possibility of salvation. And it is a concern that harbors hope as well as suspicion. Remember that white dive tourists are both rich white people, like the Dutch who colonized Papua, *and* potentially transformed ancestors waiting to reveal themselves. A key question for people in Willem's village therefore is this: are these foreigners (who can slip in and out of different skins like spirits and shamans) in Raja Ampat to give back to the Papuan people the greatest gift of all, namely, a future paradise that is the ancestral remnant of Papuan origin? Or are they really there to take that gift back to their own countries? If paradise will arise from the coral sea, then the intense, almost greedy interest that Willem and his peers see white dive tourists have in the Papuan waters is both the possible fulfillment of a mythical promise and the possibility of its obstruction.

The coral reefs of Raja Ampat are not just a site of record-breaking biodiversity. They are a crowded spiritscape. Biblical figures and ancestors occupy the Papuan seascape with many other kinds of metahumans. Predators or poisonous animals that harm people do so because they have become *faknik*, demon animals possessed by spirits (Tsing and Bubandt 2020). Large coral boulders, distinctive coastal limestone formations, and many mangroves and beaches are by the same token considered *mon*: taboo places inhabited by specific and often named spirits.

By what methods might tidalectic ethnography learn to see this coral world of the otherwise? For me, a methodological breakthrough came when I was fortunate enough—with the kind help of Raymond Corbey (2017, 2019), who has published extensively on colonialism and Papuan spirit worlds—to locate 150 drawings of spiritscapes from Raja Ampat in a Dutch archive. The drawings had been made by shamans (called *snon mon* or "spirit men") in Raja Ampat in the 1930s. They were collected by Freerk Kamma, a long-term missionary to and later anthropologist in West Papua, but had long been considered lost, destroyed by the Japanese during World War II. The discovery of some of these drawings in a Dutch archive, unorganized and undescribed, was a revelation and a methodological opportunity. On every return trip to Raja Ampat since 2018, I have spent many hours in coastal villages showing these drawings to people who know about spirit things. These are people who would have been called shamans *(snon mon)* in the past. Today, many of them reject that name for fear of being accused of not being properly Christian or, worse, of being sorcerers. Still, it is to them that people turn in case of illness, birthing problems, or other misfortune, and most villages have at least one of these knowledgeable people. Listening to their interpretations of the old drawings has become an important methodological pathway into the spirit world that also helped me see coral reefs otherwise when I snorkel. Snorkeling-with also means snorkeling with spirits. It means paying attention to which places people avoid going when spear fishing because these places are taboo *mon* places. It means to listen differently to people's accounts of seeing strange beings or feeling odd underwater. The drawings help me cultivate this attention and to snorkel the coral reef differently, because they turn out to be maps of a coral spiritscape, reminders to never completely trust one's senses—below and above water. Take the drawing in Figure 6.4, for instance. According to a reading of the drawing made collectively by a knowledgeable man and woman in 2018, it is a map of an underwater spirit place as a shaman sees it.

To the ordinary eye, the drawing appears to be an oddly adorned *korkór* or chambered nautilus shell *(Nautilus pompilius)*. But in the spirit world, as

FIGURE 6.4. Drawing by an unknown shaman of an underwater spirit place *(mon)*, circa 1930s. From the collection of Freerk Kamma held at the Utrecht Archive. Collection Raad voor de Zending, 1951–99, access no. 1102-2, inventory 6740.

the shaman knows, nothing is as it appears. The drawing reveals the shell to be the dwelling place of a witch spirit called *baradota* as the shaman would encounter it during his or her dream travels (called *datutér*). The spiraling structure of the seashell is a spiritual castle or fortress, a complex maze in which the witch spirit hides the soul *(rúr)* of the person who lies sick back home in the village. It is the task of the shaman on his or her spirit travels to enter the shell structure and negotiate with the witch spirit for the release of the soul and to return it to the patient in the village. This is a difficult and dangerous task. The naive shaman, I was told, might be tempted to enter through what looks like a door at the entrance in the top left corner of the drawing. But these are actually fish hooks designed to catch intruders. The wise shaman will instead enter through the large "head" in the top right corner. This "head" is the seat of power, referred to with the Indonesian word *benteng* (which literally means "fortress"). The flags that protect the bottom half of the spirit dwelling are also fortresses, magical forms of protection. If the shaman has less spiritual power *(baraka)* than the spirit, the shaman will also be trapped in the fortified shell of the witch. To successfully cure the sick, a shaman must be more powerful than the witch causing the illness. It is for this reason that villagers both need and fear shamans.

One gets a hint at the ultimate source of the spiritual power that makes the shaman both a successful curer and a frightful sorcerer by paying attention to the flags of the magical fortress in the drawing. They are, one will note, striped horizontally like the tricolor national flag of the Netherlands, the country that colonized West Papua until 1962. Reading the drawing alerts us to the fact that in the underwater spirit world of Raja Ampat and other parts of Papua, spirits frequently have the shape of animals but harness the power of colonialism (Rutherford 2003). Like the spiraling structure of the nautilus shell, colonial power has no outside in Papuan animism. This is because, as we saw already in the myth of Manarmaker, colonial power actually comes from the inside of the spirit world: it is because of Manarmaker's magic derived from a Biak sugar glider that Holland could colonize Papua and Indonesia in the first place!

The secret of the Papuan underwater world, and the interest of white dive tourists in its multispecies life, is the secret of this hidden, but historically denied, link between global power and animist magic. Snorkeling the coral reefs of Raja Ampat therefore means paying attention to histories of the otherwise as they transcend the boundaries between the wet and the dry. It is through these tidal histories of spirit connections that the natural history of corals as multispecies worlds and the political history of corals as colonial sites of domination are revealed. On the coral reefs of Raja Ampat, the Anthropocene becomes fully visible and audible only through the Anthropo-not-seen (de la Cadena 2015).

CONCLUSION: AMPHIBIAN METHODS

Snorkeling the coral reefs in the Anthropocene is a venture into a multispecies world that is overwhelmingly in your face, even as it stands at the brink of extinction. Bringing natural science and ethnobiological methods into an curious investigation of this world is critical. They become ways of snorkeling-with. But they are not enough by themselves. After all, the multispecies underwater worlds of West Papua are also deeply colonial and historical worlds. As I have tried to show, both colonial history and contemporary global history entangle the coral worlds of Raja Ampat intimately in the Anthropocene: in a high-CO_2 world that threatens the future of tropical coral reef as an ecosystem, Raja Ampat has become a last refuge, a last paradise that attracts European tourists by the thousands amid changing marine-ecological conditions. This novel economy and influx in turn change the sensibilities that locals like Willem bring to these reefs. A critical curiosity toward global history and planetary boundaries is necessary to an understanding of the coral reefs of Raja Ampat. But on its own, that, too, is not enough. I have therefore argued for a third kind of methodological curiosity, namely, one that helps us see the coral world otherwise. In this part of Indonesia, political history is always intimately tied up with the politics of spirits and other metahumans (see Bubandt 2014, 2019). Coral seascapes are multispecies worlds changing politically and ecologically with global history in the Anthropocene. And this change implicates the underwater spiritscapes more than one might think. The Last Paradise on Earth in Raja Ampat is simultaneously an emblem of global tourism, a specter of ecological crisis in the Anthropocene, and a hope for multispecies millenarianism that transcends a deeply asymmetrical racial and political history in Papua.

Tidalectic ethnography is a rubber boots method, a snorkeling-with that takes a resolutely multispecies perspective while framing it within histories of colonialism and capital (as Willem's interest in nudibranchs showed me).

It is also a tidal methodology that challenges the terracentrism of conventional portrayals of humans in anthropology and philosophy, while insisting that humans are far from the only amphibians (as snorkeling helped me see). And tidalectic ethnography is, finally, a methodology that insists on the importance of more-than-Western and Indigenous perspectives to take us beyond the comforts of the "mainland of the real" and teach us to see corals otherwise (as the Papuan shamanic drawings help me to do).

What I also called rubber flipper methods is an amphibian methodology that seeks to attend to the multispecies, colonial-historical, and Indigenous registers through which tidal zones are made in the Anthropocene. Tidalectic ethnography shares, in that sense, a great deal with recent calls for an "amphibian anthropology" (ten Bos 2009; Kirksey 2015; Krause 2017). The call borrows from Peter Sloterdijk the idea that humans are "ontological amphibians," animals who inhabit multiple elements—air and water— and who cannot resist moving between them. But as Eben Kirksey (2015, 17) rightly argues, Sloterdijk's notion of the "ontological amphibian" is anthropocentric: humans are not the only ones who are able to move beyond their own world into new ones. Indeed, the Anthropocene is a world of emergent ecologies where animals of all sorts are forced to move beyond the comfort zone of their own world. Kirksey "poaches" Sloterdijk's concept to suggest that being an "ontological amphibian" is a multispecies condition of life, particularly in a world of disturbed landscapes and ecological imperialism.

This multispecies and historically situated nature of ontological amphibians is also the methodological basis for a tidalectic ethnography. But as I suggested, this in itself is insufficient. For multispecies studies, in its efforts to reach out to natural science, all too often represents multispecies worlds as if they were a matter only of biological species, thereby evacuating spirits, metahumans, and a sense of the uncanny from the analysis (Bubandt 2018). Tidalectic ethnography in Raja Ampat has shown me that coral reefs are brimming as much with spirits and the uncanny as they are brimming with fish and invertebrates. Here Sloterdijk's amphibian ontology—like Brathwaite's tidalectics—has something to offer to multispecies studies, because both Sloterdijk and Brathwaite cultivate a "theory of moves" and transitions between the ontologically dry and the wet, between the real and the really made up. Sloterdijk's amphibian anthropology is, as he puts it, critical of an anthropology that interprets humans "as creatures who in the end can only

exist in one element, that is to say, on the mainland, in the so-called real"
(Sloterdijk and Heinrichs 2010, 336). Land all too easily becomes the realm
of the real, and a terracentric anthropology, as a result, easily becomes a
secular anthropology. I have tried to break with this secular tendency.

At the same time, I have also tried to break with an emerging consensus
in the "oceanic turn," namely, a tendency to Orientalize the ocean. Critical
of Continental thinking, proponents of the oceanic turn in my view tend to
represent the ocean as an exotic "other": a domain of wet ontology and the
otherwise that is disconnected and radically different from the terrestrial
world. I have tried to resist this Orientalism, insisting that tidalectic ethnography is about land–sea mediations that are both historical and ontological.
I have tried to show that in the Papuan tidal zone, reality is historically and
ontologically amphibian: maps drawn on land tell stories of dream travels
underwater to cure the sick in the village; underwater wonders attract white
tourists from foreign lands; and coral gardens are located at the beginning
of a sixth mass extinction fueled by land-based fossil fuel consumption,
even as they might also just be the place of a New Jerusalem from which
the world is made anew. Tidalectic ethnography is a method of curiosity
that follows these movements in and out of water, arguing that tidalectics is
applicable beyond both Brathwaite's West Indies and the Papuan waters of
the former East Indies. After all, we inhabit the haunted tidal zones of the
Anthropocene together.

NOTE

1. I use pseudonyms for all people in West Papua to protect their privacy.

REFERENCES

Álvarez-Romero, Jorge G., Robert L. Pressey, Natalie C. Ban, Ken Vance-Borland,
    Chuck Willer, Carissa Joy Klein, and Steven D. Gaines. 2011. "Integrated Land-Sea
    Conservation Planning: The Missing Links." *Annual Review of Ecology, Evolu-
    tion, and Systematics* 42, no. 1: 381–409.
Barth, Fredrik. 1987. *Cosmologies in the Making: A Generative Approach to Cultural
    Variation in Inner New Guinea.* Cambridge: Cambridge University Press.
Birkeland, Charles, ed. 2015. *Coral Reefs in the Anthropocene.* Dordrecht, Nether-
    lands: Springer.
Brathwaite, Kamau. 1967. *The Arrivants: A New World Trilogy—Rights of Passage/
    Islands/Masks.* Oxford: Oxford University Press.

Brathwaite, Kamau. 2000. *The Arrivants: A New World Trilogy.* Rights of Passage: Islands:Masks. Cambridge: Chadwyck-Healey.

Braverman, Irus. 2017. "The Life and Law of Corals: Breathing Meditations." In *Research Methods in Environmental Law: A Handbook,* edited by Victoria Brooks and Andreas Philippopoulos-Mihalopoulos, 458–81. Northampton, Mass.: Edward Elgar.

Bubandt, Nils. 2014. *Democracy, Corruption and the Politics of Spirits in Contemporary Indonesia.* Modern Anthropology of Southeast Asia. London: Routledge.

Bubandt, Nils. 2018. "Anthropocene Uncanny: Nonsecular Approaches to Environmental Change." In *A Non-secular Anthropocene: Spirits, Specters and Other Nonhumans in a Time of Environmental Change,* edited by Nils Bubandt, 2–19. AURA More-Than-Human Working Paper 3. Aarhus, Denmark: Aarhus University Research on the Anthropocene (AURA). http://anthropocene.au.dk/working-papers-series/.

Bubandt, Nils. 2019. "Of Wildmen and White Men: Cryptozoology and Inappropriate/d Monsters at the Cusp of the Anthropocene." *Journal of the Royal Anthropological Institute* 25, no. 2: 223–40.

Bubandt, Nils, and Anna Tsing. 2018. "An Ethnoecology for the Anthropocene: How a Former Brown-Coal Mine in Denmark Shows Us the Feral Dynamics of Postindustrial Ruin." *Journal of Ethnobiology* 38, no. 1 (online suppl.): 1–13. https://bioone.org/journals/journal-of-ethnobiology/volume-38/issue-1/0278-0771-38.1.001/Feral-Dynamics-of-Post-Industrial-Ruin-An-Introduction/10.2993/0278-0771-38.1.001.short.

Corbey, Raymond. 2017. *Raja Ampat Ritual Art: Spirit Priests and Ancestor Cults in New Guinea's Far West.* Leiden, Netherlands: C. Zwartenkot Art Books.

Corbey, Raymond. 2019. *Korwar: Northwest New Guinea Ritual Art According to Missionary Sources.* Leiden, Netherlands: C. Zwartenkot Art Books.

de la Cadena, Marisol. 2015. *Earth Beings: Ecologies of Practice across Andean Worlds.* Durham, N.C.: Duke University Press.

DeLoughrey, Elizabeth. 2001. "The Litany of Islands, the Rosary of Archipelagoes: Caribbean and Pacific Archipelagraphy." *Ariel: A Review of International English Literature* 32: 21–51.

DeLoughrey, Elizabeth. 2017. "Submarine Futures of the Anthropocene." *Comparative Literature* 69, no. 1: 32–44.

DeLoughrey, Elizabeth. 2018. "Revisiting Tidalectics: Irma/José/Maria 2017." In *Tidalectics: Imagining an Oceanic Worldview through Art and Science,* edited by Stefanie Hessler, 93–101. Cambridge, Mass.: MIT Press.

DeLoughrey, Elizabeth. 2019. *Allegories of the Anthropocene.* Durham, N.C.: Duke University Press.

Elias, Ann. 2019. *Coral Empire: Underwater Oceans, Colonial Tropics, Visual Modernity.* Durham, N.C.: Duke University Press.

Gargaillo, Florian. 2018. "Kamau Brathwaite's Rhythms of Migration." *Journal of Commonwealth Literature* 53, no. 1: 155–68.

Halpern, Benjamin S., Colin M. Ebert, Carrie V. Kappel, Elizabeth M. P. Madin, Fiorenza Micheli, Matthew Perry, Kimberly A. Selkoe, and Shaun Walbridge. 2009.

"Global Priority Areas for Incorporating Land–Sea Connections in Marine Conservation." *Conservation Letters* 2, no. 4: 189–96.

Haraway, Donna. 2016. *Staying with the Trouble: Making Kin in the Chthulucene.* Durham, N.C.: Duke University Press.

Helmreich, Stefan. 2007. "An Anthropologist Underwater: Immersive Soundscapes, Submarine Cyborgs, and Transductive Ethnography." *American Ethnologist* 34, no. 4: 621–41.

Helmreich, Stefan. 2009. *Alien Ocean: Anthropological Voyages in Microbial Seas.* Berkeley: University of California Press.

Helmreich, Stefan. 2016. *Sounding the Limits of Life: Essays in the Anthropology of Biology and Beyond.* Princeton, N.J.: Princeton University Press.

Helmreich, Stefan, and Caroline Jones. 2018. "Science/Art/Culture through an Oceanic Lens." *Annual Review of Anthropology* 47, no. 1: 97–115.

Hessler, Stefanie, and Markus Reymann, eds. 2018. *Tidalectics: Imagining an Oceanic Worldview through Art and Science.* Cambridge, Mass.: MIT Press.

Hughes, Terry P., Michele L. Barnes, David R. Bellwood, Joshua E. Cinner, Graeme S. Cumming, Jeremy B. C. Jackson, Joanie Kleypas et al. 2017. "Coral Reefs in the Anthropocene." *Nature* 546, no. 7656: 82–90.

Johannes, Robert E. 1981. *Words of the Lagoon: Fishing and Marine Lore in the Palau District of Micronesia.* Berkeley: University of California Press.

Jue, Melody. 2020. *Wild Blue Media: Thinking through Seawater.* Durham, N.C.: Duke University Press.

Kamma, F. C. 1972. *Koreri: Messianic Movements in the Biak-Numfor Culture Area.* Koninklijk Instituut voor Taal-, Land- en Volkenkunde Translations 15. The Hague: Martinus Nijhoff.

Kirksey, Eben. 2015. *Emergent Ecologies.* Durham, N.C.: Duke University Press.

Krause, Franz. 2017. "Towards an Amphibious Anthropology of Delta Life." *Human Ecology* 45, no. 3: 403–8.

Latour, Bruno, and Peter Weibel, eds. 2020. *Critical Zones: The Science and Politics of Landing on Earth.* Cambridge, Mass.: MIT Press.

Lattas, Andrew. 1998. *Cultures of Secrecy: Reinventing Race in Bush Kaliai Cargo Cults.* Madison: University of Winconsin Press.

Lewis, Martin. 1997. *The Myth of Continents: A Critique of Metageography.* Berkeley: University of California Press.

Mackey, Nathaniel. 1991. "An Interview with Kamau Brathwaite." *Hambone* 9, no. 9: 42–59.

Mangubhai, Sangeeta, M. Erdmann, Joanne Wilson, Christine Huffard, Ferdiel Ballamu, Nur Ismu Hidayat, Creusa Hitipeuw et al. 2012. "Papuan Bird's Head Seascape: Emerging Threats and Challenges in the Global Center of Marine Biodiversity." *Marine Pollution Bulletin* 64, no. 11: 2279–95.

Marx, Robert. 1990. *The History of Underwater Exploration.* New York: Dover.

McNiven, Ian J. 2003. "Saltwater People: Spiritscapes, Maritime Rituals and the Archaeology of Australian Indigenous Seascapes." *World Archaeology* 35, no. 3: 329–49.

Mentz, Steven. 2020. *Ocean.* London: Bloomsbury.

Merchant, Stephanie. 2011. "Negotiating Underwater Space: The Sensorium, the Body and the Practice of Scuba-diving." *Tourist Studies* 11, no. 3: 215–34.

Narchi, Nemer, and Lisa Leimar Price, eds. 2015. *Ethnobiology of Corals and Coral Reefs*. Cham, Switzerland: Springer International.

Neimanis, Astrida. 2017. *Bodies of Water: Posthuman Feminist Phenomenology*. London: Bloomsbury.

Ota, Yoshitaka. 2006. "Fluid Bodies in the Sea: An Ethnography of Underwater Spear Gun Fishing in Palau, Micronesia." *Worldviews: Global Religions, Culture, and Ecology* 10, no. 2: 205–19.

Picken, Felicity, and Tristan Ferguson. 2014. "Diving with Donna Haraway and the Promise of a Blue Planet." *Environment and Planning D: Society and Space* 32, no. 2: 329–41.

Pugh, Jon, and David Chandler. 2021. *Anthropocene Islands: Entangled Worlds*. London: University of Westminster Press.

Raycraft, Justin. 2020. "Seeing from Below: Scuba Diving and the Regressive Cyborg." *Anthropology and Humanism* 45, no. 2: 301–21.

Robbins, Joel. 2001. "Secrecy and the Sense of an Ending: Narrative, Time, and Everyday Millenarianism in Papua New Guinea and in Christian Fundamentalism." *Comparative Studies in Society and History* 43, no. 3: 525–51.

Rutherford, Danilyn. 2003. *Raiding the Land of the Foreigners: The Limits of the Nation on a Indonesian Frontier*. Princeton, N.J.: Princeton University Press.

Saltford, John. 2003. *United Nations and the Indonesian Takeover of West Papua, 1962–1969: The Anatomy of a Betrayal*. London: Routledge.

Sloterdijk, Peter, and Hans-Jürgen Heinrichs. 2010. *Neither Sun nor Death*. Cambridge, Mass.: Semiotexte.

Smith, Richard. 2019. *The World Beneath: The Life and Times of Unknown Sea Creatures and Coral Reefs*. New York: Apollo.

Steinberg, Philip, and Kimberley Peters. 2015. "Wet Ontologies, Fluid Spaces: Giving Depth to Volume through Oceanic Thinking." *Environment and Planning D: Society and Space* 33, no. 2: 247–64.

ten Bos, René. 2009. "Towards an Amphibious Anthropology: Water and Peter Sloterdijk." *Environment and Planning D: Society and Space* 27, no. 1: 73–86.

Tsing, Anna, and Nils Bubandt. 2020. "Swimming with Crocodiles." *Orion Magazine*. https://orionmagazine.org/article/swimming-with-crocodiles/.

Veron, John, Lyndon M. DeVantier, Emre Turak, Alison Green, Stuart Kininmonth, M. Stafford-Smith, and N. Peterson. 2011. "The Coral Triangle." In *Coral Reefs: An Ecosystem in Transition*, edited by Zvy Dubinsky and Noga Stambler, 47–55. Dordrecht, Netherlands: Springer.

Wägele, Heike, Annette Kolb, Verena Vonnemann, and Monica Medina. 2008. "Heterobranchia I: The Opisthobranchia." In *Phylogeny and Evolution of the Mollusca*, edited by Winston Ponder and David R. R. Lindberg, 384–408. Berkeley: University of California Press.

Worsley, Peter. (1957) 1970. *The Trumpet Shall Sound: A Study of "Cargo" Cults in Melanesia*. London: Paladin.

Wright, Rob. 1959. "Spearfishing in Pacific Waters." *SPC Quarterly Bulletin*, June, 43–46.

CHAPTER 7

Stickiness in a Monsoon Air Methodology

HARSHAVARDHAN BHAT

The monsoon is so much more-than-rain and is much-more-than-a-season. The South Asian monsoon as an air permeates through every aspect of life and death, livingness and form, on the Indian subcontinent and beyond. Its forms and patterns shape methodologies, and our ways constantly confront its form. As a recurring form of predominantly aqueous matter, monsoon-like experiences can be thought of as planetary. Wind reversals, convections, precipitative processes, are felt around the world, but the subcontinent is the only place where intertropical convergence happens over land (Gadgil 2018; Francis and Gadgil 2013; Gadgil, Joseph, and Joshi 1984; Sikka and Gadgil 1980), conspiring a powerful, lively conversation of the air bringing energy, matter, ocean, landscapes, and activities together inside this magnificent holding of aerial aqueous matter. So, when we theorize the monsoon as just-rain or as a volumetric outcome as resource, we are participating in a crucial essentializing trope inherited from colonial research and fetishism that cultivated extensive methodologies on inferring difference between the lively matter of the air and volumetric water for plantation economies. These processes, having written the basis of modern monsoonal description, haunt our understandings of what the air is actually doing. I argue that monsoon descriptions are so much more-than-wet and express the dynamics of lifeworlds with which the monsoon is intertwined. One must always remember that the convergence of monsoon air has wide-ranging temporalities even inside the expansive hopeful gray that blankets the sky of South Asia. Inside this perceived blanket are many-many-many worlds, co-constituting and cotheorizing its flow.

Through the stories in this chapter, I suggest a methodology of *àntu,* which is a very rough transliteration of "stickiness" (a sensing of something that qualifies as sticky) from a southern Karnataka dialect of Kannada.[1] It is a sensibility of description that often begins with the body, contemplating the felt *àntu* of summer, where perspiration is felt across lively forms, anticipating the gathering of the monsoon, which expands that sensibility of life to matters across scales, atmospheres, and strata. In the passing of the monsoon, the sensibility calls for a holding, a stickiness of living through until the next monsoon, as postmonsoon breeze completes the theorizing of one version of matter, cycle, and time. It is an organizing of working against forms of knowledge that tell us to invest in monsoonal alienation and is instead a call to dive deeper into curiosities of this stickiness that holds some of our worlds together. The monsoon drenches claims into the world beyond its seasonal temporality, and as the monsoon breathes itself into the living, it gives agency to its claims, which also trouble attempts to structure it, such as interests that attempt to forecast it or organize it for extractive planning. The use of this term of *àntu* (henceforth as stickiness) implies many possible things. Stickiness refers to a kind of clinginess, a dependency, a fondness of sorts, an undetachable form of trouble and something that is also sensed between different beings, processes, and materials. It is an embodied connection with monsoonal atmosphere and is a relational descriptor that can be used also to deploy ambiguities and uncertainties opened up by the monsoon: its weather, climates, waters, relations, and so on. I find that different forms of monsoon research and sensing rely on this methodological quality to figure with the monsoon.

Using and celebrating stickiness is also a way of operating against the invention and fetishization of so-called tropical nature and atmosphere. By recognizing this methodological quality of the monsoon-air we are inside, I am interested in how different approaches get latched on to monsoonal sway.[2] The word *monsoon* itself, which may be an inheritance from the Arabic word *mausim,* implying season, seems to have cajoled its way as an oceanic wind to broader seasonal description.[3] Histories of the state and science have also positioned the monsoon as a geopolitical and economic weather construct (Bhat 2022; Dash 2020; Coen 2018; Gadgil and Gadgil 2006). In general, I do not have a problem with the idea of the monsoon, as it is after all an ocean that moves to the sky through the change in the direction of the wind. However, from my audit of several South Asian languages, of which there are hundreds, I have not found a general "monsoon" as a hegemonic weather formation anywhere. There are many-many words (and worlds) for rains, winds, airs, and relations. One of the things that this piece exposes is how different cosmologies, approaches, materials, and processes are swayed by monsoonal stickiness. The monsoon, too, can meet its many *kalas* (times). Linking two stories from Delhi and southern Karnataka and navigating through conversations with atmospheric science, meteorology, fluid dynamics, and monsoon history, I show how the monsoon stickily intertwines itself into these stories, exposing collaborations, realizations, and methodological breakdowns.

Like the monsoon winds and waters felt on your slippers or through your shoes, methodology is a confluence of inheritances, coproductions, and inventions. *Splash.* I've never used rubber boots before. While the geography of the rubber boot aims at resisting against water, monsoonal walkways for me have been watery rituals of absorption, wash, and angling, allowing for the air and the sun to take away moisture and bring it back to life again. Rubber boots, then again, offer safety amid water when needed. Rubber boots methods can be humble companions in monsoonal worlds, but they, too, can and will be swayed through monsoonal becoming. To study the monsoon, to follow atmospheric concerns, to collaborate on it, is in many ways also about the monsoon making presence and study possible. It makes worlds possible, performing what the editors in the Introduction to this volume call a "multispecies otherwise." I am interested in conversations with different kinds of descriptions and monsoon storytellers, listening to their methods and paying attention to how they somehow, intentionally or otherwise, meander in the complex offering of life-transforming-wind. Here genealogies, capacities, affinities, technologies, all mingle in stickiness—as something that is not just a metaphor but also literal description that co-collapses into puddles, storms, droughts, anxieties, hopes, and more.

This chapter thus takes up Bubandt, Andersen, and Cypher's invitation to think "figuratively," inspired by Donna Haraway's work, which allows me to connect specific moments in my research story with the broader conceptual, material, life-altering field called the monsoon. The work of figuring (Haraway 2008) here is also linked to this realization of stickiness, which is gradually pulled out of monsoonal forms for conceptual, socioeconomic,

FIGURE 7.1. Suspension of monsoon air over/and into the Yamuna, as photographed from the Noida side of the Okhla Bird Sanctuary in Delhi NCR on August 4, 2019. Photograph by the author.

and political interests but, at the same time, has already permeated through the many ways it does ambiguously, unclearly, often undiscernibly, as it holds things together as *àntu.*[4] It is in this spirit that the chapter is divided into three parts featuring research from a broader interdisciplinary doctoral research project that was part of Monsoon Assemblages, housed at the School of Architecture and Cities at the University of Westminster. It begins with the commissioned city of Delhi as the material starting point but finds itself distributed across several locations and temporalities as the monsoon picks up those stories.[5]

AN ANTHROPOCENE-OLOGY OF AEROSOLS

Delhi, the Monsoon Assemblages case city where my inquiry on monsoon–urban relations began, had many stories about the air to offer. Having lived in Delhi for a few years prior to my PhD, I had a relationship with this air and was familiar with its capacities, which had become part of global pollution folklore. Every winter, the conditions for a thick, soft, smoggy, misty air came together over this vast urban landscape. The megacity emitted, and the air held. The region emitted, and the air gathered. During the two years I spent in the region, I developed an acute sensitivity to what can only be ambiguously described as a respiratory allergy, allergic to nothing in particular but to everything, as the air oscillated materials, making it indiscernible to expanding its temporality within bodies. I have yet to meet a person who does not have some form of respiratory symptom living in the region. It constitutes situated breath. Some days are much worse than others.

> And sometimes your eyes burn.
> Skin flakes.
> Nose leaks.
> Respiration becomes known.
> Breath becomes a relentless project.
> Everybody and everything.[6]

The opportunity to work back in the city was a welcome one, as through 2017, when I started my PhD, I had come to grips in managing breath to some degree. Furthermore, the Schedule H third-generation hybrid antihistamine that required a prescription was something that my local pharmacist would just hand out, telling me each week how lucky he was to still have these packs as the city constantly runs out of supply. The work of breath and the atmospheric assault on living in Delhi have been well studied in recent years (Srigyan 2016; Negi 2020; Sharan 2020; Ghertner 2020; Narain 2017). Every winter, episodes of agricultural burning would spike pollution levels in the region. Typical postharvest rice stubble residue would be burned in neighboring states, including Punjab, Haryana, and Uttar Pradesh, among many others in the region. Much of more than 13.5 million hectares of the Indo-Gangetic plain (Mahajan and Gupta 2009), the region in and around Delhi, follow a wheat–rice bicropping system. The green revolution through the last decades of the twentieth century, which instrumentalized the use of chemical fertilizers, commercial seed, and machines, was forced and absorbed considerably well in the region. The combine harvester, which is a machine that integrates reaping, threshing, and winnowing, leaves stubble behind, and this is burned to accelerate the time from one agricultural cycle to the next (Mahajan and Gupta 2009). This burning, not limited to but persistent during October through November, and even December, through the years of my study (due to delayed harvests), nurtured a broader blanket of smog and density throughout the sub-Himalayan Indo-Gangetic central Indian region. As one farmer I spoke to in 2018 told me, "karna padta hai," they had to do it, indicating precarious socioeconomic and circumstantial predicaments that required so, despite alternative machines and processes that have been available in recent years.[7] Interestingly, much of the stubble that is burned is that of rice. The industrial varieties of rice grown here tend to be water intensive, and monsoon waters are channeled, rerouted, drilled from the ground, engineered from river and glacial systems. The engineering of water manipulates monsoonal form.

Atmospheric science experiments since the 1990s have offered glimpses into how anthropogenic atmospheres in the subcontinent speculatively transform monsoon metabolism and becoming. The 1990s International Indian Ocean Experiment, for instance, led by Veerabhadran Ramanathan, was one such key moment, where they came to the conclusion that what they identified as "atmospheric brown clouds," essentially anthropogenic clouds consisting of several different carbons, sulfates, nitrates, ash, and dust, that they distinguished separate from what they called natural clouds, created a dimming effect that impacted precipitation and, hence, rain (Sharma, Nunez, and Ramanathan 2016). They linked the haze with reduced solar radiation absorbed by the ocean. Following this discourse, aerosols have extended spatial linkages bringing together the Himalayas, the oceans, and the monsoon (Ganguly et al. 2012; Gautam et al. 2007; Ramanathan et al. 2007a; Ramanathan et al. 2001; Ramanathan et al. 2007b; Lüthi et al. 2015). By digging through the volumetric archives of meteorology and the speculative expanse of atmospheric science experiments, the haze of these anthropogenic clouds participates in disrupting what the monsoon becomes, both as concept and as a lived reality.

Aerosols as matters in/of the air seemed to have atmospheric temporalities, and I was curious about how the monsoon swayed in multiple temporalities, collecting, remaking, and being remade by them. Gauging the aerosol from New Delhi, for instance, was an attempt at seeing how aerosols manipulate monsoon stories. While listening to my conceptual monsoon–air hypothesis, K, an atmospheric scientist from the Indian Institute of Technology, pulled out a graph sheet and drew representational images of clouds, cyclones, tornadoes, and typhoons to tell me that scale exists.[8] The scribble connected representational conditions, but he insisted that the grid enables the discipline to produce forms of certainty, to the extent science allows. The sway of aerosols seemed to be also methodologically in tune to spatial and altitudinal attunements. K explained that the science requires the grid as a scaling methodology to assess phenomena.[9] This was a basic fact for his method and methodology. More importantly, winds at different scales seemed to pick up some matter and not others, in these imaginaries. Yet, the air constantly materialized transformation. Facts change in it, just as fast as they are created and sustained. From what I understand, it is air's own methodology in sustaining its materialization. The monsoon made it particularly visible. However, the questions I had received from friends, interlocutors, and people passing by in conversation was one of transformation and insane change: a monsoon that is unrecognizable, a monsoon that is dying, a monsoon that does not rain, a monsoon that sheds mountains, a monsoon that unexpectedly drenches, and so on. There was a crisis of monsoon perception. So, as I listened to K, the questions I had were of methodological transformation: How is the monsoon transforming what we know of it? And what forms of transformation such as that of the agricultural aerosol find their entanglements creeping into the production of a different kind of monsoon future?

Following the aerosol through the winter, and into the summer, where premonsoon dust storms breeze into Delhi, I noticed how all these material timelines shifted and made themselves noticeable through a variety of agencies, such as pollination, oviparity, cultivation, and urban forestry. I was reminded of Timothy Choy and Jerry Zee (2015), who insist that the air is always occupied and suspended in multiplicities (Choy 2011; Zee 2017). By the end of June 2018, the monsoon had already onset over the region a few weeks prior to what the Met Department had predicted. The city was also witnessing a mixture of timelines where the seasonality of dust infused with the seasonality of rain. It was July 10, 2018, and as I followed reports of these so-called premonsoon dust storms through May and June, and subsequent failed forecasts, I had an opportunity to pop by the India Meteorological

Department (IMD). The IMD, which was set up in 1875 following the meteorological legacy of the British East India Company and the Raj, is currently the apex meteorological agency in India, which infamously also declares what it calls the annual onset of the monsoon. The IMD also inherited a knowledge practice of colonial weather connections that spanned the history of Company observatories set up on the subcontinent and other observatories set up by Empire in other parts of the world, such as the Caribbean, East Africa, the South Pacific, and Southeast Asia. In recent years, the agency, under its role as part of the Ministry of Earth Sciences, has also broadened its portfolio, developing meteorological communications for agriculture, heat plans for cities, and other experiments in urban adaptation in a changing climate.

Many of the popular reports about the dust storm in the media drew from NASA dispatches, highlighting the role of satellite spectral visuality in "seeing" materialities unfurl. Friends who were around for dust storms then and past spoke of it as a dirty monsoon; something that rained water and mud; something that tore away the house plants in balconies; that torrented through the gaps under the door; something that disturbed visibility, almost like that perpetual winter smog that hits breath. The dust merged with the monsoon in different ways. I was curious how these aerosols and this dust registered with the IMD. I spent some time with Dr. V. K. Soni, who works on aerosols, dust, and air pollution, among other things. He generously took me through some of their recent studies on aerosol radiative forcing that took to the air from "ground," drawing on a range of "sensors" and data collection methods that fed into mathematical translational analytics and modeling. At the terrace of Mausam Bhavan (the name of the IMD building, which translates to "Weather House"), one was confronted by the compelling view of structured gardens and government buildings from their positionality in Lodhi Road. Among the several technical objects around, he pointed to an instrument he'd been working with recently for his dust and air pollution study. I saw it moving on its circular rotor, and he told me, "This is the sky radiometer used for the measure of aerosol optical properties. It looks at the sun and measures the direct radiation and takes measurements from the sky. So from the sky it measures the diffused radiation, or sky irradiance, and using both these measurements, we calculate aerosol optical properties like aerosol optical depth, single-scattering albedo, asymmetric parameters, size distribution of aerosols, and refractive indices and use these things to calculate aerosol radiative forcing." He raised his finger and pointed out to the sun on what was a fairly clear-skied day, despite the monsoon, and said, "It's right now pointing to the sun ... and moves like that," indicating its path, and the boundary of its data sampling path. It was a sample reference to the point he was trying to make about the range of instrumental and technological ecologies that are scattered on landscapes to derive a science of the air. These objects happened to be methodological interlocutors for data, for the more that can be modeled, the more they can test and validate. Multiple technical, mathematical, and chemical translations and inferences figure in the values and properties of, in developing units, theories, and notions of, matter and force in atmosphere.

Aerosol radiative forcing is the effect aerosols (particularly anthropogenic aerosols) have on climates; aerosols absorb and distribute radiation, and radiative forcing implies its process in different parts of the atmosphere. Aerosols constitute some of the biggest mysteries for scientists, because it is unclear to them what they do to the monsoon. Although aerosols can be isolated to infer their chemical properties, the way they operate in the atmosphere opens up ambiguity and speculation for their models. Some aerosols have heating effects, others cool, but in ever evolving complex mixtures, attunements, and monsoonal sway, they drift into their own stories. Although there is agreement that they do impact the monsoon, there is little understanding of the precise ways in which they do, because aerosol mixtures suspend certainties. Furthermore, akin to clouds, which simultaneously warm and cool climates, aerosols are speculative and sticky in that they tend to draw us into ambiguities about how they change the world, despite their constant, relentless force in doing so. They complicate the relationship between fields and knowledges, calling for a variety of convergences, experiments, and speculations.

Continuing our conversation, Soni took me downstairs by an airconditioned room with buzzing servers. On a desk sat two black, rectangular boxes: a Magee scientific aethalometer and an Ecotech nephelometer. Pointing at the Magee, he explained, "Basically, air passes through this filter paper, and then relative accumulation of aerosol on this filter paper is measured using this relative attenuation of radiation, and then absorption is calculated, and from that absorption, black carbon aerosol concentration is estimated. Because it is measuring the absorption in seven different wavelengths, we can also measure the biomass concentration of aerosol using some model, and that way we can estimate how much is coming from fossil fuel burning and how much is coming from biomass burning." Moving to the Ecotech, which measures a "scattering coefficient," he explained, "We are interested in monitoring the extinction by aerosols—extinction is scattering plus absorption by aerosols. So, the scattering is basically the redistribution of energy; when radiation falls on any particle, then it redistributes the energy—*that is the scattering.*" And then pointing to the Magee, he explained, "And in case of absorption, the radiation is absorbed by the particle and reemitted by another form," hinting at a difference between the two objects. Absorption takes place with aerosols like soot and other carbons, and scattering takes place with sulfates and dust aerosols that cool. The differences between aerosols matter, he insisted. They influence world making within models, and hence expanding technoscientific material webs, such as these data collecting objects, assists in the ways models can speculate the ways aerosols potentially meet winds. Modeling aerosols is intensely hard, as each speck is inherently unique. So aerosol stories tend to attenuate the gathering of collaboration, as they have already changed the nature of relations by the time the gathering has taken place. And thus my argument here is that aerosols are sticky matters. Perhaps they show us how the monsoon and winter were closer airs, much more intimate and wound up. Perhaps monsoonal agencies played into matter, indiscriminate of their timeline. Perhaps the specificity of aerosols dissolved in the anthropogenic soup of living atmosphere, remaking it.

In 2016, a collective of global atmospheric scientists proposed for the first time the need to integrate the category of natural aerosols as part of the monsoon "system" to eventually develop methods to get closer to how aerosols complicate air (Li et al. 2016). Unlike the speculations offered by thermodynamics and theoretical modeling of warming futures, aerosols complicate how, why, when, with-what, and in-relation with what atmospheric thermals behave—from the scale of the gene and the microbe to the scale of cities, atmospheres, and the planetary. Furthermore, in contrast to popular discourse that considers the monsoon a sway of rain that brings down pollution, aerosols are found to be in complex and intensely intimate relationships with the manipulation of flow, precipitation, and variability and journeying

(Ding et al. 2015; Ayantika et al. 2021). Matter in the air changes what the air does. The aerosols gather the methodological apparatus through their very dissolution.

> The empirics heat up.
> Aerosols here with aerosols there.
> Cough. Boom. Shadow.
> The cloud bursts, and drops their toolkits to the floor.

The intensities of those 2018 premonsoon dust storms were aligned with the heat wave and the increased warming in the northwestern expanse and Tibetan plateau that also drew the monsoon into being. As K. J. Ramesh at the IMD told me in 2018, "we will have to be ready to reorganize our time schedules for the day, from school timings to office hours and congestion on the road and so on." If the Himalayas pulled the ocean into the sky, these stories of heat prior and disturbed patterns through and after spoke of multiple different monsoonal airs in the making. Delays, erraticism, strangeness, intensity, and silences have been some of the few registers of description attempting to speak to the changing qualities of the monsoon. As atmospheric science description asks us to think in planetary terms, the monsoon speaks back of the many multiples that it occupies in the same place. As Anna Tsing reminds us in the roundtable (chapter 14), Anthropocenes have place, and the story of the monsoon in this instance is also a material entanglement that needs to be read in its closeness, its inside-ness, where its very motion is brought into ambiguity and questioning by every single speck of space it meets.[10] The next section takes this further.

READING COLLABORATIVE CONVECTION

Andrew Turner, a leading monsoon scientist from the University of Reading, has been an important advisor and interlocutor to the Monsoon Assemblages project.[11] In 2016, Dr. Turner and Dr. G. S. Bhat (from the Indian Institute of Science, Bangalore), with a series of collaborators, set out on a project called INCOMPASS, which, among other things, was a campaign to develop better linkages between atmospheric fluxes just above ground and at higher altitudes, taking diverse measurements from an airborne flight facility called the Facility for Airborne Atmospheric Measurements (FAAM) (Turner and Bhat 2020).[12] The reason why this project's outputs spoke to my work was because Turner suggests that the Indo-Gangetic plain and its subregion (which surround the National Capital Region), as some of the most irrigated spaces on the planet, potentially have an impact on the regional becoming of the monsoon. It was an important moment for atmospheric science, as it connected the matters of ground with the matters of the wind as entangled forms in a methodological practice that has for the longest time relied on volumetric speculations and statistical repetition, implying a narrative of a phenomenological clock—that seasons, expected and emerging, offer themselves as processes of mere phenomena. As an inherited nod from the Continental tradition, which has been influential in contemporary weather research and writing, it's been used in the naturalization of statistical and technopolitical assumptions of the monsoon as a form of volume that can be governed in a format of time that is expected to be. The monsoon, however, has consistently taught us otherwise—that it, too, has a million other collaborations and entanglements.

Turner's thesis spoke to me of the possibility of stickiness between methodological worlds, that the monsoon troughed up north because of a careful choreography with the attached quality of monsoonal landscapes. They observe that, "despite its importance, currently there is no accepted physical explanation for the advance of the monsoon onset across India" (Parker et al. 2016, 2256). One of the key proposals was that premonsoon forms of oceanic organization encourage the moistening of mid- and low-level airs, which prepare soils for encouraging monsoonal possibilities. Parker et al. (2016) suggest that monsoonal onsets are driven to certain extents by the moisture premised just before (and even after) their onsets, hinting at a synoptic relationship between temporalities—a stickiness, if you will, that sociologically displaces the thesis of an externalized weather. Consider the bonds and "figuring" of monsoon air with the convective capacities of landscapes: moist, dry, grassy, agricultural, urbanized, industrialized; of forests, microbes, and multiple others. Stickiness permeates in enabling what we even conceptualize as the monsoon, which deepens stickiness as moisture and rain.

At an Indian monsoon workshop at the University of Leeds in February 2018, Turner generously took me through one of their posters showing the path of one of their 2016 FAAM flights from Lucknow to the edge of Rajasthan (on a higher altitude) and back to Lucknow, hovering at a lower altitude of approximately 191 meters above ground. It was a unique exercise for them that allowed for observations of lower airs and the ground. In recently published work, they disclose that soil moisture impacts deep convection. As they write, "the impact of irrigation on temperature and wind patterns implies that historical changes in irrigation are likely to have influenced mesoscale processes within the Indian summer monsoon and, perhaps more importantly, future changes are likely to do the same" (Barton et al. 2020, 2901). The intuition that also seemed to be guiding them, via what I understood from Turner, is that waters transformed on ground, below ground, and brought out of the ground develop conversations with waters above ground. Despite the monsoon being the air that brings waters to the earth, the manipulation of systems in bringing waters out into the air transforms the monsoon by converging multiple times of water. Water had time, and anthropogenic forces manipulating/encouraging their convergence in air influenced what the monsoon became. Among other conclusions, they note, "Patterns of shallow cloud correspond well to regions of low soil moisture, indicating an important role of land surface state in the development of shallow convection. Deep convection developed either in association with topography or on the dry side of soil moisture gradients" (Barton et al. 2020, 2902). This meant that dryness played a role in monsoon convection and that ecologies of aridity were just as important as the imaginary and entrenchment of wetness. Stickiness, then, in this instance can be viewed as an expanding zone of visibility where variables attach formations of air not perceived to be attached before. *Ántu.*

It can also be thought of as a formation that holds things together, like the way Dr. Rama Govindarajan (at the International Centre for Theoretical Sciences [ICTS], Bangalore) speaks from the field of fluid mechanics and thinks of the monsoon as an "underlying current," an air that is a liquid (R. Govindarajan, pers. comm., January 9, 2018). Referring to the scientific mystery of monsoonal variability within this current, she argues that "we have the fundamental belief that all this is being driven by an underlying dynamics, and the randomness we see is due to other factors sitting on top of this underlying dynamics. So, if you're going to try and capture the underlying dynamics, we have to look at the rainfall as a probabilistic manifestation of this underlying dynamics."[13] She represents this through what she defines as a "sticky family," displaying an image of the subcontinent with a patchwork of colors, some of rain and others in absence. She was the first person from the natural sciences who reminded me that the monsoon is a multidisciplinary and transdisciplinary investigation. Speaking of the monsoon, she said, "It's possibly one of the biggest problems there is to solve," and that everybody was needed and had something to offer.

The monsoon cannot be contained. And it escapes. It dismantles discipline. Constantly. It becomes the world of its own stories, and how one imagines the air therefore matters to how one encounters its stickiness, despite already being entangled in it. I learned two other things from Govindarajan at ICTS that day: first, that the foundational equations for liquid dynamics (such as the Navier–Stokes equation) that are used to model and forecast weathers don't operate well just yet with monsoonal dimensions and, despite vast parallel supercomputing capacities, pose huge mathematical challenges that might never be resolved. The other thing I learned that day was that clouds form the largest uncertainties for climate scientists and that it is not entirely understood if clouds amplify global warming or soften it. As Govindarajan and Ravichandran (2017, 271) write, "the reigning dogma is that low clouds reflect more and therefore cool the Earth's surface, whereas high clouds act like blankets to keep the heat in. Monsoon clouds are thin and tall, and do both." It is in fact one of the reasons (among many others) why "we" may not fully understand how climate change impacts the monsoon. As speculations in atmospheric science compete to claim results of the possible death or deepening erratic intensity of the monsoon, what is clear is that the Anthropocene in/of the subcontinent and its many places plays with the monsoon: its past, present, and futures. We need better constructs and ways of figuring inside monsoonal worlds.

Furthermore, the monsoon for many has also been a way to construct a relational colonial weather map: a geography within a tropical idea. One can tell through monsoon history that the monsoon is also a conceptual invention of sorts (Cullen and Geros 2020; Davis 2000; Sivasundaram 2020; Amrith 2018; Sikka 2010). It collaborates with the production of states, society, and culture, and colonial processes theorize to extract from monsoonal life— human and more-than. As a state construct, colonial observatories expanded the scale of the monsoon to the Himalayas and provided early teleconnection speculations, such as El Niño, with observatories elsewhere. This was a leap from climatological boundaries drawn by prior kingdoms and principalities. The observatories that the British East India Company set up through the late 1700s and the imperial enterprise that went on to develop them on a larger scale drew on the monsoon to enhance extraction and profitability. Following history writing, for instance, one can see the politics of developing monsoonal data as a process of accumulation, amalgamation, and appropriation of monsoonal knowledge systems to the view of the colonial apparatus and its simplification. Geology, climatology, and meteorology evolve under monsoonal surveillance because they try to figure it as it constantly undermines them with alternative stories of reality, as they fail to forecast the monsoon to extract life from it. The obsession for forecast precision might have developed scientific technologies and practices, but they also relied on planetary colonization to figure the clinical between worldmaking teleconnections like El Niño and the Southern Oscillation, which were theories developed from a small web of stations in the South Pacific, northern Australia, and Peru and Ecuador. As contemporary wisdom in atmospheric science often indicates, though it seems easier to understand the macro (the planetary), it is extremely hard to understand the granular, where issues like monsoonal "variability" are enacted. Meteorologists often say that not all drought years are El Niño years, but all El Niño years are drought years—although one can also argue that droughts are coproduced, cultivated, and deepened by enterprise and extraction, which also rely on meteorological description to validate precarity (Davis 2000). I invite the possibility of the monsoon enabling us to ask different questions of connections that are dictated by the methodology not of correlation and extractive mapping but of survival and the work for joy in changing airs. Deep investments in monsoonal description have historically emerged from modeling planetary nature–human relations to objectify, fetishize, and extract (Sivasundaram 2020; Chakrabarti 2020; Davis 2000) a theory of air, out of life webs. Probing monsoon studies and searching beyond it, I realized that better monsoonal descriptions can always be found when there is attention being paid to the monsoon as something-more, as something that constitutes the living in viscerally real and sticky ways.

Therefore, along with the critical and collaborative aspects of research and description as argued by Bubandt, Andersen, and Cypher in the Introduction to this volume, as well as by the decolonial and mixed approaches to which Tsing and Hastrup refer in the roundtable in the last chapter of this volume, I want to allude to the possibility of detunings and the monsoonal collapse of assemblages that drench us into puddles. *Detuning* is a word that I've borrowed from Natasha Myers and Shiv Visvanathan, but it's also a term that I have heard in some version growing up in South India: to un-tune, when instances for an embodied curiosity were called in or when I was asked to ignore noise or interference. It's strange the way this un-tuning or detuning can be called into operation because it's widely applicable in a range of situations: like crossing a large but shallow monsoon stream with a current, calling to detune one's fear and trust elders to cross the stream with them; tuning against a broadcast for violence; detuning, as Natasha Myers suggests, one's own perception of plant ecologies; intertwining one's tuning of monsoonal worlds through cosmological diversity and material multiplicities; un-tuning against the American geographers at a workshop who insisted that the monsoon is "just a shitload of rain"; detuning the weight of sweat of premonsoon heat as an indicator of a monsoon to come; tuning against binaries of disasters and flourishing but asking how all of them cohabit the same air; and most importantly, detuning investments against *àntu* and exploring ways through which this attachment keeps many of the things "we" know alive. Let us tune instead for ancestors that repeat themselves as something much more than "just a shitload of rain."[14] Like the "shimmer" of which Bird Rose (2017, 0G54) speaks in *Arts of Living on a Damaged Planet*—the darkest skies through the year entrench us with shimmer, even when they're silent, even when they're broken, even when they wash over and stay. A monsoon air methodology is an investment in keeping that story.

An investment in holding on to air.
A methodological oscillation of investments.
Currents that are cultivated.
Stories that are told.

A BROKEN CIRCULATION

I pivot to a story here to end on a slightly personal note. It situates a parallel yet connected inquiry that has permeated the process of my research

FIGURE 7.2. Monsoon ground. Photograph by the author.

FIGURE 7.3. Monsoon ground. Photograph by the author.

project since 2017. I perceive this note to be somewhat of an opening to the *stickiness* and *why* of monsoonal methodologies, or how the construct of the methodology is in fact constantly being rewritten by the monsoon itself. I argue that this *àntu* is a methodology that is accessible to people and living forms trying to live within a changing monsoon. Collaborations, slippages, and experiments are already being made. Our monsoonal contexts have some useful stories.

So, this requires me to take to you a moment of positionality, a place of opening to these different conversational curiosities. A site that has been important for me in coming to terms with the meaning of the monsoon has been my grandmother's place in southern Karnataka, a village at the foot of the western *ghats* (rainforest hills) toward the coastal town of Mangalore. At the outset, it is a mosaic space with dense *tota* (cultivated agricultural forest), *gadde* (fields), *todu* (streams), *gudde* (hills), and some *kadu* (forest).[15] From coconut trees, areca palm, mangoes, jackfruit, cashews, rice, pumpkins, gourds, pepper, coffee, tubers, roots, stems, leaves, flowers, flourishing grass, ferns, medicinal plants, wild berries and fruits, and even some violent rubber to the hundreds of other forms of monsoonal collaborations that inhabit the geoformations of the *tota,* the changing monsoon impacts all in different ways. My grandmother was probably eighty-five when she passed away in 2017. It was an unspeakably incredible loss. Among the vastness of the loss was a very late realization in me that a lot of what I knew about the monsoon as nature–culture was thanks to her. This led me to an exercise of writing a little archive and validating some of the stories in it through some of my family members who were also very close to her. Ajji (Grandmother) was the one who made me understand that the monsoon was not a singular season but in fact a vast temporality of multiple different winds and rains.[16] Although she never used the word monsoon, as it doesn't exist in Old Kannada, as with most other South Asian languages, the attributional term here is *mallé kala* (time of the rains), which conceptually holds both time and matter. There are multiple forms of *kala* and *rain* within *mallékala,* but *mallékala* is the general term for what I can best translate as the monsoon. She insisted that it carried the influences of many different places (and worlds) and constantly remade her world, carrying matters further. She insisted that monsoons (or just *mallékala,* "time of the rains") were in conversation with other *mallékalas* before and yet to come. For every monsoon that did not deliver the rain expected, and as each got worse, one could register a kind of intense breakdown of sadness she would exhibit about the change in the wind.

We would in occasional years when I was younger search for a particular flower that comes out just after monsoon rains. In our dialect of Kannada, this flower is called the *ammange badade kai,* which is a translation of what they call in Tulu *appege nothina kai,* which means "the hand that beats the mother."[17] It is a flower in deep yellow and red with a green-brown stem with leaves close to a ginger plant's, but slightly slender. It resembles a broken hand of sorts from its stalk and is really an intricate find when you spot one in the land. I never went searching for these myself, as they usually tended to pop up in corners of dense, wet vegetation by a stream or in places that promised the richness of crawling life. The common story about the flower was that it blossomed as a broken hand as punishment of having beaten its mother, presuming a previous life as a human or more-than-human. Ajji's version was slightly different. She suggested to me that the flower emerged with its broken hand because the world needed ways to circulate the violence being released into it. It was a gesture of peace, of some kind of reconciliation that can possibly take place with the earth. It was also something that literally emerged from the ground, was not cultivated, and was hard to find in their land. The earth had its own way of circulating these flowers back into visibility to hold the pain, which would eventually scatter or transform. They were very postmonsoon flowers in the sense that there was a very brief time when they would blossom, and they were never found in clusters, so you had to really look for them or simply encounter one, which was usually what happened. In recent years, they've disappeared, and one of the last stories I heard from her of the flower was a story of disappearance: that the capacity to hold the violence, the capacity to be reborn as the broken hand, was being broken.

My mother recently told me that the last time she spotted the flower was a few years ago by a native jackfruit tree about which I had recently written (Bhat 2020). I hear that the jackfruit tree, too, has not been producing its bloom, the flower that breaks from its stem bark during the *phala gaali* (flowering winds in Tulu) that usually come right after the monsoon, just into September and deeper into October, but had not arrived that year, in 2020.

The bark remains unbroken. The wind is missing. Here is a lesson about stickiness: its materiality crosscuts time and cross-pollinates. It holds violence and possibilities, and as *ajji* often used to suggest, one doesn't know what the wind carries, but all we can do is gather for it and do everything to let it carry from us better things. This is meant both literally and conceptually. At my grandmother's home, *ajja,* my grandfather, lately has been paying less attention to the rain stars and constellations of the *panchanga* (the local cosmological calendar) and instead follows the weather forecasts on TV and science reporting that he reads in newspapers. In a conversation with Sharada *ajji* in 2019, who is a family member who consults on matters of the monsoon via the local *panchanga* calendar, she acknowledged that her reading of the pattern of the *panchanga* has been changing with the weather: that she, too, had to consider the purpose of the consult to make judgments of whether her reading of the *panchanga* would deliver the right advice for the particular plant/crop being considered. She finds herself negotiating the fixities of the calendar system with the changing materialities of place and weather. When last I met her in 2019, she carefully took me through some of her historical records of wet seasons: moments she got wrong and moments she got right. She said, "I'm still doing better than they are," referring to state predictions. In a conversation that transpired a few hours, she said, "Didn't they after all learn that from us?" suggesting that the cosmologies of power that held weather knowledges were perhaps collected and reassembled from one system of power to another, to develop a colonial science.[18] While I have not yet been able to completely explore the implications of that claim, I sense that Sharada *ajji* is also asking me if cosmologies don't play and draw from each other. I think they do. Monsoon histories often speak of the role of upper-caste assistants and data collectors in the assembling of colonial science (Sivasundaram 2020; Amrith 2018). Sivasundaram's (2020) essay on modeling empire, for instance, asks us to consider the fact that disciplinary practices and collaboration cross-pollinated intensively through the late 1700s and beyond, citing the Madras Observatory.

My sense is that the monsoon confronted discipline then and continues to do so now. It breaks with mountains, and it breaks with cities. It dries up expectations and surprises speculation. I wonder if and how the winds that futures become translate these cosmologies and matters to circulate blooms, in some way if not the other. I wonder if there is a possibility of better conversations inside and with, despite the brokenness of changing monsoon winds. Breaking away from the cartographic and extractive impulse to appropriate monsoonal diversities and lives in the figuration of the system of local and global racial-and-its-many-advanced-industrial-capitalisms, this chapter has suggested that the antidisciplinary stickiness of the monsoon(s) give us repetitive chances not just to live but also to make space and time for reconciliation, reparation, working for the many beings that need monsoonal waters, and making futures possible. Stickiness becomes a way through which monsoonal methodologies mold, crack, and enliven disciplinary containment as they dwell in the fantasy of sense making—all this while, many sky blankets have unfolded, rained, showered, quieted, and gone making matter(s) matter.[19] It is the air within which worlds make sense that the time of *mallé* offers us recurring opportunities to make livability despite the work against it.

NOTES

1. Pronounce the *t* in *àntu* the way you say *t* in *top*.

2. "We" opens up ethical trouble, but in the interest of this section, "we" is anyone who would want to read the monsoon as something that they are inside and are some of the storytellers this piece is in conversation with.

3. There is an alternative theory of *monsoon* as an inheritance from Malay, but the Arabic inheritance is more widely accepted.

4. Thinking through figurations, I've also found Haraway's development of the term *attachment sites* at the 2018 Rubber Boots Methods in the Anthropocene workshop in Aarhus to be very helpful.

5. The Monsoon Assemblages project is led by Professor Lindsay Bremner. Visit http://www.monass.org/ to find out more.

6. I use poetics in continuity and without distinction as monsoon stories have often been gaslighted into categories of culture and/or science. Poetics helps me think through things.

7. Transliteration from Hindi.

8. K has been anonymized.

9. Although the idea of attunements has been developed over the years by scholars like Donna Haraway and Nick Shapiro, I use it rather simply here, in the sense that aerosols probably have preferences to altitudes based on conditions.

10. Thanks to Rachel Cypher for reminding me of this particular argument that Anna Tsing makes.

11. See Turner (2017) in the conference proceedings of "Monsoon [+other] Airs" organized by the Monsoon Assemblages project. Talks are archived on the Monsoon Assemblages YouTube channel.

12. INCOMPASS stands for Interaction of Convective Organisation with Monsoon Precipitation, Atmosphere, Surface and Sea and was the name of Turner and Bhat's team's field campaign. FAAM is based at Cranfield University.

13. Talk on "Raindrops, Buoyancy, the Indian Monsoon" at the International Centre for Theoretical Sciences in Bangalore (published on the ICTS YouTube page).

14. The reference to the ancestor is an acknowledgment to how my grandmother referred to it, as implying an air of origins. Just that. It is not a reference to deities or gods here, although nature cultures on the subcontinent will surely have multiple ways of perceiving ancestors and monsoons.

15. These are rough transliterations from my dialect of Kannada, just as I'd use them in a text message.

16. Her name was Honnamma, but she claimed never to like it, and hence I don't use it.

17. I believe it's a local type of wild gloriosa. However, I don't use the formal nomenclature here, to let the story be. I also want to recognize that gloriosa is said to have medical uses, and variants of it are commercially grown. Although it seems to be largely endangered in the region, it is still the state plant of Tamil Nadu, where it is called *karthigaipoo*. The flower has apparently also been the state flower of Rhodesia, although I'm not sure if there are deeper colonial and caste interconnections and implications to this story in time.

18. The quote is a rough transliteration from what was said in a local dialect of Kannada.

19. Thanks to Karen Barad's talk on "After the End of the World" at the 2018 Melbourne Anthropocene Campus, which influenced my own dealing with "matter(s)."

References

Amrith, Sunil. 2018. *Unruly Waters: How Mountain Rivers and Monsoons Have Shaped South Asia's History.* London: Allen Lane.

Ayantika, Dey Choudhury, R. Krishnan, Manmeet Singh, Swapna Panickal, Sandeep Narayanasetti, Prajeesh A. Gopinathan, and Ramesh K. Vellore. 2021. "Understanding the Combined Effects of Global Warming and Anthropogenic Aerosol Forcing on the South Asian Monsoon." *Climate Dynamics* 56: 1643–62.

Barton, Emma J., Christopher M. Taylor, Douglas J. Parker, Andrew G. Turner, Danijel Belušić, Steven J. Böing, Jennifer K. Brooke et al. 2020. "A Case-Study of Land-Atmosphere Coupling during Monsoon Onset in Northern India." *Quarterly Journal of the Royal Meteorological Society* 146, no. 731: 2891–905.

Bhat, Harshavardhan. 2020. "Stickiness of the Halasina Hannu." In *Monsoon [+other] Grounds,* edited by Lindsay Bremner and John Cook, 50–53. London: Monsoon Assemblages Project.

Bhat, Harshavardhan. 2022. "The Air of the Monsoon: In Myth, Pause and Story." In *Monsoon as Method,* edited by Lindsay Bremner, 236–41. Barcelona, Spain: Actar.

Chakrabarti, Pratik. 2020. *Inscriptions of Nature: Geology and the Naturalization of Antiquity.* Baltimore: Johns Hopkins University Press.

Choy, Timothy, and Jerry Zee. 2015. "Condition-Suspension." *Cultural Anthropology* 30, no. 2: 210–23.

Choy, Timothy. 2011. *Ecologies of Comparison: An Ethnography of Endangerment in Hong Kong.* Illustrated ed. Durham, N.C.: Duke University Press.

Coen, Deborah R. 2018. *Climate in Motion: Science, Empire and the Problem of Scale.* Chicago: University of Chicago Press.

Cullen, Beth, and Christina Leigh Geros. 2020. "Constructing the Monsoon: Colonial Meteorological Cartography, 1844–1944." *History of Meteorology* 9 (November).

Dash, Biswanath. 2020. "Science, State and Meteorology in India." *Dialogue.* http://dialogue.ias.ac.in/article/21688/science-state-and-meteorology-in-india.

Davis, Mike. 2000. *Late Victorian Holocausts: El Nino Famines and the Making of the Third World.* Reprint ed. London: Verso.

Ding, Yihui, Yanju Liu, Yafang Song, and Jin Zhang. 2015. "From MONEX to the Global Monsoon: A Review of Monsoon System Research." *Advances in Atmospheric Sciences* 32, no. 1: 10–31.

Francis, A. Pavanathara, and Sulochana Gadgil. 2013. "A Note on New Indices for the Equatorial Indian Ocean Oscillation." *Journal of Earth System Science* 122, no. 4: 1005–11.

Gadgil, Sulochana. 2018. "The Monsoon System: Land–Sea Breeze or the ITCZ?" *Journal of Earth System Science* 127, no. 1: 1.

Gadgil, Sulochana, and Siddhartha Gadgil. 2006. "The Indian Monsoon, GDP and Agriculture." *Economic and Political Weekly* 41, no. 47: 4887–95.

Gadgil, Sulochana, Porathur Vareed Joseph, and Niranjan V. Joshi. 1984. "Ocean–Atmosphere Coupling over Monsoon Regions." *Nature* 312, no. 5990: 141–43.

Ganguly, Dilip, Philip J. Rasch, Hailong Wang, and Jin-ho Yoon. 2012. "Fast and Slow Responses of the South Asian Monsoon System to Anthropogenic Aerosols." *Geophysical Research Letters* 39, no. 18: GL053043.

Gautam, Ritesh, N. Christina Hsu, Menas Kafatos, and Si-Chee Tsay. 2007. "Influences of Winter Haze on Fog/Low Cloud over the Indo-Gangetic Plains." *Journal of Geophysical Research: Atmospheres* 112, no. D5: JD007036.

Ghertner, D. Asher. 2020. "Airpocalypse: Distributions of Life amidst Delhi's Polluted Airs." *Public Culture* 32, no. 1: 133–62.

Govindarajan, Rama, and S. Ravichandran. 2017. "Cloud Microatlas." *Resonance* 22, no. 3: 269–77.

Haraway, Donna. 2008. *When Species Meet.* Minneapolis: University of Minnesota Press.

Li, Zhanqing, William Ka Ming Lau, Veerabhadran Ramanathan, Guoxiong Wu, Yihui Ding, M. G. Manoj, J. Liu et al. 2016. "Aerosol and Monsoon Climate Interactions over Asia." *Reviews of Geophysics* 54, no. 4: 866–929.

Lüthi, Z. L., Bojan Škerlak, Sang-Woo Kim, Axel Lauer, Andrea Mues, Maheswar Rupakheti, and Shichang Kang. 2015. "Atmospheric Brown Clouds Reach the Tibetan Plateau by Crossing the Himalayas." *Atmospheric Chemistry and Physics* 15, no. 11: 6007–21.

Mahajan, Anil, and R. D. Gupta, eds. 2009. "The Rice-Wheat Cropping System." In *Integrated Nutrient Management (INM) in a Sustainable Rice-Wheat Cropping System,* 109–17. Dordrecht, Netherlands: Springer.

Narain, Sunita. 2017. *Conflicts of Interest: My Journey through India's Green Movement.* Gurgaon, India: Penguin Books.

Negi, Rohit. 2020. "Urban Air." *Comparative Studies of South Asia, Africa, and the Middle East* 40, no. 1: 17–23.

Parker, Douglas J., Peter Willetts, Cathryn Birch, Andrew G. Turner, John H. Marsham, Christopher M. Taylor, Seshagirirao Kolusu, and Gill M. Martin. 2016. "The Interaction of Moist Convection and Mid-Level Dry Air in the Advance of the Onset of the Indian Monsoon." *Quarterly Journal of the Royal Meteorological Society* 142, no. 699: 2256–72.

Ramanathan, Veerabhadran, Paul J. Crutzen, Jeffrey T. Kiehl, and Daniel Rosenfeld. 2001. "Aerosols, Climate, and the Hydrological Cycle." *Science* 294, no. 5549: 2119–24.

Ramanathan, V., F. Li, M. V. Ramana, P. S. Praveen, D. Kim, C. E. Corrigan, H. Nguyen et al. 2007a. "Atmospheric Brown Clouds: Hemispherical and Regional Variations in Long-Range Transport, Absorption, and Radiative Forcing." *Journal of Geophysical Research: Atmospheres* 112, no. D22: JD008124.

Ramanathan, Veerabhadran, Muvva V. Ramana, Gregory Roberts, Dohyeong Kim, Craig Corrigan, Chul Chung, and David Winker. 2007b. "Warming Trends in Asia Amplified by Brown Cloud Solar Absorption." *Nature* 448, no. 7153: 575–78.

Rose, Deborah Bird. 2017. "Shimmer: When All You Love Is Being Trashed." In *Arts of Living on a Damaged Planet: Ghosts and Monsters of the Anthropocene,* edited by Anna Lowenhaupt Tsing, Heather Anne Swanson, Elaine Gan, and Nils Bubandt, 51–63. Minneapolis: University of Minnesota Press.

Sharan, Awadhendra. 2020. *Dust and Smoke: Air Pollution and Colonial Urbanism, India, c.1860 c.1940.* New Delhi: Orient Blackswan.

Sharma, Sumit, Liliana Nunez, and Veerabhadran Ramanathan. 2016. "Atmospheric Brown Clouds." In *Oxford Research Encyclopedia of Environmental Science.* Oxford: Oxford University Press. https://doi.org/10.1093/acrefore/9780199389414.013.47

Sikka, Dev Raj. 2010. "The Role of the India Meteorological Department 1875–1947." In *Science and Modern India: An Institutional History, c. 1784–1947,* vol. XV, edited by U. Das Gupta, 381–426. History of Science, Philosophy, and Culture in Indian Civilization 4. Delhi: Delhi Centre for Studies in Civilisations/Pearson Longman.

Sikka, Dev Raj, and Sulochana Gadgil. 1980. "On the Maximum Cloud Zone and the ITCZ over Indian Longitudes during the Southwest Monsoon." *Monthly Weather Review* 108 (November).

Sivasundaram, Sujit. 2020. *Waves across the South: A New History of Revolution and Empire.* London: William-Collins.

Srigyan, Prerna. 2016. "Delhi's Air: Histories, Technologies, Futures." MS thesis, Ambedkar University, Delhi.

Turner, Andrew. 2017. "The Indian Monsoon in a Changing Climate." In *Monsoon [+other] Airs,* edited by L. Bremner and G. Trower, 17–20. London: Monsoon Assemblages. http://monass.org/writing/.

Turner, Andrew G., and G. S. Bhat. 2020. "Preface to the INCOMPASS Special Collection." *Quarterly Journal of the Royal Meteorological Society* 146, no. 731: 2826–27.

Zee, Jerry C. 2017. "Holding Patterns: Sand and Political Time at China's Desert Shores." *Cultural Anthropology* 32, no. 2: 215–41.

CHAPTER 8

Cattle Tracks in the Dust

Riding the Margins of the Anthropocene
in the Pampas of Argentina

RACHEL CYPHER

When Ceferino needed help with the roundup, I dressed the horse he called Facundo with a sheepskin atop a rawhide saddle and pulled the girth soft around the stallion's belly, waiting for him to blow. When he gave up his air, I tugged the girth tighter and swung up into the seat, grasping the reins and pushing into Facundo's sides with my heels to steer him through the *quebracho* gate and onto the dirt road that made a wide circle around the soybean fields.

The horses clopped past the combine machine harvesting the leafless soybean stalks. From the back, it kicked up a powdered dust that clung to the edges of the eucalyptus trees bordering the sheep corral. Facundo shook his head and shimmied to the side into the soybeans, which rattled as his hind legs brushed against their spiny stalks. He was impatient and going a bit feral because he was not ridden enough. As I tried to gain control of the stallion, he jerked on the reins and bent his head to eat some weeds at the same time that Ceferino looked back, on his face the disappointment that I was not a better pupil. "Don't let him eat!" he called out.

Ceferino dug in his heels to pick up speed, and I followed him, past a row of pines under whose shade fenced llamas munched grasses and thistles, past the corral in which Martin and the other handler were weighing cattle, through two more swinging gates, around the alfalfa field strung with an electric fence until we got to the edge of what had been the forage sorghum field but was now a wide, shallow lake. Ceferino skirted the edge. The resident coots paddled, and a black-necked swan on an elevated bed of eggs nestled

FIGURE 8.1. Riding through the rye grass. Photograph by the author.

among the reeds, watching us from the side of her eye. At the edge of the lagoon was a corn stubble field where the cattle were grazing. It was November, nearly summer in the Pampas of Argentina, and the cattle were eating the last of the residue before genetically modified soybeans were planted.

Motioning for me to stay saddled near the open gate, Ceferino kicked his horse into a canter, then a gallop, whipping around the cows grazing at the edge of the field. The cattle bolted into action, startled by the thundering horse and by the yelps Ceferino hollered.

"Oooouuuuuuuuuupppp, ooouuuuuuuuuuuuuuuuuppp," he belted out as he tacked back around, causing the cattle to run away from him and toward each other, toward me. Still riding Facundo, I bent down to hold the gate open, struggling to keep Facundo in check as the cattle rushed through. But as the last cow bucked past, Facundo sprang into action before I even knew what he was doing. Suddenly I was swept up in the momentum of the herd. I lost my hat; the sun and dust gritted into my eyes and throat; I clutched the reins and pushed into the stirrups and tried not to fall off. Facundo's strong, sinewy body moved below me. I could feel the asymmetrical four beats and then the moment when all of his feet were in the air—it was a fleeting weightlessness.

We thundered along until Ceferino at the head slowed the herd to a trot, a walk, and then a slow plodding. I rearranged myself, caught my breath, tried to find my seat in the saddle. We circled around to the dusty dirt road, fenced in on either side to keep the cattle going toward the feedlot. I was in the rear, and so it was my job to make sure no stragglers remained behind. One cow stopped to pull up some weeds that were growing on the side of the road, and I directed Facundo so that he came up behind the cow, encouraging it to keep going with the herd. Up ahead, I saw Ceferino look back and wave, then pull out a pack of cigarettes. He tilted his head and cupped his hand against the wind to light one, then straightened up, blowing smoke into the Pampas air. I looked back behind us to see if my hat was there, but all I saw were cattle tracks in the dust.

Following the Introduction to this volume, where we offered to the reader an "invitation to get out and follow the more-than-human tracks," this chapter follows the more-than-human tracks of what I am calling the "cattle project." I think of cattle not as a species by themselves but as a political package or project that travels (Tsing 2005). This includes the hoof tracks of cattle and horses as well as the affect that is stirred up in their wake. Like Andrew Mathews's method of walking (chapter 1) and Nils Bubandt's method of tidalectic snorkeling (chapter 6), this chapter is a motion-based study, suggesting that a method might be riding (for motion-based studies, see also Hamilton and Taylor 2017, 115). This evinced itself because in the Pampas, horses are highly individualized, symbiotic partners, while cattle are not individuals but a project. Multispecies fieldwork always entails making distinctions like these, on one hand because we cannot attend to all species equally all the time, and on the other hand because this distinction follows the divided interests of my human interlocutors, for whom horses are partners—companions in a very real sense—while cattle are a herd, a multiplicity of objects to work with as a political project. Riding as method helped me to understand this distinction, both as participant and as observant. Riding afforded a shift in perspective, the ability to experience new soybean landscapes through a marginalized mode of travel and a mode of being in the landscape that was part of affective conquest and survival. Riding horses gave me access to affects and landscape details that were absent in other modes of movement, such as walking and driving. Only from atop a horse was I able to follow cowhands as they showed a mastery of their field as well as a loss of traction with the everyday. Only from atop a horse did I get a feeling for the cattle project, for the exhilaration and the violence, the accumulation and the restlessness, the shifts between control and its sudden loss.

The anthropologist–horse relationship offered an ethnographic extension into a more-than-human landscape. My rubber boots were the hooves of a horse, "methodological shoes," as we delineate in the Introduction, that afforded both a new sensibility and access to a particular colonial mode of landscape making. This is in part because cattle and horses were key parts of Spanish conquest, species of empire that ushered in a sweeping destruction of Indigenous environments. The cloven hooves of cattle trampled native grasses and created spaces for European weed seeds to take root, while the barefoot hooves and bodies of horses enabled new forms of astonishing mobility across Pampean flatlands. Riding the edges of this history in the plains, it was possible to see how a new form of accumulation, the soybean plantation, was squeezing out colonial ranching landscapes. In other work, I have called these ecologies of belonging, ecological formations brought by settler colonists to remind them of home (Cypher 2022). Riding transported me into the margins of these ecologies of belonging affectively, and the rancher's exclusion of cows as individual beings pulled me into the scene as a project, ongoing in the genetically modified soybeans that were destined for animal feed.

Precisely because riding had become a marginal way of passing through a landscape, it enabled me to see and feel the duality of conquest and survival within the cattle project. To ride the margin of the new Anthropocene landscapes of the Pampas was to see how ranching had made the new soy landscapes possible, even as ranching landscapes were replaced. It was a method of affection as well as disaffection, individualizing a partnership with horses while at the same time alienating cattle into a project. Others have also noted the potential for "affection" with other-than-human beings. Julie Archambault (2016, 246), for example, describes human–plant relations in Mozambique as "affective encounters," what she defines as "the transformative potential of everyday engagement with the material world." Along with others in the ontological turn, she is interested about what happens between ethnographic subjects and what happens when we count nonhumans as subjects. If I had gone this route, I might have looked at Ceferino's relationship with the horse—but I argue instead that riding itself is a method and is one example of a method I call *affective empiricism*. In the Pampas, riding horses was a method that allowed me to approach "Anthropocenic" research problems—disturbance, alien invasions, swift loss of biodiversity, anthropogenic landscape change—affectively and empirically, to feel the outlines of a world-making project even as I was pulled into their details. Affective empiricism extends multispecies sensibilities with nonhuman ethnographic partners. In this case, loving the cattle not as individuals but as a project, like the ranchers do, was critical to understanding the world-making effects of cattle in ranching.

The chapter thus gives name to the "cattle project" to denote the way in which it is not just the species themselves that create degradation and erosion but also the modes of human production and pastoralism that accompany them. This is not to say that the species does not matter. Both cattle and horses are outsized species-objects in anthropology and animal studies. All over the world, their presence has changed the face of the land. Julie Livingston (2019) describes cattle as a total social phenomena and fact in Tswana life, while Deborah Bird Rose (2004) describes how cattle horses are both settler vehicles and modes of resistance in Australia. In the Americas, Virginia Anderson (2004) has called these ungulates "creatures of empire," while Ezequiel Estrada Martínez ([1933] 1971, 154) has described cattle in the Pampas as a totem so powerful that they dominate all modes of life, lending "a style to the manner of conducting commerce, to ways of entering into agreement, to marriage patterns, to being melancholic, to loving, to dancing, and to a manner of looking at things." There is no question that cattle and horses have shaped history and that they are well-studied and well-documented companion species, not just to humans but to each other. In the Pampas, horses, cattle, and humans are together a species triad. That which they create by being with each other is a plural relation and also a potent landscape-morphing force. Such potency opens up new possibilities for ethnography, as evidenced in a recent study by John Hartigan Jr. (2020) on horse sociality in Spain. For him, ethology offered an orientation toward sociality that opened up the ethnographic subject, meaning that he could study and bring new insights to behavior between horses themselves. But, I argue in this chapter, it is specifically the riding method that highlights how the relationships and affects between humans and horses change and contribute to those between humans and cattle, a difference operationalized in a horse–cattle relationship of herding that is a political project. By riding horses, by taking seriously the horse as an ethnographic partner, it was possible to see just how powerful the cattle project was, possible to cultivate a capacity for a feeling and vision that, atop a horse, saw the world for the taking. And although riding was specific to the Pampas, this chapter makes the case for such collaboration in general. Nonhuman ethnographic partners can expand and complicate multispecies sensibilities and methodological shoes.

In what follows, I unpack the account of my morning with Ceferino to develop and control an extended inquiry into the riding method.[1] This provides an orientation to build the argument of this chapter, which is that we must employ methods of noticing multispecies affects to see political landscape effects. It is auto-exegetical, a strategy to exploit within the method I

FIGURE 8.2. Rounding up the cattle. Photograph by the author.

propose. The exegesis is often historical, following the rubber boots insight that nonhumans make history. Such a strategy also shows the importance of *curiosity,* arguing through both the format and the narrative that being curious—in this case, about what is happening in this scene with Ceferino— is necessary to a method that is empirical and affective. It ventures a small methodological task, to *feel* our way through the Anthropocene, even if we are still fumbling through the dust.

When Ceferino needed help with the roundup . . . Ceferino was the cowhand on the estancia where I lived in the Inland Pampas while I was doing field-work between 2015 and 2017. He was in his mid-forties, with bright yellow eyes, a broad face, salt-and-pepper hair, and several missing teeth that had turned brittle from drinking the highly mineral Pampas well water. He was named after the local Indigenous Mapuche saint Ceferino Namancurá, the object of a Roman Catholic cult of veneration. He had, he told me, grown up poor, sometimes going hungry as a little boy. He had already buried two of his children, which he said was against the natural order of things and that because of it, he would carry for his whole life a melancholy. He was an excellent horseman and knew everything about caring for, fattening, and killing cattle and sheep. He taught me how to ride, how to round up, how to slaughter and skin sheep with a gaucho knife and a rope.

I dressed the horse he called Facundo with a sheepskin atop a rawhide saddle . . . Facundo was a criollo horse, a breed that dated back to the sixteenth-century Andalusian stallions left by a violent and failed expedi-tion of Spaniards. Together with criollo cattle, these horses spread out into the Pampas, multiplying rapidly and destroying the bunched grasses that had not evolved to live along with hardy-hooved ungulates (Mack 1989). They were settler-colonial animals, alien invaders that transformed the face of the Pampas wherever they went.

The horses clopped past the combine machine harvesting dried soybeans. From the back, it kicked up a powdered dust that clung to the edges of the eucalyptus trees bordering the sheep corral . . . Like the horses and cattle before them, soybeans were also an alien invader, although they were far from naturalized. Genetically modified soybeans had come to the Pampas in 1995–96 and since then had come to dominate most of the arable land in those great plains. More than twenty million hectares of Argentina's fertile lands were planted in soybeans in the growing season between 2015 and 2016. Their spread was both astonishing and vast, and in the early aughts, when the price of meat plummeted due to export restrictions, farmers did the unprecedented: they ripped up forage, culled their precious herds, and planted soybeans instead. Lesser echoes of an earlier conquest remained in the scene. The eucalyptus trees were imported from Australia because they grew quickly in the Southern Cone, and the sheep continued to be a cheaper source of fresh meat for ranch hands. Just beyond the edge of the eucalyptus was a feral peach orchard with a border of Osage orange trees. Under the eaves of these trees, a few bunched grasses were the only remainders of what had once been a plain of billowing, tufted grasses.

All around us was evidence that a century before, Argentine landscapes had been self-consciously remade in the shape of European models. Cattle—anonymous, unnamed herds—had paved the way; they had trampled previous Indigenous ecologies and, in pounding the bunched grasses to dust, had made room for new settler species brought to remind settlers of home. These ecologies of belonging ranged from the everyday kitchen garden to the massive remaking of once Indigenous landscapes into that which was recognizably Swiss, German, or English. "As you go south down the coast," Bruce Chatwin once wrote during his sojourn in the Southern Cone, "the grass gets greener, the sheep-farms richer and the British more numerous. They are the sons and grandsons of the men who cleared and fenced the land in the 1890s. . . . [Today] you can find, nestling behind windbreaks: herbaceous borders, lawn sprayers, fruit-cages, conservatories, cucumber sandwiches, bound sets of Country Life and, perhaps, the visiting Archdeacon" (Chatwin 1977, 46). In the most famous of these sites, Bariloche became "Argentina's Switzerland" when its streams were plugged with trout and the streets lined with decorative wooden awnings and chocolatiers to sustain the "myth of white Argentina" (Chamosa 2008). Anthropologist Gastón Gordillo (2016, 243), following these geographic projects, argues that whiteness in Argentina is both a myth and an atmosphere, affective and geographic, "defined by the not always conscious desire to create, define, and *feel* through the bodily navigation of space that the national geography is largely European." The form of privilege that claims power over landscape imaginaries, and, in so doing, remakes the face of the land, is part of every ecology of belonging.

Facundo shook his head and shimmied to the side into the soybeans, which rattled as his hind legs brushed against their spiny stalks . . . It had been a wet year, which made harvesting delayed because it is impossible, owing to the mud, and inadvisable, because it ruins the moisture content, to harvest when it is wet. After many weeks, there had finally been a break in the rains and what Pablo the farm manager called "a blessed wind" arose that dried out some of the stalks. This field was one of the final fields that Pablo was harvesting in the combine, a massive four-ton machine that trundled through the sodden earth on hydraulics. The soybeans sucked up by the combine were fuzzy, the soybean stalks scratchy to the touch. This aspect of the soybean, along with the terrifyingly rapid spread that made it in year-over-year maps look like it was a creature crawling out into the landscape, led many local inhabitants to call soybeans *la criatura,* "the creature." They also called her the "soy queen."

The soy queen was an extraordinary creature by all counts. Seven days after planting, she would burst from the soil with tiny green cotyledons. Because of thousands of years of domestication and a century of intense breeding and selection, she would grow a thick stock strong enough to support the weight of fifty pods, each of which would be filled with two to three heavy soybeans. She would be bushy, with wide, fuzzy, spear-shaped leaves that spread to absorb the sun efficiently and flowers that self-pollinated. And she would, incredibly, live through many applications of glyphosate, a potent agrochemical lethal to all plants. Later, after she yellowed and dropped all her leaves and a combine harvested the seeds from her dried pods, she would be stored in massive plastic silos as big as an airplane fuselage. Pablo would store her in these plastic silos until the landowner was ready to sell her to an exporter in Rosario.

All around us was evidence of the soy queen's astonishing spread. That summer, for miles until the earth rounded over into the horizon, all one could see was a sea of soy.

He was impatient and going a bit feral because he was not ridden enough . . . In the midst of this sea of soy lived Ceferino and his horses. Ceferino was the only cowhand who still lived on the *estancia* and the only person who rode the five remaining criollo horses. He tried to ride the horses evenly, but he did not have enough time because he was alone on the ranch. Everyone else had moved away or into town. Where once there had been tens of horses and multiple families living in the ranch houses, now there was only him. Soybeans, mechanization, and new modes of production, including genetically modified and hybrid seeds, had made numerous farm jobs obsolete. These processes had also pushed many farmers to move cattle off the farm and into smaller feedlots and to move calf–heifer production to the west, where the land was cheaper. Ceferino thus lived alone on the *estancia* that was, like *estancias* all over the region, crumbling. The sea-foam-green walls of his room crackled and split with dampness. Underneath the chinked floorboards, a wild dog had puppies. Ladybugs multiplied and swelled in the windowsills. In the middle of his small room was a steel-legged table covered with a linoleum flower-printed oilcloth. Armadillo carcasses nested on the table alongside a revolving set of objects: an ashtray, a spray can of mosquito repellent, an empty Fanta bottle, a plastic bag with three rolls of white bread, an unsheathed gaucho knife glistening with sheep fat. He had fashioned curtains out of stained, pale blue sheets that he draped over the front window. In the place where the window frame met the wall, he stuck a small photograph of a young man on a horse, wearing the traditional gaucho outfit. The young man's eyes had the look of human eyes in old photographs. They appeared like deep pools, haunted and bare to the earth.

He trimmed the horse hooves with his own knife; he killed the sheep and sliced them up for the weekly rations between workers; he made fires and cooked his meat on the ground because the earth gave the meat flavor; he listened to "El Malevo" at dusk and smoked cigarettes while staring at the clouds. He forecast rain; he spent much of his salary on expensive boots from the *pilchero* Carlos; he was stoic and prone to anger; he taught me how to drive a tractor, how to sit still, how to ride a horse, how to stick a needle in a flailing cow, how to fall into a state of prayer in order to be unbothered by mosquito hordes; he taught me how to pray in front of the altar of the folk saint of the Pampas, Gauchito Gil, and how to trim the barefoot hooves of horses, which often go shoeless on the soft, sandy steppe of the Pampas.

The cattle project depended upon Ceferino continuing to be a worker with local knowledge taught to him by his father through a long and unstructured apprenticeship. It depended on the fact that he was Indigenous but did not identify as such. Everything Ceferino did mattered, and everything he did mattered for the cattle project. Many of the things he did felt like remnants. It wasn't that he was stuck in the past; it was that cattle fever continued to live on through cowhands. Without men like Ceferino, the cattle project would have collapsed. These men were both denigrated and revered, anonymous and immortalized, through the very symbol of their production. The structure of feeling was both mundane and tantalizing.

He jerked on the reins and bent his head to eat some weeds at the same time that Ceferino looked back, on his face the disappointment that I was not a better pupil. "Don't let him eat!" Ceferino called out . . . The injunction not to eat had primarily to do with teaching the horse that snacking during work time was unacceptable. Control over his behavior, according to Ceferino, operated by the same rule through which horse herds followed hierarchy—the horse that controls the water and food is dominant. And he who could not control his horse was scorned. Such masculine domination was intimately wrapped up in the cattle project. Men gathered power to themselves through the domination of animals and women (Adams 2015; Archetti 1999). But it did not stop there—masculinity worked to create the very landscape that we were riding through. Pulling Facundo's head back with the reins, I realized that the weeds he had bent down to eat were not particularly palatable anyway, which was why he came away so easily. I recognized them as a devious plant that the farm manager had told me was an herbicide-resistant strain of Palmer amaranth. Dotted throughout the soy field were great green bushes of Palmer amaranth. These were a new class of plants that had emerged from the fields of industrial agriculture. They had been made in a hotbed of genetically modified soybeans, glyphosate, and added fertilizers; they had been bred into existence by a plantation model that alienated soybeans into assets even as it galvanized such weed seeds into rapid evolution (Cypher 2021). Palmer amaranth was an especially troublesome weed. Like other weedy invaders, it operated as a colonizing force, one that destroyed the resource base at the same time that new methods were developed to get rid of it, including more potent herbicides. Men sought to dominate it with herbicides, to obliterate it in order to establish civilization and keep ferality at bay, even as these weeds, in response, did something extraordinary: they evolved. They evolved by sequestering toxins to live within a toxic environment. They were weedy invaders and movements of resistance saw them as heroes, throwing herbicide-resistant "seed bombs" into soy plantations (Beilin and Suryanarayanan 2017). This was paradoxical, though, because to disturb the land with more herbicide-resistant weeds was to feed into a capitalist logic that sees disturbance—volatility, in the language of the market—as necessary to accumulation. To say that the ecologies were disturbed was to say that they were alive and well.

Ceferino dug in his heels to pick up speed, and I followed him, past a row of pines under whose shade llamas munched grasses and thistles, past the corral in which Martin and the other handler were weighing cattle, through two more swinging gates, around the alfalfa field . . . The cattle, like the soybeans, did not act alone as alien invaders. They were part of a system of production, part of a multispecies assemblage that destroyed Indigenous ecologies and lifeworlds. Environmental historian Alfred Crosby (1986) has called this process "ecological imperialism." His key insight was that the conquest of the world's temperate grasslands—in the Americas, Australia, and New Zealand—was as much a biological conquest as it was a political one. Europeans took over temperate latitudes with their species, engendering a profound shift in vast regions that became "neo-European" ecologies. Cattle led the way, trampling bunched grasses, making room for European weeds, and degrading the land so completely that the Europeans were forced to find a new resource base. In addition to expanding outward, settlers began to plow under the remaining grasses and plant alfalfa and other forage species. These species held the land.

The resident coots paddled, and a black-necked swan on an elevated bed of eggs nestled among the reeds, watching us from the side of her eye . . . But these conquest ecologies were not homogenous, nor were they absolute. On the edges of the Inland Pampa, in marshes and in thorn forests, some species managed to survive. Settler extermination campaigns against both the rhea *(Rhea pennata)* and the guanaco *(Lama guanicoe)* had wiped out much of the populations, but survivors had sought refuge in a crescent-shaped region curving around the settler-dominated grasslands (Beltrán 2000). Other birds, such as the red-fronted coot *(Fulica rufifrons)* and the black-necked swan *(Cygnus melancoryphus)*, the warbling doradito *(Pseudo-colopteryx flaviventris)* and Hudson's castanero *(Asthenes hudsoni)*, survived in feral ecologies on farms. These were places that were unable to be cultivated, usually because they were geomorphologically disadvantageous. Shallow depressions, for example, tended to collect water during rainy seasons and to suck up salt from the hardpan, leaving a white crust layer on the topsoil. During the rains, it was impossible to plant because it was waterlogged, and during the dry seasons, it was often too salty for maladapted domestic plants to survive. It was just such a geomorphological shape that dominated this middle half of the farm. Because of the rains, a three-hundred-hectare lagoon had formed in this shallow depression, in which so many bird species took up temporary residence for the season.

At the edge of the lagoon was a corn stubble field where the cattle were grazing. It was November, nearly summer in the Pampas of Argentina, and the cattle were eating the last of the residue before genetically modified soybeans were planted . . . Soy is one of the latest colonizing plants in a long history of ecological imperialism. It is usually rotated with corn. The corn stubble, or *rastrojo,* was what was left after the harvest and was part of a seasonal cycle of soybean and corn production. Because the Inland Pampas was historically dry—"the whole countryside used to go flying by the window!" said the farmer with whom I worked—it was advantageous to do planting

FIGURE 8.3. Survivors in the temporary lagoon. Photograph by the author.

rotations and cover cropping and to leave leftover crop residues to provide soil protection. Farmers often turned out the cows to graze on these fields. The cows ate the remaining residues and also left some natural fertilizer. This rotation was part of a soil management strategy that sought to preserve soil structure and moisture by not tilling the land when it was time to plant. This strategy also depended on the use of herbicides to kill the weeds that would inevitably arise.

Suddenly I was swept up in the momentum of the herd. I lost my hat; the sun and dust gritted into my eyes and throat; I clutched the reins and pushed into the stirrups and tried not to fall off . . . The cattle project was not just a biological invasion. It was also affective. There was a predatory and exhilarating high that emerged from and with cattle, a being-with that, as Deborah Bird Rose (2004) has shown, was part of an old raiding culture. I got a sense for this affect in the roundup. To accompany Ceferino on morning rides, to learn how to slip the bridle over the velvety muzzle of the horse, to feel horse in my whole body during siesta when I lay splayed on the floor reeling with dust and light, was to feel the lurching high of the cattle project. This was, of course, a kind of ambivalent or wary exhilaration. The affect was related to the friendship with horses and accumulation of cattle, and as a woman, I was empathetic yet critical of the affects that drove male domination. The project depended on the subordination of women and cattle, and so I had a glimpse into what was both a triumph and a trap. I also got a sense for affects by reading Argentine books that were recommended to me by the men with whom I worked. One of these, *Pampa Grass* by Diego Newbery (1953), evokes the cattle project through an autobiographical account of nineteenth-century Pampas settlement. It was clear even through authorship how men saw landscapes through histories of times past and through their fathers: the author is considered Diego Newberry, George (Jorge) Newberry's son, because he was the one who wrote down what his father had verbally told him. One of the young managers of a local feedlot loved this book and gave it to me, not just because Jorge Newberry was a storied figure in Argentina, but also because his grandfather's family had settled in the same area and were also English. In the narrative, one could read the thrill and the violence and the way they went hand in hand.

Jorge, the story goes, was a young man from North America who came to try his luck in the Southern Cone in the 1880s. He spent years wandering around northern Argentina and falling in love with the country. He had not dared go south, since President Roca was then waging his so-called Conquest of the Desert. Several years later, over warm beer in a bar in Buenos Aires, a man told him about an English company that was selling large tracts of land to men brave enough to settle on the recently opened frontier. Jorge was intrigued, and he was hooked on Argentina. He bought the land for peanuts, hired a few scraggly gauchos, and hustled four thousand head of cattle over three hundred kilometers west to a settlement near Colonel Vallejos. He was infatuated; he was caught up in this project of colonization without thought to the violence wrought on this frontier. But the gauchos he had hired knew exactly what had happened here, and they wanted nothing to do with it. They abandoned him, and he was left with just one man trying to figure out what to do with all the cattle.

It was dusk as he and his companion built a fire and sat down to think about what to do. Off in the distance, he suddenly saw a group of people coming toward the fire. Jorge and his companion got their knives ready, even though they were clearly outnumbered. A cacique entered into the ring of firelight and, without saying anything, speared the skinned rodent hanging from his belt and put it on the fire to roast. He shared the meat and then offered himself and his men to work for one month for Jorge. They worked for one month on land that had recently been taken from them for a man who had stolen it. They built fences and erected a small rancho; they taught him where to plumb water and how to get the cattle accustomed to this region.[2] And after exactly one month, they left. Jorge describes their figures receding into the distant horizon. The tragedy of colonization was captured in their receding figures. This was tragic; it was world ripping. *Go west,* said the project, *go west and do not question why.*

We circled around to the dusty dirt road, fenced in on either side to the keep the cattle going toward the feedlot . . . Feedlots were the most recent manifestation of the cattle project in the Pampas. All over the region, farmers had thrown up fences and corralled their cattle into smaller enclosed areas, thus making room for crops like soybeans and corn. These enclosed areas depended precisely on a mode of feeding that separated cattle from grass and forage, forcing them to feed on corn, roughage, and protein pellets made from soybeans or sunflower seeds. In an ongoing metabolic rift (Foster 2000), the cattle were thus separated from the fields they might have otherwise fertilized, smooshed together in small feedlots knee-deep in mud and their own feces and fed human food to gain weight.

Up ahead, I saw Ceferino look back and wave, then pull out a pack of cigarettes. He tilted his head and cupped his hand against the wind to light it, then straightened up, blowing smoke into the Pampas air . . . The privilege that claims power over landscape imaginaries, and, in so doing, remakes the land, is sometimes hard to see, but it is always there. Whether it is in the soy harvester or the herbicide-resistant weeds, the alfalfa or the cattle herd, in every landscape, it is possible to see the way privilege and power imbue landscapes and their histories. Making certain ecologies was a powerful project of belonging and exclusion, but such world making works both ways. To focus only on the atmosphere of white Argentina is to continue to see only one kind of affective geography. This is a conundrum for Argentina in general. In Argentina, race has consistently been negated, even as society is built upon it (Gordillo and Hirsch 2003). While other countries in Latin America began in the early twentieth century to embrace a mestizo nationalism, Argentina continued to sustain the "myth of white Argentina" (Chamosa 2008). To a certain extent, this discourse continues to be promulgated as a source of pride, and even in other Latin American countries, it is cited as a source of arrogant exceptionalism. As several new histories of race in Argentina have pointed out, unfortunately, this only continues to reinforce a discourse that has no grounding in material reality, unintentionally confirming the disappearance and national erasure of Indigenous and Afrodescendant inhabitants (Alberto and Elena 2016, 3). And it is easy to see why such a discourse continues to be promulgated if one only looks at the heartland of the Pampas. But to end there would be to broadcast the same histories that are always told and to buy into progress. If we look at the edges, however, it is possible to see that the heartland is historically changing, and it is this history that produces patchiness and heterogeneity. From the remnant margins of the ranches, it is possible to see how old forms of domination are overtaken by new forms.

On the edges was where Ceferino's ancestors had survived. Eighty kilometers west of the farm, the region of refuge that had sheltered Indigenous livelihoods for three centuries rose up in the shape of a thorn forest. It was there that landscape formation and ethnic formation had gone hand in hand, where the other side of the cattle project had manifested into a thorny ecology of belonging. It was neither Swiss nor Indigenous but a conflation of the two, and in that ecological murkiness, Ranqueles and Mapuche had been able to carve out a territory that was one of the longest-held Indigenous territories in the Americas. This was a necessary and missing ingredient of the multispecies history I have been tracing. Indeed, while histories of Argentina have considered the role of settler elites and their wheat/cattle that led to the "rise of capitalism on the Pampas" (Scobie 1964; Solberg 1987; Amaral 1998), none have analyzed the way that conquest and inequality, as well as collaborative survival, require multispecies world making. Ecologies of belonging were central to the European conquest of the Pampas, as well as to Indigenous survival.

If we keep following cattle tracks, then, on the edges of the Pampas, we see a parallel story emerge. Beyond the plains, Mapuche and Ranqueles had established a massive cattle trade route up the eastern side of the Andes and into Chile. Over several centuries, this region became more forested because cattle ate the sweet beans of mesquite trees, unintentionally dispersing the seeds through their excrement along the trail.[3] As the forest rose up, so, too, did a region of refuge that was punctured only by the late nineteenth-century "Conquest of the Desert," an Argentine military campaign that sought to wipe out Indigenous resistance. Ranquel and Mapuche survivors were sequestered into reservations or pressed into menial labor on ranches (Tarquini 2010). Anthropologist Claudia Briones (2007, 99–100) describes how Mapuche, pushed south after Roca's war, now construct "different maps of meaning to interpret regional geographies and make sense of their structured mobilities" to promote "affective investments of belonging." Ceferino's grandparents were the children of survivors who had been taught to deny their ancestry to live, even as, through him, the cattle project lived on.

I looked back behind us to see if my hat was there, but all I saw were cattle tracks in the dust . . . In this chapter, I have explored how it is possible to "get out and follow the more than human tracks" through a landscape, figuratively, historically, and affectively. Through the format, which begins with a description of a roundup in which I participated one morning in the Pampas, curiosity leads to critical description of the politics of multispecies relationships. The format seeks to highlight the fact that within each glimpse, worlds are contained, worlds full of colonial legacies, ecological histories, violence, and survival. Riding horses enables the empathic extension of multispecies sensibilities and curiosity about the "cattle project." The cattle multitudes pummel the earth with their hooves and make dust and, in so doing, make worlds. But it matters who has the privilege to claim power over landscape imaginaries; it matters how cattle and horses and humans live together, how humans force cattle to reproduce in strategies of accumulation called feedlots, where they feed on corn and protein pellets, further degrading the land beneath their hooves even as all around them are the very plantations by which they are fed. These are the worlds of the Anthropocene, worlds that suffer from massive disturbance regimes, swift invasion of species, herbicide-resistant weeds, plantations, and feral ecologies (Tsing et al. 2021).

The chapter gives name to affective empiricism, a method that develops a mode of emic being-with to draw out the affective and empirical dimensions of disturbance. Riding horses in the Pampas is one method through which to both see and experience the margins and to feel the hurtling and vertigo-inducing rush of processes that have led to conquest and survival in the soy fields. It is a distinctly rubber boots method because it tries on the methodological shoes of nonhuman partners, in this case, the barefoot hooves of horses, arguing for affective empiricism as an empathic extension that begins with curiosity. Multispecies ethnography's methods can in this sense not just be "about" but also "with," complicating the already messy distinctions between human and nonhuman and complicating the traditional view of the solo anthropologist. This method requires certain kinds of noticing, requires curiosity and critical description, and can be improvised, in the sense in which I only scratched the surface in this short chapter, and also in the sense that it is generalizable to other nonhuman partners and other ecologies.

To try on other methodological shoes is to bring out the patchiness of landscapes, the entangled and interwoven histories and lifeworlds made by humans and nonhumans together. Riding horses is, after all, an outdated and marginalized practice on the Argentinian Pampas today, but that view from the margin of a practice that once reshaped the landscape to enable new forms of disturbance is what makes it the right method to draw out affective empiricism and landscape patchiness, to see ecologies of belonging. Conquest ecologies and ecologies of belonging are not, of course, limited to the Pampas. Almost everywhere in the world, it is possible to see and to feel the world-ripping cadence of ecological conquest, the rhythmic push and pull of a moody, relentless change this volume calls the "Anthropocene."

NOTES

1. Clifford Geertz invented a similar strategy in "Religion as a Cultural System"; see chapter 4 in *The Interpretation of Cultures* (Geertz and Darnton 2017). Thanks to Nils Bubandt for the reference.

2. The Spanish word is *aquerenciamiento,* verb form *aquerenciar.* The word *querencia* is metaphysical, coming from the verb *querer,* which means "to want" or "to desire." The word may have originated from bullfighting, when the bullfighter draws the bull out of his *querencia,* his safe place, to kill him.

3. Elinor Melville (1997) has traced a similar process of mesquite dispersion in the Valle del Mezquital (the Mesquite Valley) in Mexico.

REFERENCES

Adams, Carol J. 2015. *The Sexual Politics of Meat: A Feminist-Vegetarian Critical Theory.* New York: Bloomsbury Academic.

Alberto, Paulina, and Eduardo Elena, eds. 2016. *Rethinking Race in Modern Argentina.* Cambridge: Cambridge University Press.

Amaral, Samuel. 1998. *The Rise of Capitalism on the Pampas: The Estancias of Buenos Aires, 1785–1870.* Cambridge: Cambridge University Press.

Anderson, Virginia DeJohn. 2004. *Creatures of Empire: How Domestic Animals Transformed Early America.* Oxford: Oxford University Press.

Archambault, Julie Soleil. 2016. "Taking Love Seriously in Human–Plant Relations in Mozambique: Toward an Anthropology of Affective Encounters." *Cultural Anthropology* 31, no. 2: 244–71.

Archetti, Eduardo P. 1999. *Masculinities: Football, Polo and the Tango in Argentina.* Global Issues. Oxford: Berg.

Beilin, Katarzyna Olga, and Sainath Suryanarayanan. 2017. "The War between Amaranth and Soy: Interspecies Resistance to Transgenic Soy Agriculture in Argentina." *Environmental Humanities* 9, no. 2: 204–29.

Beltrán, Gonzalo Aguirre. 2000. *Regiones de refugio: el desarrollo de la comunidad y el proceso dominical en mestizoamérica.* Jalapa, Mexico: Fondo de Cultura Economica.

Briones, Claudia. 2007. "'Our Struggle Has Just Begun': Experiences of Belonging and Mapuche Formations of Self." In *Indigenous Experience Today,* edited by Orin Starn and Marisol de la Cadena, 99–122. Oxford: Bloomsbury Academic.

Chamosa, Oscar. 2008. "Indigenous or Criollo: The Myth of White Argentina in Tucumán's Calchaquí Valley." *Hispanic American Historical Review* 88, no. 1: 71–106.

Chatwin, Bruce. 1977. *In Patagonia.* New York: Summit Books.

Crosby, Alfred W. 1986. *Ecological Imperialism: The Biological Expansion of Europe, 900–1900.* Cambridge: Cambridge University Press.

Cypher, Rachel. 2021. "From the Fields of Industrial Agriculture, a New Class of Plants Has Emerged." In *Feral Atlas: The More-Than-Human Anthropocene,* edited by Anna Tsing, Jennifer Deger, Alder Keleman, and Feifei Zhou. Stanford, Calif.: Stanford University Press. http://www.feralatlas.org/.

Cypher, Rachel. 2022. "Belonging in the Pampas: Ecologies of Conquest and Survival in Argentina's Heartland." PhD diss., University of California, Santa Cruz.

Foster, John Bellamy. 2000. *Marx's Ecology: Materialism and Nature.* New York: Monthly Review Press.

Geertz, Clifford, and Robert Darnton. 2017. *The Interpretation of Cultures.* 3rd ed. New York: Basic Books.

Gordillo, Gastón, and Silvia Hirsch. 2003. "Indigenous Struggles and Contested Identities in Argentina Histories of Invisibilization and Reemergence." *Journal of Latin American Anthropology* 8, no. 3: 4–30.

Gordillo, Gastón. 2016. "The Savage Outside of White Argentina." In *Rethinking Race in Modern Argentina,* edited by Paulina Alberto and Eduardo Elena, 241–67. Cambridge: Cambridge University Press.

Hamilton, Lindsay, and Nik Taylor. 2017. *Ethnography after Humanism: Power, Politics and Method in Multi-species Research.* London: Palgrave Macmillan.

Hartigan, John, Jr. 2020. *Shaving the Beasts: Wild Horses and Ritual in Spain.* Minneapolis: University of Minnesota Press.

Livingston, Julie. 2019. *Self-Devouring Growth: A Planetary Parable as Told from Southern Africa.* Durham, N.C.: Duke University Press.

Mack, Richard. 1989. "Temperate Grasslands Vulnerable to Plant Invasions: Characteristics and Consequences." In *Biological Invasions: A Global Perspective,* edited by J. A. Drake and Harold A. Mooney, 155–79. New York: Wiley.

Martínez, Ezequiel Estrada. (1933) 1971. *X-Ray of the Pampa.* Austin: University of Texas Press.

Melville, Elinor. 1997. *A Plague of Sheep.* Cambridge: Cambridge University Press.

Newbery, Diego. 1953. *Pampa Grass: The Argentine Story Told by an American Pioneer to His Son.* Buenos Aires: Editorial Euarania.

Rose, Deborah Bird. 2004. *Reports from a Wild Country: Ethics of Decolonisation.* Sydney, Australia: University of New South Wales Press.

Scobie, James R. 1964. *Revolution on the Pampas: A Social History of Argentine Wheat 1860–1910.* Austin: University of Texas Press.

Solberg, Carl E. 1987. *The Prairies and the Pampas: Agrarian Policy in Canada and Argentina, 1880–1930.* Stanford, Calif.: Stanford University Press.

Tarquini, Claudia. 2010. *Largas noches en la pampa: itinerarios y resistencias dela población indígena (1878–1976)*. Buenos Aires: Prometeo Libros.

Tsing, Anna Lowenhaupt. 2005. *Friction: An Ethnography of Global Connection*. Princeton, N.J.: Princeton University Press.

Tsing, Anna, Jennifer Deger, Alder Keleman, and Feifei Zhou, eds. 2021. *Feral Atlas: The More-Than-Human Anthropocene*. Stanford, Calif.: Stanford University Press. http://www.feralatlas.org/.

PART III

COLLABORATION

CHAPTER 9

Marine Hitchhikers and Nested Holobionts

Is the Aquarium Trade Creating Weedy Sponge Invaders?

JOSEPH KLEIN, STINE VESTBO,
PETER FUNCH, *and* ANNA TSING

So, naturalists observe, a flea
Has smaller fleas that on him prey;
And these have smaller still to bite 'em,
And so proceed ad infinitum.

Jonathan Swift, *On Poetry: A Rhapsody*

Put a stony coral under a microscope and a landscape opens itself to you. Visible first are the polyps, the animals who make the body of the coral: you can see their fleshy mouths, which open and close to feed on passing plankton, and the smooth tissue that connects each polyp to its neighbor, shiny with bacteria-rich mucus. The surrounding substrate is pocketed with sessile organisms, which jostle for elbow room: coralline algae, fungi, tunicates, and sponges. But look closer. Beneath the polyps, the cavernous pores of the coral's calcareous skeleton belch forth all manner of organisms. Curled brittle stars, fuzzy polychaete worms, tardigrades, isopods, amphipods, rotifers, and myriad larval creatures pour out like clowns from a tiny car. And even with the microscope, you miss a profusion of life: archaeans, viruses, and protists abound. Now stuff your coral into a plastic bag and ship it across the ocean: what might come out the other side?

The live coral trade connects the far-flung reefs of Indonesia's eastern islands to the saltwater aquariums of the Global North, shuttling little bags full of aquatic life from one side of the planet to the other. Along the coral supply chain, whole microcosms of reef life are redistributed across the globe. We call these "hitchhikers," organisms that are moved by human agency but without human design, borrowing a term used by aquarium hobbyists to name the forms of life that make surprise appearances in their tanks.

Hitchhikers are everywhere—not just in aquariums. They were seeds of European grasses tucked in the guts of livestock bound for New World pastures. They were rats aboard ships in the age of sail, colonizing islands, and they are fruit fly larvae stuck to the peel of a ripe banana, taking up residence in your kitchen. They are diatoms sucked into the ballast tanks of Panamax ships, and they are *Phytopthora* spores in the soil of a commercial nursery tree. They are also novel zoonotic viruses with delayed symptom onset stowed away in apparently healthy hosts on a transatlantic flight. Hitchhiking gives name to both the organisms and to a process of feral distribution by which weedy invaders are not only spread but also born (cf. Tsing et al. 2021).

While plenty of invasions are caused by intentional introduction (consider the global nursery plant trade, European colonization of North America, or the introduction of cane toads to Australia), hitchhikers are the primary driver of invasion today (Westphal et al. 2008). This isn't surprising, given how little we still know about the living world and the diversity it contains:

FIGURE 9.1. Live coral in a Styrofoam cooler after being collected, and corals in a holding tank awaiting fulfillment of orders. Photographs by Joseph Klein.

our best estimates suggest that between one-third and two-thirds of marine species remain undescribed, representing many hundreds of thousands of species (Appeltans et al. 2012). Many species are moved without anyone noticing until far in the future, or possibly never, because many descriptions occur only *after* organisms have been widely moved around (Jarić et al. 2019). And because life is constituted by webs of relations, no creature ever moves alone; instead, we always bring a retinue of companions along for the ride.

But how to study the problem of hitchhikers? Hitchhiking is a matter of both ecology and supply chains, a topic that requires both social and natural science in equal measure. We came to the topic in a roundabout way when anthropologist Joe Klein began to study the Indonesian live coral trade with the help of Anna Tsing. They turned to biologists Stine Vestbo and Peter Funch for help in thinking through the trade's ecological impact. Lucky to find ourselves under the auspices of Aarhus University Research on the Anthropocene (AURA), together we set out to find a back door into understanding hitchhiker dynamics and their feral effects.

We began with our differences, but also with what we shared: a deep curiosity about the living world. From the beginning, this curiosity was manifested in many shared practices, including a propensity to walk slowly and look closely, trying to piece together clues and signs from the world around us. In this sense, for us, "rubber boots" is about putting yourself in a position to notice—a reminder to get up out of the armchair and into the world (Tsing 2015). But through our work together, we also learned to notice in new ways; for instance, against its caricature as a sterile space of disconnected abstraction, the anthropologists learned to see the laboratory as its own kind of "field," its own place to get dirty and look closely (after all, labs require their own footwear suited to their particular environments, not to mention the rest of the wardrobe of personal protection needed to trek safely through different laboratory landscapes).

But figuring out exactly *how* to work together in a way that took seriously the expertise and methods of both disciplines was a significant challenge. Humanists and natural scientists have drifted so far apart in their approaches to knowing the world that coming back together has taken us many years, video calls, field trips, dead ends, workshops, laboratory MacGyverisms, and oceans of patience. In fact, developing this project together has been such an important part of our collaboration that we present it here as our first "finding" in the form of a research proposal.

And so the rest of this chapter has two parts. First, we discuss questions of method and some conceptual tools that helped us to approach these problems together. The second part is a proposal for research—a blueprint that draws from our successes and failures, charting a program to study the itineraries of sponges as they hitchhike global supply chains. We offer both as a model of what is possible for collaboration between the humanities and the natural sciences as we learn to follow the intertwining of human and nonhuman histories toward a critical description of the Anthropocene.

NEGOTIATING METHOD

It turns out that the word *method* means something entirely different in biology and anthropology. In biology, a method is what anthropologists might call a "technique." Looking through a microscope; polymerase chain reactions; boiling for five minutes: each of these is a method in this sense of the term. By contrast, in anthropology, a method is a strategy for answering questions; biologists might, at best, call it an "approach." Indeed, "rubber boots" may be a method in the humanities, but it is hardly one in the natural sciences.

The definition of *method* matters because it lies at the heart of how both disciplines imagine the world and work to understand it (cf. Harding 1987; Mol 2002). Anthropologists—with our wildly generous view of method and proclivity for asking big questions—often bite off more than we can chew from a biologist's perspective. Consider what happened to us in our equivocations of method. From the beginning, the four of us were interested in the redistributive power of the coral supply chain, that is, how it might be unintentionally moving organisms from one place to another. Of central interest to the anthropologists was *Symbiodinium*, the genus of symbiotic photosynthetic dinoflagellates that live within the tissue of the corals. We had come across an exciting study arguing that a weedy strain of *Symbiodinium* from the Indo-Pacific had infected corals of the Caribbean. The invading *Symbiodinium* was outcompeting local symbionts because of its heat tolerance in a time of global warming, but simultaneously slowing down colony growth (Pettay et al. 2015). While the interaction between coral and

Symbiodinium has usually been regarded as a mutualism with equal benefits for both partners, researchers had begun to show that coral–*Symbiodinium* relations are much more variable, falling on a spectrum from mutualism to outright parasitism (Lesser, Stat, and Gates 2013; Baker et al. 2018). In the Caribbean case, we may have been witnessing something closer to the latter—a weak mutualism at best. Here was a seemingly clear case of ecological disruption caused by hitchhiking coral associates, and so the anthropologists asked, why not study the ecological effects of the transport of *Symbiodinium* in the global coral trade?

Several things, indeed, were wrong with this ambition—and they can be understood in relation to method in the biological sense. Biologists don't imagine they can study just anything: it takes many years, a great deal of expertise, and expensive, specialized equipment to learn about just one kind of organism. No one in Peter Funch's laboratory had an expertise in coral or *Symbiodinium*. But even drawing in an expert was not enough. The late coral scientist Ruth Gates from the University of Hawai'i, no stranger to bold experimentation, generously spent four days consulting with us as we tried to find ways to open up our questions and attach them to "method." It was exciting, and it was terrible. In 2015, when we met, the baseline biological knowledge just did not exist for the questions we wanted to ask. At the time, researchers were only beginning to work out the systematics of different *Symbiodinium* clades and their various functional relationships with coral; nothing was settled enough to begin asking questions about long-distance transport. Indeed, the article we read about the Caribbean turned out to be a surprising result of ordinary research, not the product of a clever experiment. Thus, too, no experimental setup could possibly handle the complexity of our question. We were left with a good idea with no way to study it: there was no "method."

We had several such false starts in our work together. Yet, through each attempt, we gained valuable insight on the "how" of collaboration, and along the way, we developed key insights and tools that began to cumulatively build from each other. In what follows, we explain three.

1. Nested holobionts: The living world is composed of a series of nested holobionts. For decades, corals have been paragons of symbiosis, central to the theoretical advances that have helped us to see a living world populated by organisms that not only interact with one another but are themselves physically composed by and dependent on a bewildering diversity of life (Haraway 2008; McFall-Ngai et al. 2013). As model organisms in this research, corals have shown us how to loosen our obsession with the individual as the primary unit of biological knowledge, helping the sciences wake up to the old truths (never forgotten by Indigenous thought) that all things come into being through webs of interaction, and that each individual is always already a multitude, an assemblage, a network (cf. Todd 2016; TallBear 2017).

In this new biology, organisms come into being in relation to a set of others; together, this relational group of "bionts" acts as a "holobiont," that is, a composite unit of ecology and evolution (Guerrero, Margulis, and Berlanga 2013; McFall-Ngai et al. 2013). The term seems to have emerged from nineteenth- and twentieth-century research on lichens. Lichens are composed of various component "bionts," including algal *phycobionts* (also called *photobionts* for their photosynthetic abilities) and fungal *mycobionts,* which come together in a *holobiont* composite. The holobiont concept was later extended by Margulis (1991) and has since become closely associated with studies of coral symbiosis following the model proposed by Rohwer et al. (2002).

This coral holobiont model provides a framework for understanding how corals thrive in nutrient-poor waters by relying on nutrient cycling and exchange between corals, endosymbiotic *Symbiodinium,* and other associated microorganisms. And while most research has focused on the role of microorganisms in this assemblage, La Barre (2013) helpfully proposes the term *extended holobiont* to further include other species associated with the structure of the host—which in the case of corals includes epifauna, meiofauna, and cryptic organisms that live in and on the coral and its substrate.[1] Although our earlier attempts to study *Symbiodinium* had attuned us to thinking about corals as assemblages, we lacked basic information about this extended holobiont and could only speculate about what species might survive arduous transportation.

In their lab in Denmark, Vestbo and Funch worked with undergraduate student Frederik Skovby Felding to set up a simple experiment (see Figure 9.2). They ordered "live rock" from a commercial supplier in the Netherlands, sourced from an Indonesian coral reef. "Live rock" is calcium carbonate

FIGURE 9.2. Diverse organisms colonizing the aquarium from the "live rock" experiment at Aarhus University, Denmark. Photograph by Stine Vestbo and Peter Funch.

(aragonite)—mostly reef substrate—taken from the ocean and put into a saltwater aquarium to purposefully introduce wild marine organisms into the closed system, as well as provide three-dimensional structure to an empty tank. They placed the live rock in a simple aquarium in the lab at Aarhus University, extracting the fauna hiding within for examination. From the rock came at least twenty different animal species and three protists that had survived transportation from Indonesia through the Netherlands to Denmark; the tank was soon replete with algae, sponges, and other blooms of marine life (Figure 9.2). Here was proof of concept: clearly the live coral trade could be a vector for transmitting organisms from one marine habitat to another. While many tropical species would not survive radically different environmental conditions, some clearly could. If these hitchhikers could survive weeks of rough handling in transport, oxygen deprivation, fluctuations in temperature and salinity and light, and moving from tropical ocean to Danish university, perhaps they could survive introduction to new marine habitats, generating potential for invasion.

But the experiment did something else as well: through the simplicity of its method, it generated new conceptual worlds. Consider a coral: it is partially composed of its own microbiome—along with fungi, protists, and viruses closely associated to it. However, its extended holobiont also includes cryptic fragmentary sponges that may be growing on or within its porous substrate. These sponges in turn contain their own microbiomes, and possibly their own extended holobionts, including organisms that in turn have their own microbiomes, and so on. From here, we learned an important lesson that would shape our research going forward: the living world is composed of a series of nested holobionts. When we think about the transportation of corals and their companions, we must think about it in this framework.

2. Disturbance taxonomies: Human disturbance transforms nested holobionts. With a small grant secured for preliminary explorations, Funch and Vestbo joined Klein in the eastern part of Sulawesi, Indonesia, where he was doing research on the live coral industry. The premise of the trip was to acknowledge "method" in the biological sense: what might we find that we could actually study? Our team started studying microscopic meiofauna—a diverse assemblage defined by size and consisting of animals small enough to pass through 1-millimeter sieves, but too big to pass through a sieve with holes about 0.04 millimeters in diameter. Funch had spent much of his career studying the diversity of meiofauna, including describing two new phyla (Cycliophora and Micrognathozoa); if anyone had the methodological expertise to help us see what kinds of meiofauna were moving with coral, it was Funch.

In the city of Kendari, we visited coral companies, borrowed a bit of lab space, and, with help from our hosts, set about trying to develop protocols for meiofaunal extraction (see Figure 9.3). Meiofauna are very comfortable in their homes deep within the coral substrate, and to make them visible, we needed to get them to emerge. There are a few ways to do this, such as shocking them with a splash of freshwater or relaxing them with carbonated water, forcing them into the surrounding water. When the water with extracted meiofauna is filtered through a fine sieve called a "mermaid bra," the meiofauna are concentrated and can be examined under a microscope, photographed, fixed, and then placed on microscopic slides in glycerin. Now this was "method"! Still, things never work quite as well as one hopes—and we hit roadblocks in our efforts, especially being unable to locate the correct chemicals that would allow us to make proper fixations of the meiofauna.

With limited time and access to the necessary equipment to finish the meiofauna project, the team pivoted and began to examine the corals with stereomicroscopes and to document as many organisms as we could. From these practical steps, a new world opened. Even with the extreme stress of transport and limited laboratory conditions, the corals were awash with life. Brittle star arms poked out from the coral calcium carbonate skeletons, while crustaceans swam across our fields of vision, followed by annelids, tardigrades, and all manner of creatures. The substrate itself was pocketed with sessile organisms trying to claim their inch of territory—coralline algae, dozens of tunicates, and sponges. Sponges! This moment under the microscope

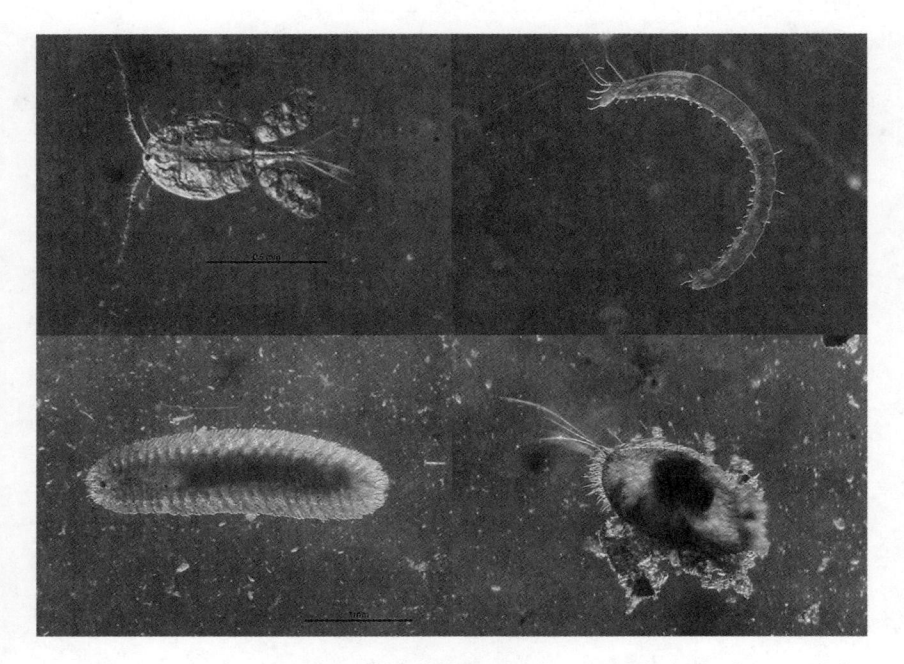

FIGURE 9.3. Meiofauna extracted from live coral samples. *Clockwise from top left*: a female copepod with egg sacks (Arthropoda); unknown polychaete (Annelida); an ostracod or "seed shrimp" (Arthropoda); a scale worm (Polynoidae, Annelida). Photographs by the authors.

changed everything: sponges were another area of expertise of Peter Funch and his laboratory, and our ornamental corals were covered in them (see Figure 9.4). Funch's laboratory had already invested in the expertise and equipment to pay attention to sponge sytematics. Sponges could allow us access to a "method."

Once we began looking for them, we found cryptic hitchhiking sponges everywhere—including in public aquaria both in Denmark and in California, where we collected samples of nineteen different species that seemed to have been introduced to the aquaria by an outside source—possibly the ornamental coral trade, which was used to supply these institutions. These sponges were not considered desirable organisms, fit for display, by the aquarium staff. Instead, they were cleaned out and removed. Still, they flourished in both display and backstage tanks, growing across every available substrate.

Our project pushed the limits. Instead of identifying sponges in relation to their native places and ecologies, our interest in sponges *traveling inside*

FIGURE 9.4. Unidentified encrusting sponge covering part of the white coral skeleton on the right-hand side of a *Blastomussa* spp. colony. Photograph by the authors.

of corals brought us to a strange subset: sponges that could survive the conditions of human transport. In contrast to more classic taxonomical forays, ours might take advantage of depauperate ecologies to figure out what survives. We began to consider research tracking how the aquarium trade might not only be moving these sponges around the world—but be accidentally selecting for the hardiest, most weedy sponges, which bring along their own suite of microorganisms as they travel.

We were on our way to developing "disturbance taxonomies," that is, methods for learning about what survives within human disturbance, such as long-distance transport. Put differently, we came to understand that human disturbance simplifies and recomposes nested holobionts. To move to this insight was a major breakthrough for our work in common. Here was a topic that needed both natural scientists and humanists and where methods in each sense of the term might find traction with each other. In turn, these insights into nested holobionts and disturbance taxonomies pushed us to think in new ways about supply chain research as we followed hitchhiking sponges.

3. Relational supply chain analysis: Supply chains generate and redistribute disturbance taxonomies. How might we approach the trade that carries sponges hidden inside of corals? Natural scientists, however well meaning, have not had the time or means to investigate the supply chains that carry potentially destructive new species around the world. Social scientists likewise have not known enough about nonhuman organisms to explore the nondesigned ecological effects of supply chains. Neither natural science nor social science literature helps us understand what actually happens within relations among humans and nonhumans in a supply chain.

Anthropologists working on supply chains have offered important insights, significantly transforming the supply chain literatures that emerge from economics. Instead of limiting questions of value to price, anthropologists have discussed value as a problem of the making of people and things in relation (e.g., Rofel and Yanagisako 2018; Tsing 2015). In contrast to economists, who view supply chains as networks within already-settled conventions of value (see discussion of this problem in Bair 2009), anthropologists have shown how supply chain nodes can link quite different world-making projects, allowing transactions between people with entirely different motivations and ways of life (e.g., Tsing 2015). In understanding why a fisherman in a small village in Sulawesi might procure corals that are eventually sent to Denmark, for example, such "relational supply chain analysis" seems essential. Supply chains make it possible not just to link far-flung parts of the world but also to create the articulations that make people bother to pursue one livelihood or another.

Such analysis is already "relational" in mapping the social interactions through which people and products are made *in relation.* But this insight can be extended: what about the social relations of corals and sponges? Thinking through the supply chain literature, our team stumbled into something to offer: a relational analysis without the assumption of human exceptionalism. How might we understand the constitutive social relations of the trade if we explore them through the relations of corals and sponges?

Four features of Klein's fieldwork suddenly took on new relevance. First, the aquarium trade in Indonesia is highly capitalized in comparison to other wildlife economies. Where a local man can go into the forest, find a snake, and sell it to a dealer, a coral harvester needs an expensive boat, difficult-to-obtain permits, diving equipment, a warehouse, and more. The coral trade is controlled by bosses, rather than being dispersed. A study of the coral supply chain from the perspective of corals and sponges requires attention to the institutional features of the political economy, rather than merely the practices of procurement, storage maintenance, and shipment.

Second, the coral harvesters are not fully proletarianized but instead rotate across many livelihood options. They are constantly looking at places and marine products not just for the job they are doing but for the possibilities of other livelihoods, which may or may not be sanctioned by their boss or the state. They slip multiple endeavors into the shadows of their coral procurement. This means that coral harvesting must always be studied in relation to the promise of the species assemblages around any given coral, even as these fluctuate by season, location histories, and even the luck of just being there.

Third, no coral maintenance infrastructures operate as closed systems. The capitalization of the trade has not resulted in technically advanced facilities; the money is needed for boats, permits, and connections. Instead, rickety warehouse buildings draw water from the closest polluted bay and return it, potentially full of disturbed biodiversity. The corals and sponges that survive this treatment must have a high tolerance for living in human messes.

Fourth, state regulators are often more interested in words than deeds. The gap between de jure and de facto in wildlife trade regulations is so wide that a study examining only laws on the books would entirely miss the actual conditions and practices of coral maintenance and shipment. A quarantine for coral shipments might just mean leaving the boxes in an office for a few days before dropping them off at the airport. It is only through immersion in the practices of the trade that a researcher can hope to enumerate the disturbance conditions through which some corals and sponges thrive, while others die.

Hitchhikers showed the social scientists new ways of understanding supply chains—no longer just as techniques of rendering disparate worlds commensurate for capital but as processes of ecological disturbance and redistribution. The supply chain was not only selecting for weedy sponges but had become a mode of dispersal, funneling them into new habitats across the world. Indeed, reimagining the coral trade through the shifting associations of sponges, humans, and corals offered our team a new set of research priorities. We determined to explore each node of the supply chain in relation to its constituent political economy as well as its personnel, its practices as well as its regulatory apparatus, its workers as well as its bosses, and its water circulation as well as its list of organisms. This is relational supply chain analysis as if hitchhikers matter.

Our team took several years to get to the point of even being able to write a research proposal together. What follows is an excerpt from one proposal in process. To appreciate it as a humanist, consider some of its goals. First, we show how approaches developed from different disciplinary homes might come together in better descriptions of the world. Our proposal brings together what we've here called disturbance taxonomy and relational supply chain analysis. It also carefully attends to the varied meanings of method, allowing them to fill each other with new potential. Second, we make relations among organisms a topic for anthropologists, even as we make global trade a topic for biologists. Anthropocene disturbance is a problem for us all, and a few weedy sponges show the way.

IS THE AQUARIUM TRADE CREATING WEEDY SPONGE INVADERS?

Our group was so curious about the sponges popping up at the distal end of the live coral supply chain that we began to ask, is the aquarium trade creating—and then spreading—weedy sponge invaders? We wondered what a hitchhiking sponge could teach us about the subtleties of Anthropocene disturbance regimes, and we wanted to better understand what kinds of organisms can survive as hitchhikers as well as how supply chains generate specific patterns of disturbance.

Hitchhiking sponges also posed bigger question about the inadvertent ways that Anthropocene disturbance generates new forms of resilience, resistance, and adaptability through intense selective pressure. When we blast a landscape or douse our hands in sanitizer, we never succeed in total destruction; what remains are those able to survive the disturbance (cf. Seeberg 2021).

Every supply chain is a machine of specific disturbance and selection; through the conditions of its ordinary functioning, it subjects hitchhikers to an iterative suite of pressures and therefore produces selective effects. In all cases, some life will make it—you could even launch a tuft of moss into the vacuum of space and safely bet that at least a few tardigrades would survive (Jönsson et al. 2008). We expect that the general result of all this pressure is to select for organisms that will tolerate fluctuating conditions that might kill many other forms of life. This is the general hypothesis we seek to test with this project.

Here is our plan in its most basic outline: using relational supply chain ethnography and molecular methods, we will characterize the disturbance of the supply chain and track which sponges are able to hitchhike successfully along the chain. To do this, we will follow coral fragments as they move across the chain from collection on the reef to a series of holding tanks to international airports to their final destination in both public and private saltwater aquariums. At each stage, we will both ethnographically study disturbance regimes produced by the supply chains and collect samples for molecular and morphological analysis. This combined approach will allow us to track coral-associated sponges, tracing which associations hold stable, which new associations emerge, and which ones are snuffed out by the stressors of transnational transport.

The potential implications of the project are several. First, we are interested in better understanding how supply chains like the live coral trade can serve as both selectors and vectors of potential marine invasive organisms. Second, we will contribute important basic knowledge to the taxonomy and systematics of sponges (Porifera)—currently among the least described major phyla of animals (van Soest et al. 2012).[2] In particular, by studying weedy and resilient sponges, we may identify new model organism species amenable to laboratory culture and research. Third and perhaps most significantly, we hope to build a more fundamental understanding of the stability and resilience of nested holobiont associations to a broad range of anthropogenic disturbances.

BACKGROUND

A cacophonous symphony of planet-crossing markets and commodity chains has rearranged ecosystems in recent decades, as biological invasions and other forms of nonnative species introductions throw food webs into chaos, reduce biological diversity, and rapidly recast ecological niches and functions in unexpected and surprising ways (Dunoyer et al. 2014; Molnar et al. 2008; Rosenfeld 2002; Ruiz et al. 1997). Although introductions and invasions have been increasingly common since the advent of European colonization and the genocidal ecological imperialism of the so-called Columbian exchange (Crosby 1986), researchers have shown that the accelerating pace of global trade since the mid-twentieth century has become the key driver behind today's rapid growth of invasions (Drake and Lodge 2004; Lockwood, Hoopes, and Marchetti 2013; Westphal et al. 2008). Indeed, human-driven trade and exchange are so effective at species introductions that we are often challenged to differentiate between "native" and "introduced" species, particularly for understudied taxa (Carlton 1996). And when examining documented extinction events, invasive species pressure is either the first or second leading cause of extinction worldwide (Clavero and García-Berthou 2005). Though no biome has been unaffected, the oceans have been particularly vulnerable, with marine biological invasions causing radical alterations to ecosystem health (Occhipinti-Ambrogi and Savini 2003; Sardain, Sardain, and Leung 2019; Ruiz et al. 1997; Ruiz et al. 2000).

Researchers have shown how anthropogenic redistributions of organisms, both intentional and unintentional, have intensified biological invasions and radically transformed coastal ecosystems (Bax et al. 2003; Carlton 1996;

Hixon et al. 2016). However, even within marine biological invasion research, the vast majority of attention is focused on introductions made through industrial shipping, including transport in ballast water and fouling on ship hulls. By contrast, very little attention is paid to the global aquarium industry, despite its demonstrated success as a vector for virulent species introductions (Padilla and Williams 2004; Rixon et al. 2005; Semmens et al. 2004; Strecker, Campbell, and Olden 2011). Indeed, the aquarium trade is responsible for one-third of the aquatic species on a list of the one hundred worst invasive species compiled by specialists (Lowe et al. 2000).

Many factors influence the ability of a nonnative species to survive and reproduce in a new environment, including the biodiversity of the new habitat and abiotic factors like temperature, but perhaps the most critical factor is what invasion biologists call *propagule pressure,* or the number of individuals introduced and the number of introduction events (Lockwood, Hoopes, and Marchetti 2013). Although invasion is more likely when the habitat is suitable to the invader, in many cases, propagule pressure may be a more significant factor; that is, a constant flow of propagules greatly increases the odds of invasion (Simberloff 2009; see also Lockwood, Hoopes, and Marchetti 2013; von Holle and Simberloff 2005).

Sponges have sometimes been considered "poor invaders" because of limited dispersal capacities as sessile invertebrates with short-lived larvae and thus have rarely been examined as potential threats for invasion (Guardiola, Frotscher, and Uriz 2016). However, recently, several well-established cases have highlighted successful sponge invasions with serious ecological implications (Ávila and Carballo 2009; Guardiola, Frotscher, and Uriz 2016; Vicente et al. 2020), with known introductions dating back decades (e.g., de Laubenfels 1950). For example, the orange keyhole sponge, *Mycale grandis,* has become a weedy sponge invader across the harbors and bays of Oʻahu and Maui in Hawaiʻi, likely introduced from its native range in the Indo-Pacific by military vessels and container ships. Thriving in this new environment, the orange keyhole sponge outcompetes local sponges and native coral species alike (Vicente et al. 2020).

This isn't surprising, given that sponges often make cunning hitchhikers— not least because they are able to regenerate from incredibly small fragments thanks to special pluripotent cells called *archaeocytes.* Even a small fragment consisting of a few cells clinging to a live coral could grow into a new adult sponge. More incredible still, even a brutally dismembered sponge is able to reaggregate its body; if you force certain species of sponge through a sieve, its individual cells are able to reconstruct a full-grown sponge like a self-assembling puzzle (Curtis 1962). These points indicate that even the mechanical removal—such as scraping with a knife—common within the coral trade might actually *increase* sponge presence in the supply chain through dispersal of fragments.

Sponges are both incredibly diverse and incredibly understudied; with approximately nine thousand described species and vast numbers remaining undescribed (van Soest et al. 2012).[3] Like corals, sponges themselves make their living in association with a suite of microorganisms, which may compose up to 40 percent of the sponge's biomass (Taylor et al. 2007). Indeed, sponges host a bewildering diversity of microbial symbionts that likely play roles in immune response, nutrient cycling, photosynthesis, and other roles, though very little is known about these symbionts (Hentschel et al. 2012). Moreover, some species of sponges are tolerant of fluctuating environmental conditions, which would increase their likelihood of successful establishment in new habitats (e.g., Webster et al. 2011).

Given these factors, along with the fact that a single sponge fragment could found an entire population through regeneration, and asexual and sexual reproduction, we find a worrying confluence of factors that could make sponges resilient hitchhikers and potential invaders. Accordingly, we expect to find high numbers of cryptic sponges traveling along with live corals, a hypothesis corroborated by our own preliminary observations of nineteen cases of "outbreaks" of unintentionally introduced sponges in three public aquaria in Europe and the United States. We hope that this project will not only help us better understand sponges as potential invaders and hitchhikers but also demonstrate the broader unexpected ecological impacts of global trade, and of the aquarium trade in particular, as a potential pathway for marine biological invasions.

METHODS AND EXPERIMENTAL DESIGN

It takes a rethinking of method to track how the supply chain generates disturbance and transforms coral–sponge associations in the live coral economy. In the rethinking, we hope to recombine ethnographic and biological methods, bringing together relational supply chain analysis with DNA metabarcoding and morphological analysis to understand how sponges are transformed in the crucible of the supply chain.

As we follow corals, we are interested in tracking the diverse forms of association and relation among corals, sponges, and humans. For example, at the open-water reef sites of coral collection, histories of human use and anthropogenic destruction matter as much as quantitative data on depth and water temperature. In this framework, biological methods like high-throughput DNA metabarcoding become tools of social analysis, making visible the forms of relation and association that surround corals and sponges. Likewise, stalwart methods of anthropology, such as participant observation, become important sensors for analyzing disturbance regimes. Building from this understanding, our group has put together a design to both research the disturbance of the supply chain and track coral–sponge associations in flux as they follow the many curving paths from reef to tank.

Our relational supply chain analysis means following corals with special attention to the interplay between anthropogenic conditions and the multispecies assemblages of corals, sponges, and other organisms. What matters here is not simply passive movement through disparate nodes but the ways transportation-as-disturbance reshapes the potential forms of relation between these multispecies assemblages. Toward this end, our approach is to track coral holobionts and their movement from the moment they are plucked from the reef, dropped into a basket, stashed in a holding tank, and packed into small baggies for transport, recording data on environmental conditions, including temperature, light, and salinity, as well as carefully recording structural details, such as diagraming the flow of water through holding tank facilities or the length of time spent in transit. All these details will matter for understanding the regimes of disturbance to which the corals are subjected and for later interpreting the molecular data we produce.

Critical to understanding these disturbance regimes are the confluence of cultural, economic, and historical forces from which the supply chain emerges. For instance, long-standing processes of proletarianization and dispossession have driven millions of people from subsistence coastal livelihoods into market-based relationships, providing abundant labor reserves made up of skilled waterfolk; yet these marketized livelihoods also build from culturally specific frameworks of patronage in which laborers are indebted to their bosses, creating cycles of exploitation that drive collection practices.

Simultaneously, the same processes of industrial coastal development that have severely impacted nearshore coral reefs have also degraded the quality of water recirculating through coral holding tanks, producing a kind of doubled disturbance that may further impact nested holobionts. Our approach is to work to understand how these forms of disturbance emerge from social, cultural, and historical patterns to shift holobiont assemblages.

But what about "method" in the biological sense? How to harness the tools of natural science to follow and tell these hitchhiker stories? To get there, our attention to disturbance regimes will be paired with molecular methods, including DNA metabarcoding, at each stage in the supply chain. Because organisms regularly shed DNA into the surrounding environment, including into the water column, researchers are able to detect the presence of hundreds of species through these remnant molecular fragments. By sequencing relatively short, highly conserved regions of DNA or even the whole mitochondrial genome, researchers are able to develop "barcodes" that correspond to specific species. Many of these barcodes are already publicly available in online genetic databases. However, because much marine life— especially microscopic invertebrates and microorganisms—remains undescribed, in our work, we will also likely be developing new barcodes for yet undescribed species. This method of DNA metabarcoding will give us a strong sense of the number of different species within different taxonomic groups present at each stage of the supply chain.

At each site, water samples will be sampled in triplicate on sterile filters for DNA metabarcoding. Following collection, samples will be stored in buffer at subzero temperatures until further analysis. Research by Collins et al. (2018) has estimated a forty-eight-hour window for the degradation of extraorganismal DNA beyond detection in marine aquarium environments. This implies that eDNA samples collected at intervals of forty-eight hours or greater would provide a relatively high-fidelity snapshot of living species present at time of sampling. By repeating this process at each stage of the supply chain, we will be able to observe how the coral holobiont shifts in response to the disturbances of the supply chain, with particular focus on sponges.

Because holobionts are composed of all types of organisms, including bacteria and eukaryotes, we will conduct DNA metabarcoding analysis using appropriate bacterial and eukaryote primers in order to characterize both the coral microbiome and the extended holobiont. We expect the extended coral holobiont and the microbiomes of corals and sponges to become increasingly uniform along the steps in the supply chain and, at the final stage of the trade, to consist mainly of robust and generalist organisms that can withstand changing environments. Although here we will focus on sponges, our pilot studies indicate that the list of species that are part of the coral and sponge holobionts would include a range of crustaceans, halacarids, annelids, gastropods, bivalves, and nematodes, as well as bacteria, protists, fungi, and various algae. It is possible that such an analysis could identify species that have already become invasive, thus providing a potential explanation of transport and introduction. Metabarcoding of water samples will give insight into which part of the supply chain coral and sponge holobionts could gain and/or lose associated organisms and symbionts.

In addition to sampling water, when sponges are observed on the substrate of sampled corals, specimens will be fixed for standard taxonomic identification and ultrastructural studies following the protocols from Wörheide and Hooper (1999) and Willenz and Hartman (1989). We will identify hitchhiking species of sponges based on morphological characters, including spicule morphology and body architecture (Hooper and Van Soest 2002), combined with DNA barcoding (Wörheide, Erpenbeck, and Menke 2007). Sponges are poorly known taxonomically, and we expect to find several new species. In the aquaria, the final stage of the trade, we expect to find many tropical sponges native to the Indo-Pacific, indicating a high probability of introduction as hitchhikers.

Through this unusual combination of relational supply chain analysis and disturbance taxonomies, we hope to better understand the supply chain as a machine of disturbance and to identify what forms of life are selected for and distributed. We hope not only to shed light on supply chains as selectors and vectors of weedy species dispersal but also to model what is possible for collaboration between social and natural sciences as we work to understand the innumerable ways that Anthropocene processes are reshaping multispecies interactions on our planet.

A SLOW CONCLUSION

Hitchhikers in global commodity chains are part of a structure of perpetual invasion by which ecosystems are subjected to endless pressures from weedy invaders. This is a legacy of European colonialism, the hegemony of free trade, and neoclassical economies of scale that promote a planetary economic order organized around shipping anything and everything—a fantasy land where production is imagined to be materially decoupled from place. While circuits and patterns of trade have linked the world for centuries or millennia, the differences today are speed, distance, volume, and the unrelenting and ever-increasing frequency by which organisms are moved. Changing the story requires a fundamental rethinking of how we make and consume things.

Indeed, hitchhikers have remade our world—but how does one know, see, and study them? We argue that doing so requires the tools of both social and natural science—here both anthropology and biology—and the invention of new tools to be shared between them. We have tried to show some of the emergent approaches and methods that have allowed us to follow just one kind of hitchhiker—the humble sponge. In particular, we turn to relational

FIGURE 9.5. Unknown anthozoan polyp growing on orange sponge retrieved from a coral sample. "And these have smaller still to bite 'em, / And so proceed ad infinitum." Photograph by the authors.

supply chain analysis and disturbance taxonomies as two ways that have helped us get inside the problem of hitchhiking sponges by focusing on the relations between sponges, their hosts, and the anthropogenic disturbance of the supply chain. These tools work because they recognize the divergent ideas of what counts as "method" between different disciplines, and they incorporate both. They create a new ground of shared assumptions from which unstable objects like "hitchhikers" come into view.

However, this is all much easier to say than to do. Here we have tried to show the reality and the promise of slow work in common; three key things kept us going. The first has been a willingness to continue working together *without results* as we debated our focus, followed tantalizing leads, and searched for that set of questions that might best respond to our respective expertise. The second has been an openness to the serendipities and surprises of research and the fortitude to follow these surprises, allowing them to transform the project into something new. The third has been genuine curiosity about worlds beyond the purview of our own training and experience, worlds that we have tried to open up for one another through years of friendship and collegial patience.

An inconvenient conclusion of ours is that these collaborations must be "slow," in the sense advocated by Isabelle Stengers (2018). For too long, the hypercompetitive academy has told us that every moment spent on a project without generating and publishing results, no matter how shallow or rushed, is a moment wasted. Critical inquiry, ethical collaboration, community work: all must yield to the compulsion to produce. Slow science is a modest way to push back against the hallucinatory worship of progress and productivity, against the relentless and dogmatic paradox that we have forever been running out of time.

A call to slow down can feel indulgent or wasteful in a moment when the world burns—yet this is exactly when careful attention, listening, thought, and arts of noticing become obligations, responsibilities. In a world where markets have conditioned us to speak more than we listen, without slowing down, we risk drowning in empty noise. It is at these moments that the freedom to muck around—purchased at great expense by a commitment to taking one's time—can open new worlds. In the slowness, you might even encounter a sponge busy hitchhiking the Anthropocene, waiting to take you for a ride.

Notes

This research would not have been possible without funding from the Danish National Research Foundation through the project Aarhus University Research on the Anthropocene (AURA). The authors thank Per Andersen, Annie Brandstrup, Nicklas Bisbo, Tomas Cedhagen, Sarah Bickel Flensburg, Frederik Skovby Felding, Josephine Goldstein, and Josefine Callesen Madsen for their assistance. We are grateful to our hosts in the Faculty of Mathematics and Natural Sciences at Universitas Halu Oleo University in Kendari, Indonesia, including Dr. La Ode Amaluddin, Andi Septiana, La Ode Abdul Fajar, and Sahrir. Thank you to Prof. I Gusti R. Sadimantara and Dr. La Ode Kadidae at Halu Oleo University's Office of Global Partnerships. Funding from the National Science Foundation, the Fulbright-Hays Program, and the UCSC Center for Southeast Asian Coastal Interactions was also essential. Thank you to Dr. Benny Baskara, Oce Astuti, Danial Kapitoi, and Dr. Wa Iba for assistance in Kendari. Thank you to the American Indonesian Exchange Foundation, RISTEKDIKTI, Dr. Tony Rudyansjah, and the University of Indonesia's Faculty of Social and Political Sciences for sponsoring Klein's ethnographic research in Indonesia.

1. In biology, the term *cryptic* can have several meanings. Commonly it is used as a synonym for "hidden" or "unknown," but it can also be used to describe camouflaging features, for example, "cryptic plumage in a bird to minimize predation." The concept "cryptic species" is used to describe higher than expected genetic diversity in species with similar morphology. Such undetected crypticity can confound biological research and is often only detected through analysis of molecular data.

As we use it here, *cryptic* refers both to the fact that sponges are often hidden inside the porous coral fragments and to the fact that morphologically similar sponges grouped together could turn out to be distinct species if molecular data were available.

2. The Sponge Barcoding Project is working to identify DNA signature sequences for the phylum Porifera (sponges). See https://www.spongebarcoding.org/.

3. See also the Sponge Barcoding Project, https://www.spongebarcoding.org/.

References

Appeltans, Ward, Shane T. Ahyong, Gary Anderson, Martin V. Angel, Tom Artois, Nicolas Bailly, Roger Bamber et al. 2012. "The Magnitude of Global Marine Species Diversity." *Current Biology* 22, no. 23: R996–97.

Ávila, Enrique, and José Luís Carballo. 2009. "A Preliminary Assessment of the Invasiveness of the Indo-Pacific Sponge *Chalinula nematifera* on Coral Communities from the Tropical Eastern Pacific." *Biological Invasions* 11, no. 2: 257–64.

Bair, Jennifer, ed. 2009. *Frontiers of Commodity Chains Research*. Stanford, Calif.: Stanford University Press.

Baker, David M., Christopher J. Freeman, Jane C. Y. Wong, Marilyn L. Fogel, and Nancy Knowlton. 2018. "Climate Change Promotes Parasitism in a Coral Symbiosis." *ISME Journal* 12, no. 3: 921–30.

Bax, Nicholas, Angela Williamson, Max Aguero, Exequiel Gonzalez, and Warren Geeves. 2003. "Marine Invasive Alien Species: A Threat to Global Biodiversity." *Marine Policy* 27, no. 4: 313–23.

Carlton, James T. 1996. "Biological Invasions and Cryptogenic Species." *Ecology* 77, no. 6: 1653–55.

Clavero, Miguel, and Emili García-Berthou. 2005. "Invasive Species Are a Leading Cause of Animal Extinctions." *Trends in Ecology and Evolution* 20, no. 3: 110.

Collins, R. A., O. S. Wangensteen, O'Gorman, et al. 2018. "Persistence of Environmental DNA in Marine Systems." *Communications Biology* 1, no. 185 (2018). https://doi.org/10.1038/s42003-018-0192-6.

Crosby, Alfred W. 1986. *Ecological Imperialism: The Biological Expansion of Europe, 900–1900.* Cambridge: Cambridge University Press.

Curtis, A. S. G. 1962. "Pattern and Mechanism in the Reaggregation of Sponges." *Nature* 196, no. 4851: 245–48.

de Laubenfels, Max Walker. 1950. "The Sponges of Kaneohe Bay, Oahu." *Pacific Science* 4, no. 1: 3–36.

Drake, John M., and David M. Lodge. 2004. "Global Hot Spots of Biological Invasions: Evaluating Options for Ballast-Water Management." *Proceedings of the Royal Society of London, Series B: Biological Sciences* 271, no. 1539: 575–80.

Dunoyer, L., L. Dijoux, L. Bollache, and C. Lagrue. 2014. "Effects of Crayfish on Leaf Litter Breakdown and Shredder Prey: Are Native and Introduced Species Functionally Redundant?" *Biological Invasions* 16, no. 7: 1545–55.

Guardiola, Magdalena, Johanna Frotscher, and Maria-J. Uriz. 2016. "High Genetic Diversity, Phenotypic Plasticity, and Invasive Potential of a Recently Introduced Calcareous Sponge, Fast Spreading across the Atlanto-Mediterranean Basin." *Marine Biology* 163: Article 123.

Guerrero, Ricardo, Lynn Margulis, and Mercedes Berlanga. 2013. "Symbiogenesis: The Holobiont as a Unit of Evolution." *International Microbiology* 16, no. 3: 133–43.

Haraway, Donna J. 2008. *When Species Meet.* Minneapolis: University of Minnesota Press.

Harding, Sandra. 1987. "The Method Question." *Hypatia* 2, no. 3: 19–35.

Hentschel, Ute, Jörn Piel, Sandie M. Degnan, and Michael W. Taylor. 2012. "Genomic Insights into the Marine Sponge Microbiome." *Nature Reviews Microbiology* 10, no. 9: 641–54.

Hixon, Mark A., Stephanie J. Green, Mark A. Albins, John L. Akins, and James A. Morris Jr. 2016. "Lionfish: A Major Marine Invasion." *Marine Ecology Progress Series* 558 (October): 161–65.

Hooper, John N. A., and Rob W. M. Van Soest, eds. 2002. *Systema Porifera: A Guide to the Classification of Sponges.* Vol. 1. Berlin: Kluwer Academic/Plenum.

Jarić, Ivan, Tina Heger, Federico Castro Monzon, Jonathan M. Jeschke, Ingo Kowarik, Kim R. McConkey, Petr Pyšek, Alban Sagouis, and Franz Essl. 2019. "Crypticity in Biological Invasions." *Trends in Ecology and Evolution* 34, no. 4: 291–302.

Jönsson, K. Ingemar, Elke Rabbow, Ralph O. Schill, Mats Harms-Ringdahl, and Petra Rettberg. 2008. "Tardigrades Survive Exposure to Space in Low Earth Orbit." *Current Biology* 18, no. 17: R729–31.

La Barre, Stéphane. 2013. "Novel Tools for the Evaluation of the Health Status of Coral Reefs Ecosystems and for the Prediction of Their Biodiversity in the Face of Climatic Changes." *InTech, Topics in Oceanography,* 127–55.

Lesser, M. P., Michael Stat, and R. D. Gates. 2013. "The Endosymbiotic Dinoflagellates (*Symbiodinium* sp.) of Corals Are Parasites and Mutualists." *Coral Reefs* 32, no. 3: 603–11.

Lockwood, Julie L., Martha F. Hoopes, and Michael P. Marchetti. 2013. *Invasion Ecology.* New York: Wiley.

Lowe, S., M. Browne, S. Boudjelas, and M. De Poorter. 2000. *100 of the World's Worst Invasive Alien Species: A Selection from the Global Invasive Species Database.* Vol. 12. Auckland, New Zealand: Invasive Species Specialist Group.

Margulis, Lynn. 1991. "Symbiogenesis and Symbionticism." In *Symbiosis as a Source of Evolutionary Innovation,* 1–14. Cambridge, Mass.: MIT Press.

McFall-Ngai, Margaret, Michael G. Hadfield, Thomas C. G. Bosch, Hannah V. Carey, Tomislav Domazet-Lošo, Angela E. Douglas, Nicole Dubilier et al. 2013. "Animals in a Bacterial World, a New Imperative for the Life Sciences." *Proceedings of the National Academy of Sciences of the United States of America* 110, no. 9: 3229–36.

Mol, Annemarie. 2002. *The Body Multiple: Ontology in Medical Practice.* Durham, N.C.: Duke University Press.

Molnar, Jennifer L., Rebecca L. Gamboa, Carmen Revenga, and Mark D. Spalding. 2008. "Assessing the Global Threat of Invasive Species to Marine Biodiversity." *Frontiers in Ecology and the Environment* 6, no. 9: 485–92.

Occhipinti-Ambrogi, A., and D. Savini. 2003. "Biological Invasions as a Component of Global Change in Stressed Marine Ecosystems." *Marine Pollution Bulletin* 46, no. 5: 542–51.

Padilla, Dianna K., and Susan L. Williams. 2004. "Beyond Ballast Water: Aquarium and Ornamental Trades as Sources of Invasive Species in Aquatic Ecosystems." *Frontiers in Ecology and the Environment* 2, no. 3: 131–38.

Pettay, D. T., D. C. Wham, R. T. Smith, R. Iglesias-Prieto, and T. C. LaJeunesse. 2015. "Microbial Invasion of the Caribbean by an Indo-Pacific Coral Zooxanthella." *Proceedings of the National Academy of Sciences of the United States of America* 112, no. 24: 7513–18.

Rixon, Corinne A. M., Ian C. Duggan, Nathalie M. N. Bergeron, Anthony Ricciardi, and Hugh J. Macisaac. 2005. "Invasion Risks Posed by the Aquarium Trade and Live Fish Markets on the Laurentian Great Lakes." *Biodiversity and Conservation* 14, no. 6: 1365–81.

Rofel, Lisa, and Sylvia J. Yanagisako. 2018. *Fabricating Transnational Capitalism: A Collaborative Ethnography of Italian-Chinese Global Fashion.* Durham, N.C.: Duke University Press.

Rohwer, Forest, Victor Seguritan, Farooq Azam, and Nancy Knowlton. 2002. "Diversity and Distribution of Coral-Associated Bacteria." *Marine Ecology Progress Series* 243 (November): 1–10.

Rosenfeld, Jordan S. 2002. "Functional Redundancy in Ecology and Conservation." *Oikos* 98, no. 1: 156–62.

Ruiz, Gregory M., James T. Carlton, Edwin D. Grosholz, and Anson H. Hines. 1997. "Global Invasions of Marine and Estuarine Habitats by Non-indigenous Species: Mechanisms, Extent, and Consequences." *American Zoologist* 37, no. 6: 621–32.

Ruiz, Gregory M., Paul W. Fofonoff, James T. Carlton, Marjorie J. Wonham, and Anson H. Hines. 2000. "Invasion of Coastal Marine Communities in North America: Apparent Patterns, Processes, and Biases." *Annual Review of Ecology and Systematics* 31, no. 1: 481–531.

Sardain, Anthony, Erik Sardain, and Brian Leung. 2019. "Global Forecasts of Shipping Traffic and Biological Invasions to 2050." *Nature Sustainability* 2, no. 4: 274–82.

Seeberg, Jens. 2021. "Bombarding Microbial Life with Antibiotics Creates an Explosion of Drug Resistance." In *Feral Atlas: The More Than Human Anthropocene*, edited by Anna L. Tsing, Jennifer Deger, Alder Keleman Saxena, and Feifei Zhou. Stanford, Calif.: Stanford University Press. http://www.feralatlas.org/.

Semmens, Brice X., Eric R. Buhle, Anne K. Salomon, and Christy V. Pattengill-Semmens. 2004. "A Hotspot of Non-native Marine Fishes: Evidence for the Aquarium Trade as an Invasion Pathway." *Marine Ecology Progress Series* 266: 239–44.

Simberloff, Daniel. 2009. "The Role of Propagule Pressure in Biological Invasions." *Annual Review of Ecology, Evolution, and Systematics* 40: 81–102.

Stengers, Isabelle. 2018. *Another Science Is Possible: A Manifesto for Slow Science*. Hoboken, N.J.: Wiley.

Strecker, Angela L., Philip M. Campbell, and Julian D. Olden. 2011. "The Aquarium Trade as an Invasion Pathway in the Pacific Northwest." *Fisheries* 36, no. 2: 74–85.

TallBear, Kim. 2017. "Beyond the Life/Not-Life Binary: A Feminist-Indigenous Reading of Cryopreservation, Interspecies Thinking, and the New Materialisms." In *Cryopolitics: Frozen Life in a Melting World*, edited by Joanna Radin and Emma Kowal, 179–202. Cambridge, Mass.: MIT Press.

Taylor, Michael W., Regina Radax, Doris Steger, and Michael Wagner. 2007. "Sponge-Associated Microorganisms: Evolution, Ecology, and Biotechnological Potential." *Microbiology and Molecular Biology Reviews* 71, no. 2: 295–347.

Todd, Zoe. 2016. "An Indigenous Feminist's Take on the Ontological Turn: 'Ontology' Is Just Another Word for Colonialism." *Journal of Historical Sociology* 29, no. 1: 4–22.

Tsing, Anna Lowenhaupt. 2015. *The Mushroom at the End of the World: On the Possibility of Life in Capitalist Ruins*. Princeton, N.J.: Princeton University Press.

Tsing, Anna, Jennifer Deger, Alder Keleman, and Feifei Zhou, eds. 2021. *Feral Atlas: The More-Than-Human Anthropocene*. Stanford, Calif.: Stanford University Press. http://www.feralatlas.org/.

van Soest, Rob W. M., Nicole Boury-Esnault, Jean Vacelet, Martin Dohrmann, Dirk Erpenbeck, Nicole J. De Voogd, Nadiezhda Santodomingo, Bart Vanhoorne, Michelle Kelly, and John N. A. Hooper. 2012. "Global Diversity of Sponges (Porifera)." *PLoS ONE* 7, no. 4: e35105.

Vicente, Jan, Andrew Osberg, Micah J. Marty, Kyle Rice, and Robert J. Toonen. 2020. "Influence of Palatability on the Feeding Preferences of the Endemic Hawaiian

Tiger Cowrie for Indigenous and Introduced Sponges." *Marine Ecology Progress Series* 647 (August): 109–22.

Von Holle, Betsy, and Daniel Simberloff. 2005. "Ecological Resistance to Biological Invasion Overwhelmed by Propagule Pressure." *Ecology* 86, no. 12: 3212–18.

Webster, Nicole S., Emmanuelle S. Botté, Rochelle M. Soo, and Steve Whalan. 2011. "The Larval Sponge Holobiont Exhibits High Thermal Tolerance." *Environmental Microbiology Reports* 3, no. 6: 756–62.

Westphal, Michael I., Michael Browne, Kathy MacKinnon, and Ian Noble. 2008. "The Link between International Trade and the Global Distribution of Invasive Alien Species." *Biological Invasions* 10, no. 4: 391–98.

Willenz, Philippe, and Willard D. Hartman. 1989. "Micromorphology and Ultrastructure of Caribbean Sclerosponges: I. *Ceratoporella nicholsoni* and *Stromatospongia norae* (Ceratoporellidae: Porifera)." *Marine Biology* 103: 387–402.

Wörheide, Gert, Dirk Erpenbeck, and Christian Menke. 2007. "The Sponge Barcoding Project: Aiding in the Identification and Description of Poriferan Taxa." *Porifera Research: Biodiversity, Innovation, and Sustainability* 28: 123–28.

Wörheide, Gert, and John N. A. Hooper. 1999. "Calcarea from the Great Barrier Reef. 1: Cryptic Calcinea from Heron Island and Wistari Reef (Capricorn-Bunker Group)." *Memoirs of the Queensland Museum* 43, no. 2: 859–91.

CHAPTER 10

Anthropological Sensations

A High Arctic Travelogue

KIRSTEN HASTRUP, JANNE FLORA,
and ASTRID OBERBORBECK ANDERSEN

This chapter takes off from a collaborative research project, involving anthropologists, biologists, and archaeologists in Northwest Greenland (2014–18). The ambition was to explore the relations between animal resources and human communities around the North Water Polynya in Avanersuaq (the Thule region) in a long-term perspective. In summer 2014, twelve members of the research group traveled together through the landscape in a collaborative effort to understand it (Figure 10.1). Distinguishing analytically between *tracks, trails,* and *traces,* this chapter examines our way into a physically and intellectually demanding landscape of joint research. *Tracks* are seen as tangible remains of past history, as in ruins, artifacts, records, or landscapes that we may tread. *Trails* are identifiable routes taken by humans or animals in a particular terrain, which we may map in various ways. *Traces* are ephemeral imprints of humans and animals in a volatile landscape, inviting inquiry into ways of knowing.

Before we can move into the region, we shall briefly present it. In the High Arctic, the changing seasons are very pronounced. During the long winter, temperatures drop down to between minus twenty and forty degrees Celsius. The land- and seascape lay frozen, and precipitation comes in snow. When the sun returns on the horizon in mid-February, after months of absence, the temperature may still drop. From late April, the sun no longer sets, yet temperatures still stay far below the freezing point. Sometime in May, temperatures occasionally rise above freezing, and ice and snow start melting, if never disappearing from sight, as glaciers, inland ice, and huge icebergs are still visible all over the region. In June, the sea ice begins to fill with puddles, the small cracks that could be spanned by sledges now become impassable, and sledges are replaced by motorboats. In late June or July, the ground on land becomes soft (at least at the upper layer), grass and mosses green the landscape, and most days, the temperature stays above the freezing point. Precipitation may come as rain, the sea ice on fjords and sounds breaks, and the water opens for boating. In October, the sun sets again, polar darkness creeps back in, and moving around becomes increasingly difficult.

The seasons set the frame for movement within the region and demand particular attention to the shifting materialities of the landscape and to clothing and footwear, again each having its history, spatiality, and practicality. We use this rather facile observation as a takeoff into reflections on some of the challenges that we met with in an interdisciplinary research project in the Avanersuaq (Thule) region in Northwest Greenland, inhabited by some 750 people, mostly Inughuit hunters and their families, but also people from outside with other local functions. The project gathered archaeologists, biologists, and anthropologists in a concerted effort to study the dynamic relations between the living resources and the hunting practices around Pikialasorsuaq (the North Water Polynya), an arctic oasis where multiple species thrive and upon which human and other life in the region depends (Hastrup, Grønnow, and Mosbech 2018). Interdisciplinarity was seen as key to new knowledge, agreed to be necessary in times of rapid changes—both in ecosystems and in society—yet as it happened, there was some discord regarding the larger vision of our work; the "empirical" had very different portents, even as we found ourselves sharing the landscape. In this chapter, we investigate some of the productive frictions between ways of thinking and practicing our sciences throughout the duration of the collaborative research project. In the first field season (2014), a group of twelve traveled together in the landscape to learn from each other, while also looking ahead to future cross-fertilization of concepts and insights.

While sharing the landscape on the surface of it (literally), we rarely shared *sensations*, seen as a kind of bodily perception and implicit understanding of the surface upon which we walked. The disparity was spurred by the points of departure in distinct disciplines, each with its own call for results. We were literally disciplined to take different positions in the environment even as we moved around together. While we had hoped to cultivate the differences productively by sharing our insights, this proved more difficult than expected, because whatever common goals we might have had, the methods remained distinct and not readily scalable. It was difficult, for instance, to attune to the trusted techniques for measuring and documentation of which biologists and archaeologists make use; while easy to understand as such, these techniques never got beyond the surface—literally. Although we would often make truth claims beyond positive knowledge, as we gathered sensations along the paths pursued—whether by travel or conversation—these were not always appreciated. In the travelogue presented here, we reflect on our experiences in the landscape, and in the conclusion, we return to further discussion of interdisciplinarity. We would like to admit from the outset that despite sincere aims of collaboration, engagements may turn out as incompatible—or at least surprisingly difficult to combine into one coherent picture (see Hastrup, Flora, and Andersen 2016). In this chapter, we therefore set out to trace the productive overlaps and challenging disparities between the ways in which our three fieldworking disciplines came to know environmental

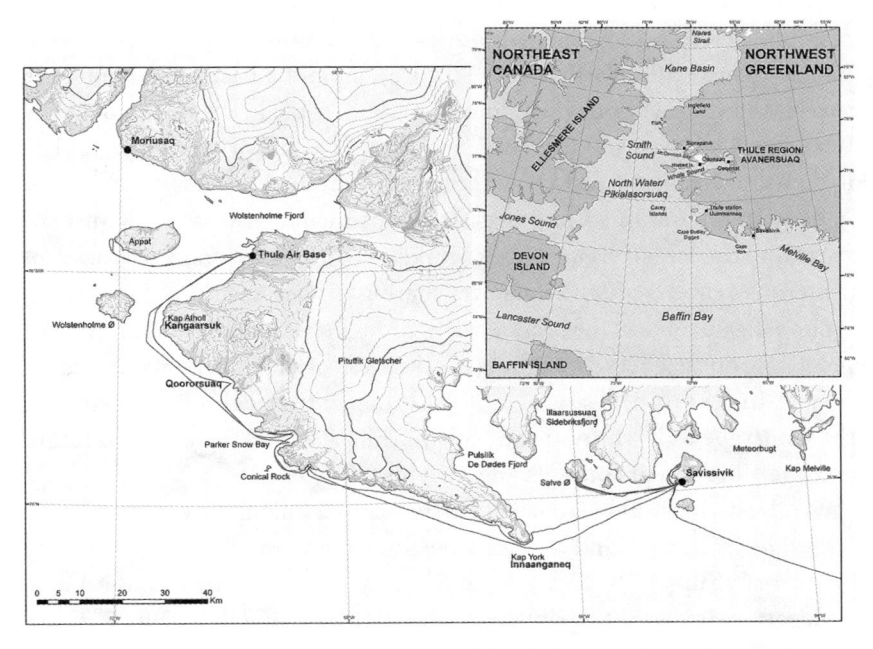

FIGURE 10.1. The region and sailing route of the fieldwork of 2014. Map and route produced with the sailing log. Printed with permission, The Now Project.

relations through different methods. Modes of sensing—through a variety of devices or through the disciplined human body—are crucial in this regard.

Tracks: Falling into Step

While moving north along the coast from Innaanganeq (Cape York) in our expedition vessel, we had what was possibly the most productive meeting of the disciplines. We were able to land in a number of sloping valleys that were only accessible from the sea, as they were closed off by the glacier on top. One of these valleys was Little Auk Valley, so named by the biologists and full of promise to all of us; its official name is Qoororsuaq (Large Valley). Two of the biologists had been there before to study the little auks, but this time, two paleoecologists had been invited along to take a closer look at the layers of turf and soil between the grass and permafrost to make a time estimate for the making of soil in the first place. It was summer, the valley was beckoning, and the landing from our inflatable motorized dinghies in the middle of the light summer night was promising, if daunting. We were not to camp there, so we could travel light in our almost identical hiking boots and with our hiking gear, including food, caps, VHF radios, and always some rifles, should one meet a polar bear or be attacked by a muskox. Safety measures dictated that no one should walk off on their own, so we divided into parties—each with a rifle. If anything, rubber boots methods in this case were equally distributed between the disciplines; they were formatted by the specific landscape we were to share with our collaborators from other disciplines.

Once we had safely landed on shore, we divided into two parties. One of us (J.F.) stayed back with the paleoecologists at the bottom of the valley, where the turf should be turned up and samples of the different layers taken, and to assist one of the biologists in locating and retrieving their photo traps, which recorded annual arrival and departure of little auks. Elsewhere, the paleoecologists would take out whole peat cores, which would eventually be analyzed in Germany for precise dating of individual layers. This would determine the beginnings of life in the valley—and as the procedure was repeated in other places, it eventually gave substantial evidence that the region had opened up for life some forty-five hundred years ago (Davidson et al. 2018)—when the last ice age retracted a little, even in the High Arctic. There was (and still is) a glacier on top of this valley, now retreating at an alarming pace along with the ice cap and reminding us of the landscape as a living force. Some forty-five hundred years ago, the coastal lands were apparently laid bare, allowing little auks to move in. Their droppings fertilized the valleys, and soon foxes entered to prey on the small birds, as did (and do) the seagulls and hawks (Mosbech et al. 2018). As grasses grew, life-forms multiplied, and at some point in time, humans entered from what today is known as Canada, maybe following in the steps of muskoxen and finding a landscape with easy access to food in the months of little auks. There was plenty of water in the stream running down through the valley from the glacier on top, and meltwater was trickling through the grassy slopes, often affecting our stride in other valleys on the coast; this added to the peculiar sense of collaborative moments (see Hastrup 2018a).

Two of us (A.O.A. and K.H.) chose to accompany biological and archaeological colleagues on a walk up through this valley. This time it was also populated by a flock of some sixty muskoxen—descendants of a small herd of seven animals translocated from Kangerlussuaq (Søndre Strømfjord) in West Greenland to the Kangaarsuk (Cape Atholl) area in 1986 (Boertmann et al. 1992). The walk up through the valley on grassy paths under which meltwater trickled was an extraordinary experience. The swirling birds, the small foxes, the running water, and the huge and enigmatic muskoxen, with whom we had to negotiate our path, made a lasting impression on us. The more-than-human world, in which we became immersed, changed us, and our running conversation with our colleagues from archaeology and biology made us "see" stones, grasses, animals, and even humans in a new way. We knew from our reading of older ethnographies on the region that elderly people would sometimes sojourn here to fend for themselves during summer, while younger and more agile relatives would attempt more demanding journeys and hunts elsewhere on the coast.

One of the biologists, who had spent earlier summers in the valley studying the little auks and setting up the photo traps, knew the location of the long-abandoned shelters, and the archaeologists in the party immediately recognized them for what they were. Seeing the actual remains of the tiny stone shelters, in use until the mid-twentieth century, when patterns of settlement and mobility changed in the region due to the placement of the American airbase in mid-district, located in the upper valley close to the stream of clear, running water, and now coming to *know* the place, was edifying. The light gear that these early, elderly dwellers would have carried with them—nets to catch birds, spears for fish, some cooking utensils—we mostly had to imagine, but that was easy, once we were there. The more-than-human literally materialized as we walked up through the valley, and so did the more-than-anthropology—sustained by archaeological and biological knowledge and fertilized by our own encounters and conversations with inhabitants in the settlements, who rarely came into the valley now because it had become inaccessible by sledge owing to the dwindling ice. Yet earlier human activity was still palpable, and for a time, it connected the disciplines.

As Lee and Ingold (2006, 67) suggested, "walking does not, in and of itself, yield an experience of embodiment, nor is it necessarily a technique of participation." Both embodiment and participation presuppose some kind of attunement to a particular landscape, and this was part of our shared experience in the Little Auk Valley. We were not only walking together; we also felt that we were heading in the same direction. In this particular landscape, we fell into step with each other across disciplines, as we met with past and present, with muskoxen and soaked turf. We were literally in it together; and we looked alike—in a uniform Arctic dress code and with waterproof leather hiking boots. Between us, we sensed the particular "ecology of materials," where materials are defined not as inert and passive materiality that is acted upon, but rather as "*matter* considered in respect of its occurrence in processes of flow and transformation" (Ingold 2012, 439). In a High Arctic ecosystem, as the one described here, different materials still reside and circulate, each offering different possibilities that come to be realized in relation to others, be they humans, animals, weather, ice, footwear, scientific instruments, or data—retrieved from a photo.

The main point here is that in this valley (and elsewhere along the coast), collaboration brought us at least temporarily on the same footing, as we sensed it. The fact that all three disciplines are field sciences (at least partially) and the fact that we walked together made a new relation emerge. Different advanced technologies also mediated the collective understanding of the place, via our natural science colleagues in particular. Paleoecologists came to know the landscape and its dynamics through peat cores that were extracted with drills to be carefully shipped to Denmark, then to Berlin for analysis. The biologist of spatial ecology and food webs visited the valley to extract the memory cards of on-location cameras, programmed to take photographs all year round to register the arrival, population dynamics, and departure of little auks. The archaeologists drew remarkably detailed sketches of the arrangement of rocks in tent rings and fireplaces, measuring them and registering all human-made structures with GPS. These practices of sensing and registration affected our own ways of seeing the landscape and generated a fertile field of reflection. It could be said that a truly "collaborative moment," enabling us to see the place through several registers at the same time, occurred (see Hastrup 2018a).

Although the bulk of the traveling on the inaugural North Water expedition was done by ship, whenever we walked through landscapes to identify and register tent rings, soot houses, meat caches, and old traps in abandoned settlements, or climbed talus slopes to put data loggers on little auks, or to witness local procedures for capturing them in their nets, it gave us a sense of heading in the same direction and of participating in each other's disciplinary practices. Walking together was a determining factor in enabling participation and sensing the landscape. The Little Auk Valley served up tangible remains of past habitation, mostly temporary; the human and non-human tracks through the landscape were evident, including the tracks of introduced muskoxen, disturbing thousands of years of moss layers of the landscape and turning these into soil and grass. Our anthropological sensitivity expanded as we fell into step with our colleagues from the other disciplines, working our way from the sea upward toward the ice-clad hinterland along the tracks of bygone (human) dwellers and present (animal) occupants. Here the tracks crossed each other, and the disciplinary methods that would often separate our thinking met and connected us.

The walk in the valley and the thoughts it engendered made it increasingly obvious that engaging with other disciplines enhanced our understanding of multispecies landscapes; we also learned that the distinct disciplinary languages and ambitions were not readily scalable. Yet, the disjunction itself was productive in that it underscored the dynamics of multispecies landscapes and opened new vistas. In the vein of Donna Haraway (2008, 244), it showed how the "open beckons; the next speculative proposition lures; the world is not finished." In the High Arctic, hunting, eating, and clothing may be read in different ways, but they all testify to deep interspecies relationships; these are not given but historically crafted. As Anna Tsing (2015, 142) suggested, interspecies relations draw evolution back into history because they depend on the contingency of encounter. They do not form an internally self-replicating system. Instead, interspecies encounters are always events, "things that happen," the units of history. Events can lead to relatively stable situations, but they cannot be counted on in the way self-replicating units can; they are always framed by contingency and time.

Our encounters with the many species and histories in the Little Auk Valley (and beyond) became even more complex as our conversations incorporated three kinds of thinking. It is unclear to us how much our colleagues saw our comments and observations as a contribution to *their* disciplines, but we felt surely on the track of deep knowledge. We could easily see how people had lived there; the affordances offered up by the landscape were tangible, and events of interspecies encounters were still forthcoming.

To end this section on tracks, as we choose to call them here, we shall move from their past remains on land to their being found in cyberspace. In a recent article, Flora and Andersen (2017) pose the question "Whose track is it anyway?" with reference to the specific collaborative endeavor within the North Water Project, involving themselves, two biologists, a GIS expert, and seventeen occupational hunters from the region who agreed to use specially programmed GPS devices to track their hunting routes, register their animal sightings and catches, take photographs, and so map their hunting activities for a full year. The collected material would show the whens and wheres of particular game and hunting events. The result is fascinating and exceptionally rich in documenting the tracks of both humans and their prey and many aspects and details of hunting that neither anthropologists nor biologists had foreseen, such as the preparation of hunting equipment, the role of the family in hunting and food-sharing practices, or the intricacies of women's work (see Andersen, Flora, and Johansen 2017; Flora et al. 2018). But the question still lingers: whose track is it? While scientists (including anthropologists) devised the project (known as Piniariarneq, or "Hunting Trip"), translated the results, and made them visible on maps, the hunters had made the trips, hunted the animals, and sent their pictures.

Complex collaboration is one thing, the question of "whose tracks" another. "Who owns the data, and who is to ensure that data is not abused after the

FIGURE 10.2. Far beyond *anthropos*. A muskox skull left from the past in the Little Auk Valley (Qoororsuaq). Photograph by Astrid Oberborbeck Andersen.

collaboration has ended, and the data is published? Who is accountable when the data circulates into realms over which neither we nor our biologist or hunter collaborators have any control?" (Flora and Andersen 2017, 113). This is where we inevitably lose track and are left with the imprint of sensation.

TRAILS: ON FOOTING WITH WALRUS

Trails do not have to be permanently visible in the landscape to be lasting. As Aporta (2009) argues, trails in the Arctic are nonpermanent features of the landscape. Trails disappear when the sledge tracks are covered by snow after a blizzard and when the ice melts at the end of each spring. Spatial itineraries, however, remain in people's memory and materialize again when the next trailbreaker makes the trip (Aporta 2009, 131). We extend this notion to the tagging of walrus with satellite transmitters to follow their migration during summer; the satellite offered a temporary trail, a kind of memory, provided the transmitters did not fall off the walrus, as they eventually would. This is where we shall start, again allowing the interdisciplinary to surface in practice, in this version, however, with additional disciplines accompanying the making of new knowledge.

In 2015, two of us (J.F. and A.O.A.) joined one of our biologist colleagues on a walrus-tagging excursion. He had worked in the region for decades and specialized in monitoring sea mammals, tracking their migration routes and foraging patterns with satellite transmitters. Walrus have inhabited Smith Sound and Pikialasorsuaq for millennia. Among biologists, much effort has gone into surveying and estimating the current and historical abundance of Atlantic walrus in general and the Baffin Bay stock in particular (Born, Heide-Jørgensen, and Davis 1993; Stewart, Kovacs, and Acqaurone 2014). To make abundance estimates, biologists engage with walrus in various ways: conducting aerial surveys and counts, collecting samples for studies of age distribution and reproduction, using information on numbers of catches and losses, and applying methods of population modeling. These estimates feed into procedures of assessing population size and dynamics and, crucially, of making the scientific recommendations to politicians whose task it is to implement quotas that allow a sustainable hunt. Since then, from 2017, the walrus hunt has been certified as sustainable in Smith Sound. The point is that the biological surveys become part of the political domain of decision-making and management. This implicates the biologists and affects their relationship with the hunters. Through the practices of tagging and counting, biology and implementation become companions and collaborating disciplines; in the process, walrus are made to move along another kind of trail, where they emerge as a different form, with particular kinds of values and qualities ascribed to them, as species, as stock, as quota, or as number (Andersen, Heide-Jørgensen, and Flora 2018).

The walrus-tagging expedition was part of the project's effort to learn about each other's modes of knowledge production. The biologist was interested in obtaining information on walrus migration between Greenland and Canada. This is of relevance for assessing the status of the stock and its exposure to hunting and of phenological changes related to new patterns in ice cover. The anthropologists were interested in human practices and uses of walrus and hunting grounds and how Inughuit negotiate and make sense of the shifts and changes in their surrounding land-, ice-, and seascapes.

FIGURE 10.3. Hunter throwing his customized harpoon to tag the walrus with a satellite transmitter in Uummannap Kangerlua (Wolstenholme Fjord) in May 2015. In this moment, the trail of the walrus moves into a digital and biological register. Photograph by Janne Flora.

The work of tagging is carried out in the habitat of the walrus, when these animals forage on the Greenlandic side of Pikialasorsuaq, and while fjords and sounds are still covered with sea ice, around which drift ice moves, tightens, and releases with the currents and winds; this makes the work dependent on the participation of hunters from the area. Inughuit hunters are experts in traveling the land-, ice-, and seascapes throughout the year, in tracking down and approaching walrus, and in harpooning them and thus fastening the transmitter securely in their thick layer of skin and blubber. Five hunters participated in the expedition. Although they were essential for the success of the endeavor, some of them joked afterward that they got no meat out of their efforts, despite their many catches; they were paid, of course, in accordance with rates agreed with the Greenland Institute for Natural Resources. Significantly, they were the ones who had directed the biologists' attention toward the fjord in which the tagging took place, and in the process, both hunters and biologists learned more about the stock of walrus in the area.

However, their own—and their dogs'—hunger for meat was not satisfied. Having confirmed the number of walrus in this particular fjord, some of the participating hunters returned once the tagging was over—this time with the purpose to hunt walrus. Despite traveling along the same trail, and well knowing that both activities involved tracking walrus and harpooning them, the tagging of walrus differed radically from the hunting of them, not least because the hunters this time were joined by their families. The anthropologists, who did not participate in this hunting trip, later saw the trail materialize as "data," as one of the hunters had brought his GPS along to track and film the hunt—not least the hunt in which he guided his son through harpooning a walrus. This was a special event of participation in the hunt and sharing their observations.

Walrus hunting takes place in two distinct seasons, between January and June and in October through November—each season offering its own possibilities and challenges for the hunt. The seasons require particular techniques for hunting—by sledge on land-fast sea ice in winter and spring, by boat in (partially) open waters in summer, and on foot on the thin new ice in the autumn (see Hastrup et al. 2018). In accordance with both local hunting rules and the latest Greenlandic legislation for walrus hunting, walrus should be harpooned before they are shot. The harpoon is tied to a floater (*avataq*), preventing the walrus from sinking once harpooned. When caught and hauled up on ice or land, the walrus is flensed, its parts separated and shared among the participating hunters. From here the trail of the walrus traverses between people and purposes—as meat for human consumption, as dog food, as trophy, as a narrative of how the hunt took place, as a catch to be reported to the authorities, or, indeed, as samples for biological studies. This social trail of an Inughuit walrus hunt is radically different from the biological one, and between them, they make visible an anthropological trail, to which we shall return later. For the biologist, the tagging itself was only the beginning of the scientific walrus trail, as one might call it, in which he could follow the individual movements of the tagged walrus digitally, almost in real time, on a web-based platform. These would later be translated and become data in a scientific publication (Heide-Jørgensen et al. 2017).

A few weeks after the tagging project, the two anthropologists found an opportunity to join two hunters on a hunting trip to Inglefield Land in the northern part of Pikialasorsuaq. The two hunters wanted walrus meat and possibly a polar bear for both meat and skin. The journey started on dog sledge to the ice edge, where a twenty-foot boat had already been brought and had been used for narwhal hunt earlier in the spring. Sledges and dogs were left there—well fed and close to fresh water, while the human party continued northward by boat. Somewhere, halfway to our destination, two walrus emerged out of the water and were pursued, until the boat suddenly turned around: "our walrus is farther north," one of the hunters explained, in a way that cautioned us not to question how he was so sure. Thus we continued northward along the coast, passing several places that had been named after walrus body parts. Collective memory is tied to places throughout the area, and these particular places became relevant on our trail in much the same way that oral narratives may achieve their purpose and meaning in the present (Cruikshank 2000). In other words, the invisible trails of the walrus hunt are transported and stored in both memory and matter; these days, they could also be stored in GPS data, photographs, and film. Yet, Claudio Aporta's (2009, 132–33) observation still holds:

> Trails are physically ephemeral as snow tracks, and not permanent visual features of the landscape (as highways, roads and streets in other geographies). Since Inuit did not traditionally use maps, trails can only be shown through

an exploration of people's memory (both individual and collective, present and past), or through the mapping of actual travel.

With our narration of a summer walrus hunt, we want to emphasize walrus hunting as one among many possible trails that underpin hunting livelihoods in Avanersuaq. On each hunting trip, such trails are performed and reproduced. Sometimes new trails emerge as old ones disappear. Knowing where the walrus can be found, traveling to reach them, and recognizing walrus of different shapes—as living animals (potential food) and as places that are named with walrus body parts—are activities loaded with cultural significance and expertise. Identifying the walrus as a proper catch, then catching it and transforming the entire animal into parts, preparing them to become different foods and materials that are shared in the community, are practices through which the trail, memory, and, indeed, hunting as livelihood are materialized, remembered, and relevant in the present.

Returning to the anthropological trail, ours was an attempt to understand an engagement with hunters and biologists as a meeting of three "disciplines" that could be studied. We cannot overlook the differing positions of the hunters via-à-vis the scholars. In the first case, the hunters were assigned the vital position of logistics operators and harpooners, on which the biologists depended for the immediate need of tagging and producing scientific articles and recommendations; meanwhile, the anthropologists could observe, inquire, and, later, ponder. The tagging expedition occurred alongside another trail of disagreement between hunters and biologists, who, despite their overlapping interests and collaboration on various projects, fundamentally disagree about hunting quotas and the science on which these are based. Walrus tagging, including the hunters' participation, contributes to this knowledge, thus creating a mismatch in the reciprocal perceptions and positions of hunters and biologists, which in some ways reflects the mismatch between anthropology and biology as we experienced it throughout our collaborative efforts.

At the core of this mismatch is the methodological approach, which, from the biological position, enhances the position that scientific results can be yielded and extracted from their surroundings and that the experiment, using a similar blueprint, can be repeated in the future by other scientists.

Ethnographic practice and fieldwork in turn are always influenced by the anthropologists' positioning in the field, their respective outlooks, and the field and set of circumstances from which an ethnographic collaboration grows (cf. Flora and Andersen 2018). In anthropology, there is no methodological blueprint to be executed, no experiment to be repeated, and no final results to be extracted from the field. What counted to us as valuable collaboration with collaborators in different fields (scientists as well as hunters) rarely counted as little more than loose talk to our natural science colleagues.

The two episodes of walrus hunting described were pursued by hunters, but with different interests and purposes. In both cases, the boating toward walrus can be read as trails—physically and visually ephemeral, yet firmly imprinted on the memory and its unwritten maps. Though unconventional, the walrus-tagging exercise was of high interest to the hunters; not only were they well paid for their efforts but they also had a vested interest in inspecting well-known hunting grounds. The ways in which this trail is imprinted in the memory—as "data" in biology and as (possible) catch in the hunting communities—differ fundamentally. For the biologists and anthropologists, the trail also settled as memory in their observations, bodies, and notebooks. It continued "trailing" into future fieldwork moments and observations and eventually found a place as memory for the scientific disciplines through their respective publications, thus also laying out a trail for future studies and examinations to follow. Although three "disciplines"— biology, hunting, and anthropology—took part in the walrus-tagging mission, physically following the same trail, there is another sense in which the three modes of thinking followed and left behind separate trails. Paying attention to such trails, and to how they format knowledge production in different ways, we suggest, is necessary in collaborative and interdisciplinary endeavors if we are to understand how fieldwork encounters are in themselves productive of relations (beyond the "results" that come out of them). It is the privilege of anthropology that our object of knowledge can be stretched and our attention scaled according to matters of concern of different actors in the empirical contexts in which we work (Hastrup 2013). This is not the case for hunters or biologists, whose objects of knowledge interest in the field are more immediate; this makes the premises for collaboration and interdisciplinarity different. Hunters, as persons and as a discipline, are mainly interested in securing their livelihood and that of their families. As described,

a lot of knowledge and expertise go into this. The biologist, in turn, is confined to producing biological data about walrus populations, which may also
be used in monitoring and managing walrus as resources. Neither hunters
nor biologists are necessarily interested in letting go of these positions to
the extent that their collaboration negates their original purpose. Despite
the shared curiosity in walrus and exchange of information, their different
interests and epistemological properties limit the potential for a shared trail
in any collaboration. Attending to the qualities and effects of different trails
is a methodological and analytical strength of anthropology—always set on
uncovering the multiplicity of trails and their implications.

Traces: Sensations on the Way

The third anthropologist (K.H.) had conducted a number of monthlong
fieldworks in different seasons from 2007 onward. Until 2014, she had always
been on her own, so to speak, and depending entirely on the hunters' goodwill to take her out. This would imply having her on the sledge when moving about and taking responsibility for her safety when camping on the ice
edge with other hunters, and not least teaching her what to do and what
not to do on the ice. The first lesson was never to jump off the sledge until
the hunter signaled that it was OK to do so. Even if the driver jumps off to
walk in front of the dogs, this may be to make sure that the ice carries. The
newcomer cannot read the ice and may disappear through it; traces left by
others on their way toward the ice edge may look massive but never last. By
such episodes, one gradually learns that the highway of the ice is far from
solid or predictable. Sledging itself is always a kind of experiment with ice,
and one to be shared: meeting another sledge on the ice far away from the
village, one would normally stop and exchange observations on the state of
the ice, animals, and dogs. Presently, the local ice conditions are increasingly
destabilized owing to the general impact of the (Anthropocene) warming of
the sea. This, of course, is a major issue for the hunters.

Dogs deserve a particular mention; they have been essential for life in
the region since the immigration of the historic Thule-culture Inuit around
1250 c.e. (Hastrup 2017). Apart from dogs and sledges, the Inuit brought
kayaks and boats, enabling people to hunt whales and other sea mammals.
But let us concentrate on the dogs that now came in packs dragging heavy

sledges, sometimes with boats on, as we heard from recent experience. The new technologies enabled long-distance hunting on the sea ice and made quite an impression on newcomers. One of them was John Ross, the first European to break new contact with Inughuit in 1818, when he succeeded traversing Qimusseriarsuaq (the Melville Bay) after centuries of packed ice that had isolated a small group of people living in the high north. Ross was interested in their dog sledges and wanted to study the dogs in the manner of a naturalist making an inventory of life in the region. Ross (1819, 133) writes, "The dogs which are the only animals that have been domesticated by the Arctic Highlanders, are of various colours.... They are of the size of a shepherd's dog, they have a head like a wolf, and a tail like a fox; their bark resembles the latter, but they also have a howl like the former." What is striking here is the ambiguous classification of the dog; though domesticated, it still bears the mark of wildness. They howled like wolves and were not to be entirely trusted.

In present-day Avanersuaq, dogs are still very prominent. The barking, the howling, the smelling, the sudden emergence of young dogs from under the houses built on stilts reaching down in the permafrost, and the temporary placement of a whelping female dog in a special doghouse along with her pups, she being tied up, of course—all this forms an underlife in the towns and settlements. The sounds of dogs may signal approaching bears or changing weather, and they are essential for sledge driving to the ice edge by Pikialasorsuaq, again supplying lots of meat. The trace of sledges is also always a trace of dogs, and both are evidence of human passage. The more-than-human is always within sight. This also applies to the winter hunt, which is relatively restricted in the darkest months. Seals may be caught in nets, spanned out between two holes in the ice relatively close to land, from where one may see hunters go out on the ice with lamps to assess the state of the net and its possible content. Again, we note how traces may be hard to see but easily sensed. This also goes for the polar bear often heard in the vicinity of the town (or its dump)—heard by the tethered dogs, that is, and their barking sounds recognized by the people. This is when all children are called in, as is a stray ethnographer.

The sea ice is (for a while yet) the primary hunting ground, and it attracts hunters whenever it is stable enough to drag out the sledge and when the light is back. The challenge right now is the receding sea ice, making hunting on

FIGURE 10.4. Hunting camp by the ice edge in spring. Photograph by Kirsten Hastrup.

the ice more restricted and depriving hunters of one of the most cherished seasons, the spring and early summer. Whether going for *uuttoq* hunting (seals basking in the sun beside their breathing hole and approached from behind a shooting screen) or bigger game, such as narwhal or walrus, the hunter mostly directs his dogs from the sledge. He has two instruments for this: the voice and the whip. He commands, sometimes loudly, when to go left and when to go right or when to speed up and prepare for a long jump over a rift in the ice. As an untrained passenger, one gradually learns the particular calls well enough to brace oneself for the maneuver. Eduardo Kohn (2013, 144–45) speaks of a "trans-species pidgin" in the case of the Amazonian Runa with whom he worked and suggests that it both instantiates and blurs the distinction between people and dogs. This is an evocative term, and while in the Inughuit case, the pidgin may appear less elaborate than seems to be the case in Amazonas, there is definitely a kind of two-way communication going on, upon which the success of the hunt may hinge; the dogs seem to have a degree of ice literacy of their own in addition to what they are told. On top of that, some hunters may have a secret language with their dogs, and there is a rather elaborate vocabulary when it comes to bear hunting in particular.

Let us return to the ice as a surface and as part of a more-than-human world. In many ways, we perceive the ground kinaesthetically (Ingold 2010, 125). Our bodily sensations tell us about the world we are traversing; with time and experience, they set as a "muscular consciousness" (Hastrup 2018b). Although the sea ice may look smooth from, say, a plane, it is actually quite rough, having possibly refrozen after a storm that broke up the first new ice, while the drift ice from glaciers may have occasioned other irregularities. The passenger behind the hunter driving the sledge must look out, lest the feet dangling over the side should hit an ice floe. The sensation of irregularity along the path settles in the body, and after a while, one just knows when to watch out.

Over several years with many trips and camps on the sea ice, one gradually learns what is necessary for feeling at home, or at least safe, on the ice, by which the hunters live (or have lived until recently). The anthropologist also learns to hear the sounds that accompany the drive, be they a conversation between hunter and dogs, cracking in the ice, calving glaciers, or swirling birds, who are known to congregate over the open water, to where one may be heading. It is a particular kind of wayfaring, which in ice-clad regions is very much a matter of weaving a strand of movement into the apparent stillness of an unlimited world (see Ingold 2010, 128). Dogs and sledges may leave unmistakable impressions on the ground, at least when the snow has newly fallen or the route is just opening up for passage due to changed ice conditions, but even they are ephemeral.

Although one may know a promontory or an island just from the shape of it, the road toward it is never straight. One must decode small topographical cues offered by winds and birds and by people one meets. The point is that one is not so much moving *upon* the ground as making a way *within* a comprehensive world of substances, weather, and wind. The traces of past movement may stand out, but not necessarily for long. Ingold (2007, 43) has suggested that a "trace is any enduring mark left in or on a solid surface by continuous movement." This fits the traces on the ice (as well as in the grassy valleys in the region), where human traffic in terms of just six to eight hunters heading in the same direction makes a veritable (if short-lived) "sledge road," as when Pikialasorsuaq becomes accessible in spring and sea mammals are rumored. In this part of the Arctic, where movement depends on sledges, the traces left stand out as lines in the snow. As footprints of human movement, they are far from permanent, however. Ingold (2010, 129) writes about such footprints,

> Yet precisely because soft surfaces do not readily hold their form, footprints tend to be relatively ephemeral. Snow may be covered by further falls or may eventually melt away, sand may be sculpted anew by the wind or washed by the tide, mud may be dissolved by the rain, and moss or grass may grow over again. Footprints thus have a temporal existence, a duration, which is bound to the very dynamics of the ground to which they belong: to the cycles of organic growth and decay, of the weather, and of the seasons.

This point makes a lot of sense in the Arctic, where, indeed, the snow and the ice are likely to have either melted or been covered by new snow before the next movement. Yet, the cover is stable enough in certain periods for the ephemeral impression of the singular sledge to become one of many, the route becoming a trace in the ground. Traces are cumulative only until the ice melts or a snowstorm covers them up. One could extend this ephemerality to anthropological knowledge more generally, seen as "fluffy" by one of our biological collaborators, who would have liked to organize a shared database, but also acknowledged by ourselves as never cast in stone—in contrast to archaeology, for instance—but recalled in sensations.

The preceding deliberations on traces serve not only to highlight their temporary nature but also to underscore the multiple nonhuman forces that impinge upon fieldwork and the sense of place. It does not consist in a simple conversation between anthropologists and their interlocutors, however important that is. It is also a conversation with a landscape that always "entails a relationship between the foreground and background of social life" (Hirsch 1995, 3). This is the stuff of anthropological sensations— embodied multispecies and cross-cultural histories that settle in one's outlook on the world, an outlook that again is always more than just seeing. Anthropologists in the field take part in a long history of life in the region and in an unbounded collective, not restricted to the present, but also including bygone times and other scholars. This sense of participation is possibly what distinguishes anthropologists from other scientists, while connecting them closer to the people living there and offering up their lifeworlds in practice and in conversation—sometimes seeing their world in a new way themselves. In doing so, the hunters (and their families) become sophisticated partners in fieldwork.

CONCLUSION: OUT OF THE COMFORT ZONE?

The anthropologist in the field—any field, not only in the Arctic—is part of a larger collective. Everything—animal, herbal, historic, prehistoric, material, watery, slippery, storied, or silenced—plays into the anthropological sensation of the field. The emphasis may be individual and take shape from particular interests, but the field itself, as lived and understood, is inadvertently more-than-human and cannot be presented as one integrated, long-lasting place. Although we do not want to reduce anthropology to sensation, we have wanted to underscore the fact that ethnographic knowledge itself emerges as relational. It is built on meetings and movements with people in the chosen field, conversations across different disciplines, knowledge registers, and personal interests. When we highlight "sensation," it is not a simple matter of seeing or sensing something. Nor is it a matter of feeling. It is an acknowledgment of the interplay of forces, materialities, temporalities, collaborations, and intellectual ambitions that become "known" through their embodiment in the anthropologist as fieldworker. This interplay is what makes the anthropological field, even as it includes other points of view—including companion scholars from biology and archaeology, in our case. Biologists and archaeologists as persons also sense the landscape, of course. But as we experienced, biological sensing becomes valid mainly when it is carried out by sensing devices that augment the biological sensing body: satellite transmitters and wildlife cameras. Concerning the archaeologists, their sense of the field was closely related to visible structures of a certain age. On the strength of our anthropological and interdisciplinary experiences, we would like to suggest that our field is always "more-than-anthropological." Anthropological thinking depends on collaboration with people who think about the world through other registers—be they based in other experiences or in other scientific fields.

To conclude, our experience in the Northwest Greenlandic landscape underscored the historical, political, and ecological connections across different times and scales and made us realize the degree to which anthropology already engages with multispecies landscapes and a plurality of temporalities and viewpoints. In the particular collaborative effort that framed the research project, our sensation of the landscape and its more-than-human inhabitants was greatly enhanced. Our working together, and to some extent sharing notes with our colleagues, in addition to participating in local hunters' work and outlooks, enhanced our understanding of the living landscape, emerging from moving ice, immigrant birds, and imported muskoxen—to mention just a few of the elements that made the place. For us, the anthropologists, the temporary sharing of the ethnographic field did not simply lead to common sensing but also made us realize that, between us, we saw the landscape, its inhabitants, *and* ourselves differently, if still recognizably. This was part of the pleasure of walking together and of sharing the anthropological sensation—including the sense of enrichment that was owed to the cross-disciplinary conversations that the North Water Project entailed, if brought to bear differently on the individual disciplines.

NOTE

We thank our colleagues on the NOW-Project (http://www.now.ku.dk/) as well as our collaborators and interlocutors in Avanersuaq, without whom our research would not be possible. The NOW-Project (2014–17) was funded by the Carlsberg Foundation and the Velux Foundations; we gratefully acknowledge their contributions. Thanks also to the editors of this volume for poignant and productive comments on the first draft of the chapter, enhancing our argument.

REFERENCES

Andersen, Astrid O., Janne Flora, and Kasper L. Johansen. 2017. *Piniariarneq: From Interdisciplinary Research towards a New Resource Management.* Aarhus, Denmark: University of Copenhagen and Aarhus University. http://now.ku.dk/documents/Rapport_Tresprog_FiNAL_WEB.pdf.

Andersen, Astrid O., Mads Peter Heide-Jørgensen, and Janne Flora. 2018. "Is Sustainable Resource Utilisation a Relevant Concept in Avanersuaq? The Walrus Case." *Ambio* 47, no. 2: S265–80.

Aporta, Claudio. 2009. "The Trail as Home: Inuit and Their Pan-Arctic Network of Routes." *Human Ecology* 37: 131–46.

Boertmann, David, Mads Forchammer, Carsten Riis Olesen, Peter Aastrup, and Henning Thing. 1992. "The Greenland Muskox Population Status 1991." *Rangifer* 12, no. 1: 5–12.

Born, E. W., M. P. Heide-Jørgensen, and R. A. Davis. 1993. *The Atlantic Walrus (Odobenus rosmarus rosmarus) in West Greenland.* Copenhagen: Museum Tusculanum Press.

Cruikshank, Julie. 2000. *The Social Life of Stories: Narrative and Knowledge in the Yukon Territory.* Vancouver: UBC Press.

Davidson, Thomas A., Sebastian Wettereich, Kasper L. Johansen, Bjarne Grønnow, Torben Windirsch, Erik Jeppesen, Jari Syväranta et al. 2018. "The History of Seabird Colonies and the North Water Ecosystem: Contributions from Palaeoecological and Archaeological Evidence." *Ambio* 47 (suppl. 2): 175–92.

Flora, Janne, and Astrid O. Andersen. 2017. "Whose Track Is It Anyway? An Anthropological Perspective on Collaboration with Biologists and Hunters in Thule, Northwest Greenland." *Collaborative Anthropologies* 9, no. 1–2: 79–116.

Flora, Janne, and Astrid O. Andersen. 2018. "Taking Note: a kaleidoscopic view on tor, or three, modes of fieldnoting." *Qualitative Research* 19, no. 5: 540–59.

Flora, Janne, Kasper Lambert Johansen, Bjarne Grønnow, Astrid O. Andersen, and Anders Mosbech. 2018. "Present and Past Dynamics of Inughuit Resource Spaces." *Ambio* 47, no. 2: S244–64.

Haraway, Donna J. 2008. *When Species Meet.* Minneapolis: University of Minnesota Press.

Hastrup, Kirsten. 2013. "Scales of Attention in Fieldwork: Global Connections and Local Concerns in the Arctic." *Ethnography* 14, no. 2: 145–64.

Hastrup, Kirsten. 2017. "Dogs among Others: Inughuit Companions in Northwest Greenland." In *Dogs in the North: Stories of Cooperation and Co-domestication,* edited by Robert J. Losey, Robert P. Wishart, and Jan Peter Laurens Loovers, 212–32. London: Routledge.

Hastrup, Kirsten. 2018a. "Collaborative Moments: Expanding the Anthropological Field through Cross-Disciplinary Practice." *Ethnos* 83, no. 2: 316–34.

Hastrup, Kirsten. 2018b. "Muscular Consciousness: Knowledge Making in an Arctic Environment." In *Pre-textual Ethnographies: Challenging the Phenomenological Level of Anthropological Knowledge-Making,* edited by Tomasz Rakowski and Helena Patzer, 116–37. London: Sean Kingston.

Hastrup, Kirsten, Janne Flora, and Astrid O. Andersen. 2016. "Moving Facts in an Arctic Field: The Expedition as Anthropological Method." *Ethnography* 17, no. 4: 559–77.

Hastrup, Kirsten, Bjarne Grønnow, and Anders Mosbech. 2018. "Introducing the North Water: Histories of Exploration, Ice Dynamics, Living Resources, and Human Settlement in the Thule Region." *Ambio* 47: 162–74.

Hastrup, Kirsten, Astrid Oberborbeck Andersen, Bjarne Grønnow, and Mads Peter Heide-Jørgensen. 2018. "Life around the North Water Ecosystem: Natural and Social Drivers of Change over a Millennium." *Ambio* 47 (suppl. 2): 213–25.

Heide-Jørgensen, Mads Peter, Janne Flora, Astrid O. Andersen, Robert E. A. Stewart, Nynne H. Nielsen, and Rikke G. Hansen. 2017. "Walrus Movements in Smith Sound: A Canada–Greenland Shared Stock." *Arctic* 70, no. 3: 308–18.

Hirsch, Eric. 1995. "Introduction. Landscape: Between Place and Space." In *The Anthropology of Landscape: Perspectives on Place and Space,* edited by Eric Hirsch and Michael O'Hanlon, 1–30. Oxford: Clarendon Press.

Ingold, Tim. 2007. *Lines: A Brief History.* London: Routledge.

Ingold, Tim. 2010. "Footprints through the Weatherworld: Walking, Breathing, Knowing." *Journal of the Royal Anthropological Institute* 16, no. 1: S121–39.

Ingold, Tim. 2012. "Toward an Ecology of Materials." *Annual Review of Anthropology* 41: 427–42.

Kohn, Edouardo. 2013. *How Forests Think: Toward an Anthropology beyond the Human.* Berkeley: University of California Press.

Lee, Jo, and Tim Ingold. 2006. "Fieldwork on Foot: Perceiving, Routing, Socializing." In *Performing Fields on Shifting Grounds: Locating the Field—Space, Place and Context in Anthropology,* edited by Simon Coleman and Peter Collins, 67–86. Oxford: Berg.

Mosbech, Anders, Kasper L. Johansen, Thomas A. Davidson, Martin Appelt, Bjarne Grønnow, Christine Cuyler, Peter Lyngs, and Janne Flora. 2018. "On the Crucial Importance of a Small Bird: The Ecosystem Services of the Little Auk *(Alle alle)* Population in Northwest Greenland in a Long-Term Perspective." *Ambio* 47 (suppl. 2): 226–43.

Ross, John. 1819. *Voyage of Discovery, made under the orders of Admiralty, in his Majesty's ships Isabelle and Alexander, for the Purpose of Exploring Baffin's Bay, and inquiring into the probability of a North-West Passage.* London: John Murray.

Stewart, Robert E. A., Kit M. Kovacs, and Mario Acqaurone. 2014. "Introduction: Walrus of the North Atlantic." *NAMMCO Scientific Publications* 9: 7–12.

Tsing, Anna Lowenhaupt. 2015. *The Mushroom at the End of the World: On the Possibility of Life in Capitalist Ruins.* Princeton, N.J.: Princeton University Press.

CHAPTER 11

Becoming Disturbed
in Disturbing Landscapes

*Methodology and Epistemology in
Anthropocene Wastelands*

MEREDITH ROOT-BERNSTEIN, FILIPPO BERTONI,
NATALIE FORSSMAN, *and* KATY OVERSTREET

Y ou will need rubber boots to get through the Anthropocene, because as you zigzag between infrastructures of accumulation and extraction—archives and industrial farms, mining pits and dumps—you are always sinking in and losing your footing. Go to the field with a backpack large enough to hold all the appropriate types of footwear. A thermos is also recommended. At first, during our time with Aarhus University Research on the Anthropocene (AURA), rubber boots emerged as an eminently practical concern (Figure 11.1). This was because of the particular territory of our field site, which had attracted our attention in the first place: its history of extraction had left behind a scarred landscape made of sand dunes, waterlogged open pits, and geological instability, a landscape where we literally got lost, sidetracked, and stuck in the mud time and again. Rubber boots allow you to venture into vague and ambiguous territory: waterlogged soil, flooded ditches, boggy expanses with unknown capacity to bear weight. They also are the vectors of a strong grip that the mud exerts on you, sucking you into its logic and its immediacy, getting you stuck. The Anthropocene that you find in apparently innocuous places like the middle of Jutland may look boring, but it is not: it is an unstable and shifting terrain, a Bermuda triangle of collapsed intentions and constant reconfigurations, out of which spills a ruderal growth of unearthed histories, complex feelings, and strange observations.

Here we reflect on one of the core conceptual and methodological tools with which we grappled throughout the AURA project: disturbance. Disturbance, in the ecological sense, is an event that transforms and rearranges biotic material. Disturbance is something that humans do, which contributes to structuring the world (Root-Bernstein and Ladle 2019). Here we propose that disturbance can be understood not only as an unreflective transformation of the environment but also as a deliberate, deliberative method for experimenting on, playing with, and learning about the affordances and structures of the world.

As we see it, methods are not simply practices that take place somewhere called the "field." Methods enact the dialogue between empirical or

FIGURE 11.1. Coauthor Meredith Root-Bernstein wearing rubber boots at BKL field site. Photograph by Meredith Root-Bernstein.

phenomenological being-in-the-world and epistemology. We play with our idea of disturbance as method by acknowledging that disturbance as method can be disturbing—it can, and even should, disturb our ideas of what methods are and our existing methodological practices. We bring this meta-disturbance to the fore by structuring our chapter like a series of ecological disturbances—a discontinuous set of patches of altered affordances and surprises within a background matrix of case study presentation and analysis. Our first patch is about deer and the deer-hunting forest structure. Our second patch is about uncovering paleohistory. Our third patch takes place at a pig farm. It may be too early to see everything that will flourish in these patches, but we give some indications of what our disturbing and being disturbed generate.

THE BACKGROUND CONDITIONS

In this chapter, we ask both how disturbance shapes multispecies landscapes and how to ourselves *be disturbed in our methodological expectation*—to cultivate the capacity to overcome the banal (Swanson 2017) or "mild" (Brichet 2016) character of the Anthropocene apocalypses. As Bubandt, Andersen, and Cypher describe in the Introduction to this volume, the former brown coal mines have a long history of disturbance in the ecological sense; Søby was the site of a major extraction project especially from World War I to the 1970s. By 1970, as mining became unprofitable, the area was left as a wasteland, scarred by deep, sandy pits that quickly became waterlogged and unstable. Not only did mining processes disrupt and make anew this swath of central Denmark but landownership, ecological succession, cultural heritage, and sociotechnical relations also overturn the territory, in fundamentally banal, everyday forms (Swanson 2017). None of this made BKL, as we called the field site following a Danish term for brown coal fields, particularly special.

One of our explicit methodological goals was to understand BKL as an Anthropocene site from a multispecies perspective, which implies disturbing disciplinary boundaries. However, what this meant in the field, in practice, and as method was unclear: How to abandon an anthropocentric stance, and how to study other species, without just actually doing ecology? Did we need to develop totally original field practices? Methods in the absence of an interpretative framing have no meaning: a key question for us was thus where to find a framework of original meaning making for our field practices.

Here we use the ecological disturbance concept to make sense of what we were doing. An attraction of the disturbance concept is that it is regularly used in the ecological literature to mean both bad, destructive things that humans do to ruin landscapes and life on Earth and natural events with a range of unique outcomes, including increasing biodiversity (see Root-Bernstein and Ladle 2019). It thus operates across the ecological divide between the "natural" and the human, the good and the bad, inherent in the debates on the Anthropocene that our group was trying to navigate. Disturbance ecology can be simultaneously used to resist and reinforce the idea that humans are somehow external to nature, the sole and unique cause of the order of the world (Willig and Presley 2018). In what follows, we interrogate this claim of human power and sole causality not in the general but in the specific, asking whether our own particular practices of disordering and rearranging, observing and sensing, are the sole determinants of what we learn from methods. We claim that methods are not only the outcome of human research practice in the field but interactions co-constructed by multiple actors and multiple species.

Patch I: Human–Deer Behavioral Disturbances

Multiple times across the years, we (N.F., M.R.-B.) booked a university car, gathered available AURA colleagues, agreed on who was both willing and legally allowed to drive, booked a local hotel, and made the several-hours' drive out to BKL. We stopped along the way to buy a large amount of food at a supermarket and sandwiches at a bakery. Arriving in the late afternoon or evening, we unpacked the car, installed ourselves, and set about cooking, drinking, and discussing the research project. In the morning, we put on our warm, waterproof outdoor clothes and our hiking or rubber boots and packed our backpacks with thermoses, notebooks, and cameras. We drove to particular bits of the forest. Some of us had a better mental map of BKL than others; it was a relatively uniform landscape, and somehow it was easy to get discombobulated on the few roads. We once got the van stuck in some mud. We walked through the landscape observing random things, followed deer tracks and trails, and tried to imagine spaces that would be interesting to deer, discussing concepts of deer interestingness among one another. We placed camera traps and track-ing cards (thick paper with charcoal bands so that animals crossing them would leave charcoal footprints). To place camera traps effectively, we had to locate appropriate positions, distances, angles, and heights, to capture frequent deer activity. We attempted to choose promising positions mainly by placing the cameras near deer trails and other animal tracks or other sites of deer activity, such as the edges of clearings and places within forest cover with bark stripping. Once we had decided where to put the cameras, we moved like deer to check that the angles and heights of the cameras were adequate for register-ing animal movement. When we chose places to put tracking cards, we also speculated about what the tracking cards looked like to the deer and other animals and whether they would step on them.

These may resemble scientific practices, but they lack the structure of scien-tific methods. They are not anthropological methods either. Is it an art project? Is it a children's game? What can these practices possibly be good for?

In addition to our seemingly inexplicable and playful modes of discuss-ing, tramping, and placing objects across BKL, on several trips structured like the generic field trip described in the preceding paragraphs, over sev-eral years, we walked around with and talked to hunters. The red deer hunters at BKL with whom we spent time had a far deeper experience with materi-alities and temporalities of disturbance. Hunters and landowners spoke of relations among the smells and sounds of human bodies, the machinery of vehicles and weapons, the vegetation of closed forest and open clearings, and the sensory acuity of deer and their social arrangements through the seasons. As we walked and spoke with them, they frequently justified their quiet steps through the forest, their long waits in their hunting blinds after a shot was fired, or their vehement resistance to mushroom pickers enacting their legal right to forage along the roads of their property holdings in terms of a strict imperative not to "disturb the deer."

Primatologist Thelma Rowell (1987) argues that particular environments suggest and guide us toward watching and understanding more-than-human socialities through the environment's material and sensory affordances. Rowell reflects on the intimate zigzagging between the structure of meth-ods, the affordances for knowledge that methods render available and select among in the world, and thus the epistemologies to which methods contrib-ute. But, in a multispecies move, she sees this zigzagging as something that other species are also doing. For example, in her reflections on the intersections between multispecies ecologies and the humans who enter these ecologies as behavioral observers, Rowell observes how red deer in the Scottish moorlands "allowed the first modern study of mammalian breeding behavior" (Clutton-Brock, Guinness, and Albon 1982) because of "the excellent visibility in the moorland habitat," which allowed researchers to observe social relations among deer in the wild with unimpeded vision (Rowell 1987, 656). She notes, however, that Clutton-Brock and coauthors express doubt as to whether the mating system they observed and recorded in such detail exists in red deer's original forested habitat of woodlands or thickets (656). In other words, both what human observers understand about being deer and what deer understand about being deer are shaped by the open moorland. Closed forest habitats may allow deer to be deer differently but also condition the affordances available to others for learning about what being deer means in such habitats. One of the big, but perhaps underappreciated, intellectual disturbances to ecology and biology has been the recognition that deer are many things and that this variance represents alternative, nongenetic livelihoods, strategies, and modes of life (Hebblewhite and Merrill 2009; Putman and Flueck 2011). Rowell (1987) helps us understand that methodological limitations on observation mean ecologists only see some of the ways of being deer.

The deer we were able to observe with our field methods seemed to be ephemeral beings well versed in avoidance (Forssman and Root-Bernstein 2018). While we could detect their trails, and scattered traces of spoor, tracks, disturbances, and browsing activity, the deer never left tracks on the tracking cards, and neither did any other animal. On our camera traps, the observed deer were flickers, fragments, and vague smudges in the dark. If we were hardcore positivist empiricists unable to take a science and technology studies perspective, we could conclude one of two things: (1) deer at BKL are a kind of ghostly being or (2) our methods are a failure as they did not render deer as we had been told deer should appear to us, either by hunters or by biological literature. Before coming back to think about this perplexing choice of conclusions, let us look at a third set of disturbances that we observed in the hunting forest.

The land managers at BKL wished both to shape the landscape into a dichotomized set of clearings and forests to shape deer behavior for optimal hunting experiences (Forssman and Root-Bernstein 2018) and to produce timber. But the deer, and the way the trees responded to deer behaviors, did not always cooperate and often disturbed the well-laid plans of the hunters and land managers. In winter 2015, some members of the AURA "red deer team"[1] walked through Fasterholt plantation with Bo, a land manager for two of the largest landowners in the area.[2] Bo described his methods for managing vegetation in Fasterholt as we walked. In particular, we discussed the tensions of managing Fasterholt both for production forestry and for wildlife. Bo and other land managers around BKL were interested in wildlife primarily for hunting, not for biodiversity as such.

Bo led us along a human-made ditch that ran in a straight line through the coniferous plantation forest. Bo stopped to point to some trees where the bark had been stripped, in some cases around the entire tree. Much of the bark stripping was at eye level, but some was further down the tree. Bo told us that this area of the forest had enough damage from deer that it was no longer profitable to harvest the trees because they could be used only for wood chips. He looked around and said, "You can't find a tree that is not damaged," to emphasize the extent of deer bark stripping, browsing, and antler rubbing in this area.

Earlier in the day, Bo and Mikkel, a professional forester who also worked with the Fasterholt landowners, had shown us an area where young trees had been heavily browsed and stripped by red deer. They had pointed to a tree that had been stripped of bark. On the tan of the bare tree was a large black spot. Mikkel described how deer browsing and bark stripping made the young trees vulnerable to fungal rot.

In some cases, deer disturbed trees so much that they would kill trees in an area and prevent new growth, creating clearings, even while deer preferred the cover of dense forest. We were shown this by Bo and Mikkel, but it is also standard ecological knowledge: browsers, especially if the density is relatively high, can damage adult trees, slow woody plant growth, create dwarfed shrub habits, or reduce seedling establishment. This kind of ecological explanation, however, depends in no way on any imputation of behavioral motivations and, even less, affect. The only feeling that has worked its way into aspects of the animal–plant interaction literature is fear of predators (cf. the "landscape of fear"; Laundré, Hernández, and Ripple 2010). But our interlocutors disturbed this affectively narrow understanding of deer–tree dynamics, and we stepped into the gap that this opened. While landowners at BKL attributed the sometimes extensive destruction wrought on vegetation by deer as "boredom," while meandering up and down the weird pits and slopes, in and out of the quiet clearings and pine stands of BKL, we speculated that activities like bark stripping and antler rubbing might indicate that deer "preferences and desires exceed" the "monotony of lives where they feel safe and are well fed by frequently cut grass and sometimes-daily deliveries of sugar beets and oat silage" (Forssman and Root-Bernstein 2018, 81).

To attempt to find methods to allow us to interrogate this speculation, we developed the field practice described in the beginning of this patch, meshing aspects of multispecies ethnography and ecology. In a multispecies ethnography mode, we intended to use existing ecological studies of the research site to identify places that were "interesting" to the deer (see Despret 2004). Maria Dahm, a master's student in biology, had already assessed deer presence at the site through a number of indices: feces, tracks, signs of bark stripping, gnawing and biting, and camera traps. Interpreting these signs through statistical analysis, she found that deer preferred areas with higher densities of trees or tall vegetation and areas far from buildings; however, plenty of variation in evidence of deer presence could not be explained (Müller et al. 2017). Following Maria's recommendations and our interviews and interactions with landowners and land managers (Forssman and Root-Bernstein 2018), we focused on deer trails, meadows, *hochsitz* (hunting seat) sites, and forest edges. But reinterpreting these objective landscape features as "interesting places" implied tinkering with and going beyond a distanced, objective ecological fieldwork practice. To speculate about the emotional valence of a landscape, or the emotional motivations of the deer, is to invite affective empathy into a subjective framing of the research question, in an anthropological mode (e.g., Stodulka, Selim, and Mattes 2018). These interesting places would then double as quadrats for an ecological phase that would look at the collaborative or interactive place making of deer along with other nonhuman species in their sharing of habitat structures over time. That, in short, is what we thought we were doing when we were doing our field methods.

As any scientist knows, the entire intrigue of methods lies in the gap between what you think you are doing, what your readers and critics claim you were actually doing, and what the study object interprets you as doing. The evidence we obtained signals a gap—a disturbance in the mesh of knowledge—that is fertile ground for the establishment of some ruderal interpretations. We need not conclude that deer are ghosts nor that our methods are failures. The camera trap videos provided temporally acontextual snapshots that began and ended in the midst of things (in media res, like this patch). Just as the reader needs an intellectual framework to understand our methods, we needed ideas about interestingness to interpret the deer behaviors that we glimpsed acontextually.

Thus, through the very incompleteness and lack of traction of our interventions, we were able to learn something about the structure of their disturbance regime. Trees, deer, and humans are not convivially carrying out a negotiated agreement or stable "disturbance regime" of coexistence so much as strategically disturbing one another's plans, distorting, ruining, and reorganizing each other's ideas of an interesting landscape. The failure of our methods to (re)produce the expected richness of standard forms of evidence showed us how the deer disturb our plans (our methods) by having their own plans that exceed any simple account of them. We were all (deer, research team, hunters) improvising, being destructive, and generally trying to find a delicate balance between fervently attending to the other actors and scrupulously ignoring them, willing the others to render themselves comprehensible (Forssman and Root-Bernstein 2018). Our speculative, playful, subjective, tentative methods allowed us to glimpse fragments of ways and aspects of being deer that cannot be seen through standard disciplinary, objective, reproducible field methods.

Epistemological Disturbance

Disturbance is an integral concept in ecological thought, as it played an important role in the early debates about why species exist in particular combinations and structures in particular places (Johnson 2017). Early definitions relied on an intuitive idea: "disturbance was understood as a rare and transitory event that destroyed or interrupted the natural development or succession of biotic communities" (Johnson 2017, n.p.). Later definitions of disturbance more broadly included any disruption to resources, substrates, or the physical structure of the habitat (Pickett and White 1985). Species that are specifically adapted to these transitory but common conditions are known as ruderal, or, in common parlance, as weeds.

In contemporary ecological theory, disturbance is rarely explicitly linked to models of ecological change but is enmeshed in clouds of concept boxes and process arrows that cannot generate clear hypotheses or be experimentally falsified (e.g., Vellend 2020; Pulsford, Lindenmayer, and Driscoll 2016). These new theoretical syntheses are themselves resilient to epistemological disturbance: they are so complex and opaque that they cannot be disproven. They capture everything and reveal nothing. The only prediction-generating theory of disturbance remains the aging intermediate disturbance hypothesis (IDH; Grime 1973; Horn 1975; Connell 1978; Figure 11.2). This graphical model predicts that disturbances that are neither very small nor very large

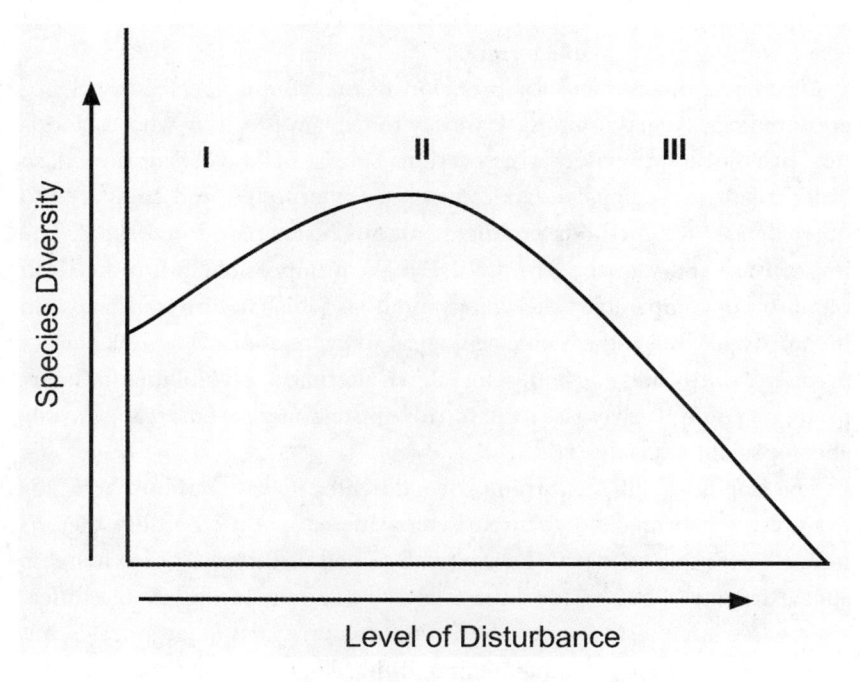

FIGURE 11.2. The intermediate disturbance hypothesis, which takes the form of this diagram. The y axis is, in this version, labeled "species diversity," and the x axis is labeled "level of disturbance." The three parts of the figure, marked I, II, and III here, refer to different outcomes: II is the famous intermediate disturbance that maximizes biodiversity. Image from Sciencerelatedusername, CC BY-SA 4.0, https://creativecommons.org/licenses/by-sa/4.0, via Wikimedia Commons.

(hence "intermediate") enhance species richness. The size of the disturbance is imputed post hoc by seeing where biodiversity peaks. This theory is generally unable to make correct predictions and thus has been challenged for decades, but without an alternative being proposed (e.g., Waide et al. 1999; Mackey and Currie 2001; Shea, Roxburgh, and Rauscher 2004; Willig and Presley 2018).

One way to frame this theoretical failure is to suggest that the range of relations and processes considered in the IDH is too diverse to successfully generalize about. More specifically, what variables are supposed to be represented on the x and y axes have been interpreted in multiple ways (see Figure 11.2; Grime 1973; Horn 1975; Connell 1978; Wilkinson 1999; Shea, Roxburgh, and Rauscher 2004; Michalet et al. 2006). What ecologists are trying to describe when they talk about disturbance is thus, in a profound sense, shifting and unstable terrain.

Of course, the demand for precision in meaning is specific to science epistemologies: we do not ask Haraway to explain precisely what is and is not "trouble" and precisely which actions consist of "staying" or how close "with" really is. Ecological data can only be interpreted, and compared to other data, if the methods producing all bits of data are isomorphic. The imprecision and vagueness of the IDF make it impossible to translate into replicable or comparable field ecology methods, which in turn translates into incomparable, unsynthesizable ecological interpretations. The IDF cannot precisely control and efficiently close down alternative possibilities for interpretation and thus gives rise to an unruly epistemological landscape unsuitable for scientific analysis.

The IDF is equally disturbing if we decenter its assumptions. Why, for example, is intermediate disturbance considered "good"? Because there is no inherent scale to the IDH, empirically small disturbances also increase species richness, but by too little to make any human-observable difference, while large disturbances also have similar impacts at large scales, but quite slowly, relative to human temporalities. There is thus a spatiotemporal framing to this concept, centered around human capacities and needs. Disturbance is best when it forms a "disturbance regime" (a pattern of disturbance as a baseline state [e.g., Cardinale et al. 2005; Clark and Johnston 2011]) within that narrow range of comportment where it contributes to the maintenance of a stable, optimized, human-scaled order. A goal of our methodological experimentations was thus to attend to and highlight that multispecies ecologies also do disturbance in all the most disturbing, non-normative ways: interactive, multitemporal and multilocated, heterogeneous, incoherent, illegible, contentious, and messy.

Patch II: Among Geo-epistemic Disturbances

BKL is full of large pits, generally squarish. They are full of water. We walk around the pits. We look for mushrooms. We sketch things. Some of the pits have sides that are very abrupt; others have slopes going down to them that can be climbed. There may be an old refrigerator next to one of them, or a scattering of pinecones, or a newt. The water may be reddish. Some have docks for boats. Some are bordered by sand, some by dense pine plantations, some by grass and paths. We talk to local people about the pits. They tell us that some have been stocked with fish. Some may be anoxic. At one point, some of the fieldwork team puts on impermeable gear and, kitted out with buckets and other scientific apparatus, wade into some shallow pit borders and take water samples. We measure something or other. We measure pH. We take photographs. This rapidly becomes intensely uninteresting. There is no question to orient our field methods, no framework on which to cobble together something that would resemble something more than anecdote.

Disturbance as method first occurred to me (F.B.)—naughtily—not while I was in my rubber boots, playing with the field methods for BKL as I was supposed to be. The idea appeared to me in my own office. One of the first documents we were given as an introduction to the BKL project was a scanned copy of a geological description of the area. This mentioned Miocene swamps, ancient flora, and geological changes in landscape, but it didn't look much further than the "facts" of science: it did not explore how these "facts" had been accumulated. As someone with a training in science studies, I found this disturbing. I looked for more connections and references to the work that grounded what we were to use to ground our AURA field site. So, disturbance appeared to me as I was digging through old scientific surveys, in the form of a grainy photo in a report addressed to the Danish Geological Society. In the photo, the exposed wall of an open brown coal pit is the main subject, occupying almost the whole scene. Framing the rock formation were, below, a raft floating on the waterlogged bottom of the pit and, above, the profiles of a few students, busy taking notes and visually joined to the raft through two long rope ladders extending along the machine-cut, steep wall of the mine. AURA was not the only project that had been attracted by the disturbances caused by the brown coal mining: in 1970 and 1971, a group of scientists and students from the Geological Institute of Aarhus University had ventured to the area before us. Their interest in the disturbance of mining was different, though. To them, the freshly dug pits of the recently abandoned mines were opportunities to easily access the sediment profiles at depth, and at no additional cost. As the two leading professors concisely put it in their annual report to the Danish Geological Society:

> field training courses have focused on the lignite beds in Fasterholt; in A/S Carl Nielsen's pit, a survey was carried out in 1970, immediately before the trench was submerged, and in 1971 the outcrops were analyzed in the partially waterlogged Damgaard's southern pit.[3] (Larsen and Friis 1973, 120)

The two sedimentological training programs, taking place as soon as the mining stopped, proved fruitful, as the exposures left by mining offered a great chance for geologists to cheaply and easily collect precious samples. The scientists involved in the training projects brought fossilized remains of plants back to the institute, where they sieved, washed, and carefully analyzed and characterized the pieces and traces of fossilized wood, pollen, seeds, fruits, and leaves they managed to collect. These trips resulted in the description of three important floras characterizing the Middle Miocene in the region: the "Damgaard," "Søby," and "Fasterholt" floras. This little, mundane, and even marginal nugget of history of science caught my attention.

As all AURA members joined in the shared experimental fieldwork in BKL, we had tried to read up on the specificities of this unconventional Danish field site. And as the geological landscape and history of the area emerged as a particularly salient characteristic of the site (as we learned thanks to the work of one of our master's students, Mathilde Højrup, on instability in BKL), I had zealously studied the documents and studies available on the geological past of the area. But these floras, and the Miocene environmental histories of which they offer glimpses, had appeared almost as mere facts, objective realities of how the territory of BKL had looked in the past. Finding out about the training field trips in the 1970s and tracing their own historical dependence on specific disturbances—in this case, the two most recent open pits, dug and abandoned at the margins of the current site of the dump—reminded me of how disturbances actively shape what we know, and what comes to count as nature, as the past, present, and future of a specific landscape. It was the disturbances caused by mining that had made possible the deep geological knowledge of the area's transformations; but it was also previous geological excavation of the area that had identified the deposits that made mining for brown coal possible.

Another, similar dynamic emerged as I followed on the development of the geological characterization of BKL's past. The flora traced through these remains was mobilized to envision other, broader disturbances that preceded the anthropocenic brown coal landscape and that sedimented into contemporary Central Jutland. The minute traces of fossilized plant remains allowed an image of the past to congeal. The North Sea, during the Tertiary, extended inland, in a large basin that opened between the vast landmass that includes the contemporary United Kingdom and the Scandinavian Peninsula and covered a large part of northern Continental Europe. The region of Central Jutland was moving between the seafloor and the coastline, following changes in temperatures and ice cover, climate, seasons, deposition of sediments, erosion, tectonic uplifts, and continental drift. These processes will have acted as disturbances to the flora and ecosystems of Central Jutland, shaping a set of environmental conditions and structures lived in, and mutually shaped by, the flora and other species, in a *longue durée* version of the same story as in patch I.

The three sets of plant remains helped characterize this multispecies détente as taking the form of a warm temperate-subtropical swampy delta, shaped by the ebbs and flows of the area during the Miocene. As leaves, wood, and other parts of the plants fell to the waterlogged forest floor, they were prevented from completely decomposing by the anaerobic and acidic conditions. With time, these partially decomposed materials formed peat, and—as it was buried under the sediments transported to the area as the coastline advanced—the peat was compressed to form the lignite or brown coal layers that characterize the area.

Following the specific disturbance of the Damgaard's southern pit, excavated in the second half of the 1960s and abandoned in 1970, other disturbances appeared, folded inside the very same pit. The ebbs and flows of the ocean, the deposition of sediments, the partial decomposition afforded by the swampy environment, the transport of plant material along with the current, the pressures of sediment turning moss into peat and peat into brown coal—all these events may be represented as disturbances to other ongoing processes that they interrupted. The traces of old disturbances bring into being new ones, and geological strata and epochs echo each other in often subtle and almost unnoticeable ways. There is no original landscape.

Notice how this claim is radically different from the IDH. The IDH claims that material-transformation events perturb a system away from a baseline because disturbance essentially wipes out all pertinent ecological conditions: disturbances take you back to zero every time. There is thus no accumulation of structure or order and no historicity or history in the IDH vision of disturbance ecology. The IDH fails to imagine a baseline that is shifting, multiple, accumulating, and interactively constituted.

This slippery and accumulative quality of disturbance interrupted my repose on the side of the field, when my boots were happily exchanged for comfortable slippers in my study as I pored over archival material. Like the geological reading of disturbances as closely interrelated and complex series of events, learning from disturbance as method in this case means attending to the specificities of how disturbance shapes the conditions of how and what we know and simultaneously of how our knowing disturbs the world constantly. Our knowledge practices cannot be separated from disturbances and their dynamics but always need to be excavated in their complex relations.

It is also the case that to find what is interesting about a field situation, one cannot politely tiptoe around; one needs to disturb arrangements, dig things up, and see what may be hidden or how the system reacts. This played out as methods when, with Colin Hoag and Nils Bubandt, we decided to focus part of our ethnographic fieldwork on AFlD Fasterholt, a recycling facility situated by the brown coal beds. Originally established in 1979 as a landfill, the dumping site—to which we familiarly referred as "the dump"—was chosen because of the devastations already brought to the area by mining (Figure 11.3). But, as we soon understood was critical, it was conditioned both on the presence of other pits in the ground and on the absence of a pit in that particular spot: a disturbance history shaping the form of another disturbance. As the Ministry of Environment stated, "the deposition of waste in the area was considered justifiable, because the groundwater and the water in the lakes and streams were already heavily polluted by chemicals following decades of brown coal digging" (Miljøcenter Aarhus 2009, 7). Furthermore, and perhaps ironically, while the disturbance of mining in the area justified the choice, the ground underneath the landfill was appropriate because it had not been excavated, and the intact brown coal and clay layers were supposed to help in preventing the leaching of pollutants from the landfill. Elsewhere (Hoag, Bertoni, and Bubandt 2018), we showed how the historical disturbance of mining animated the flourishing of marginal forms of life and life-forms at the dump. Importantly, disturbances in the dump were not just a matter of industrial archeology and environmental history: they

FIGURE 11.3.
Filippo Bertoni
at "the dump."
Photograph by
Pierre du Plessis.

were very much what the present ecologies of the dump were all about. As the site currently focuses on composting, a large part of the recycling process at AFlD Fasterholt was about controlling and managing disturbance to transform waste into nutrient-rich soil by exploiting microbial succession and taking over their metabolisms.

Wearing rubber boots can only be understood in the context of wearing slippers, and vice versa. The choice of subject (e.g., the dump), the reification of certain things as objects taken for granted (the ground, the vegetation), the interpretation of the significance of subjects and objects (what do pits tell us?), and so on are methodological acts just as much as taking water samples or sketching pinecones are. We could not ask interesting questions of the landscape that the abandoned mining pits formed, or what the absences of pits might mean, without disturbing the framing of our initial fieldwork through digging up archival material underlying the assumptions on which we based those initial field explorations. The pits were not boring, but their interestingness relative to our multispecies landscape questions was not yielded to us through naive empirical observation.

A Recent Ecological Innovation in the Concept of Disturbance

Although not rising to the level of a novel theory making testable hypotheses, one new conceptualization of ecological disturbance deserves our attention because it has implications for disturbance as method. Jentsch and White (2019) claim that disturbance events are multidimensional or multifactorial transformations of biomass and are thus always inherently pulse/disturbance events—where a pulse is the ecological term for a beneficial increase in some resource. All disturbances simultaneously provide some resources for some actors, while reducing other resources for other actors. They alter the balance of multiple resource distributions across space and time. Ruderal species, after all, do not thrive in an absolute lack of resources but are those species that are able to requalify death, disorder, and excess as "pulses" of material that can be recycled and valued. If we use Jentsch and White's pulse/disturbance conceptualization of ecological disturbance to think about disturbance as method, we can ask what intellectual resources are created, but perhaps overlooked, when we alter or reinterpret practices.

PATCH III: THE MANIFESTATION OF AFFECTIVE DISTURBANCES

In fall 2019, I (K.O.) visited a pig farm not far from BKL with my colleague Inger Anneberg. We went to the farm in the morning and arrived at just about nine, as we had agreed with the farmer. This allowed us to go to kaffepause with the farmer, Troels, and his employees. They all removed their farm clothes (coveralls) when they left the barn and took off their shoes. They came into the farmer's kitchen, and we all sat down at the table for coffee, bread, butter, jam, and liver paste. One farmworker is Ukrainian, the other Romanian. Troels took us into the barns afterward. We entered a small hallway, where we could either put on sanitary foot coverings or wear some of their rubber boots. After we put on sanitary coveralls, we went to the farrowing unit. Troels walked us through and talked about some experiments he was doing with rearing arrangements. Later, we followed workers as they did castrations (Figure 11.4).

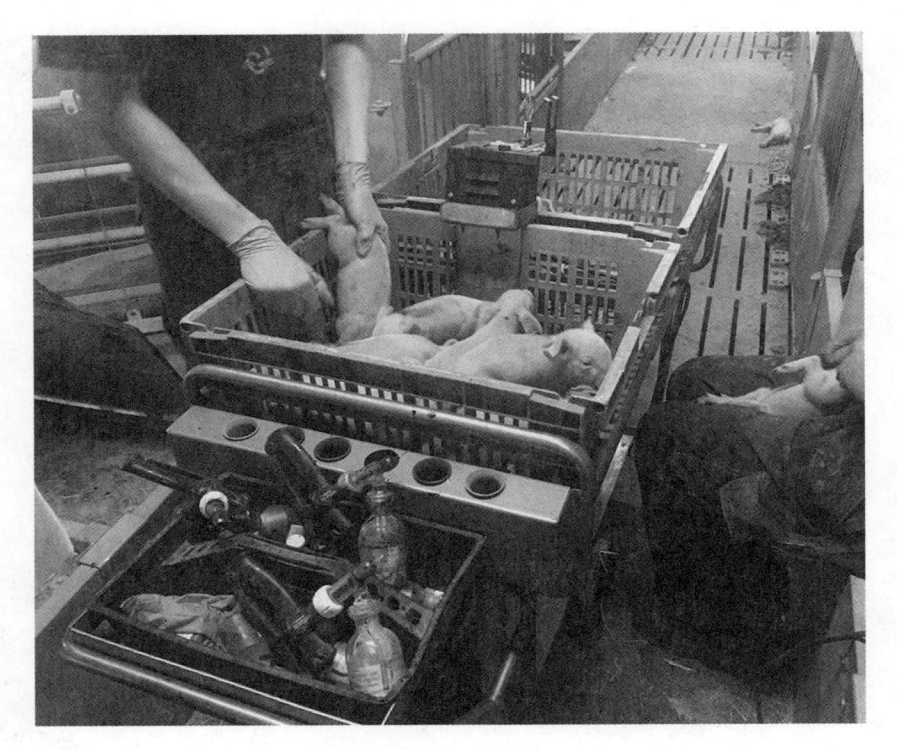

FIGURE 11.4. Observing piglet handling by workers at the pig farm. Photograph by Katy Overstreet.

The young worker from Romania allowed us to follow her as she insemi-nated pigs and then helped out in the farrowing section, where sows give birth and nurse piglets until they are old enough to be weaned. Lining the walkways between the enclosures were the dead bodies of newborn piglets. I saw that one piglet was still moving and struggling on the side of the walkway. I pointed out that the piglet was alive. The farmworker walked over and, with little surprise or change in manner, picked up the piglet, pointed out to me that the piglet's head was oddly shaped, and then, holding it by the legs, swung the piglet hard against the cement floor. She did so almost without pausing in what she was saying. She looked again at the piglet, hit it again against the floor, and then replaced it along the edge of the walkway. Moments later, she plucked a living piglet from an enclosure and put the piglet in my arms, cooing to the piglet in an affectionate way. While I held the warm and grunting piglet, she described how she loves working with the piglets because they are "very cute."

To say that I was disturbed by the disjuncture of that moment would be an understatement. But I was not disturbed by the actions of the Romanian farmworker. I knew that she was well respected by the farmer and her col-leagues for her work and her patient handling of the pigs. Indeed, her way of killing the piglet is standard in the industry and is largely held to be the humane way to kill piglets under five kilograms (Grist et al. 2018). Yet the clear lack of emotion in that act and the bucket filled with bloated piglet carcasses and flies outside the door to the farrowing unit demonstrate the normalization of this kind of death.

Attention to affect and affective disjuncture is an ethnographic field method that has gained attention in recent years (Thompson 2014; Kund-sen and Stage 2015; Stodulka, Selim, and Mattes 2018). One of the benefits of directing attention toward emotion in this way while in the field is that emotions indicate what is important, what matters, what is interesting to think about further. We have already seen this twice in the other patches in this chapter: our difficulties in interpreting deer emotions and our possible or partial failure in being interesting to deer and our boredom in an under-interpreted and undercontextualized field of "givens" consisting of abandoned mining pits. Of course, the field of the pig farm is not in BKL, and the visit was not during the BKL fieldwork period. We argue that it was, nonetheless, part of our methods of understanding BKL. It was the making-sense clue that allowed us to revisit our fieldwork practice in memory and write this chapter. If the slippers phase of methods that we discussed in patch II looks to the historical contextualization of the site and comes before (because it creates or allows) the development of interesting field methods, the sanitary-shoe-coverings phase of fieldwork comes after and allows us to reinterpret our own field methods. The Bermuda triangle of methodology, that space between what you think you were doing, what your critics claim you actually did, and what the study object interpreted you as doing, is not a fixed and eternal hole in the landscape of epistemology. One can also, and indeed perhaps must, reinterpret one's own actions in the field in light of later actions in other fields, adjusting the contours of the unknown, the unknowable, the mistaken, and the unrecognized.

Although it was unpleasant to witness the dispassionate killing of a piglet, being disturbed by death is not what we are arguing for. Instead, this aspect of disturbance as a methodology means attending to the normalization of the present to defamiliarize, contextualize, and make space for an otherwise, even while cultivating a deep understanding of how this moment and place came to be. In the case of pig farming in Denmark, the killing of piglets such as this one is connected to a national-corporate project aimed at leading the world in pork productivity.

Denmark is one of the leading exporters of pork in the world and is considered to be a gold standard in pork production in terms of environmental, animal welfare, and hygienic standards. As part of the national and industry push to lead in pork production, breeding farms in partnership with SEGES (Landbrug og Fødevarer) have developed "hyperprolific sows." Hyperprolific sows gestate more piglets than they can feed. In Denmark, this means that conventional sows give birth to an average of just over seventeen live piglets, but sows have only fourteen teats. Furthermore, many piglets are "undercooked," as the Romanian farmworker put it; larger litter sizes tend to mean that many piglets are born premature or with abnormalities that prevent them from thriving. And these piglets, as the law dictates, should be killed humanely through a hard smack on a cement floor.

This might be called the national-corporate production of vulnerability. Danish pigs, and industrially bred and enclosed pigs more generally, are vulnerable to numerous diseases and injuries. These vulnerabilities necessitate intensive forms of biosecurity that in some contexts dictate—through corporate hegemony—the social parameters of human workers' lives (Blanchette

2015). In Denmark, this vulnerability is the justification for a controversial border fence across the border between Denmark and Germany. In 2018, the Danish Parliament approved the construction of this fence to prevent the entrance of wild boars and thus African swine fever into Denmark. Denmark completed the border fence in December 2019 despite criticism by wildlife advocates and parties concerned over the implications for human migrants. In a *New York Times* article on fence construction, Bent Rasmussen, the project leader and a forest manager with the Danish Nature Agency, is quoted as describing his job in this way: "my job is to disturb the wild boar as much as possible and humans and other wildlife as little as possible" (Sorensen 2018).

Although there are no pig farms in BKL, the biosecurity needs of pig farms thus shape wildlife and ecologies of all of the Jutland peninsula, including BKL. At one point, I went on a common hunt at BKL, accompanied by another research team member, and we followed along with a man hired by the landowner to do upkeep. We were talking about eating the deer and how bad the stags taste in mating season. He mentioned that he wished there were wild boar to hunt, because they taste so good, but that the pig agricultural industry is so powerful in Denmark that it will never be allowed. The hunting forest is thus shaped by the absence of wild boar (a species moreover known for its disturbance activities that structure habitats; Barrios-Garcia and Ballari 2012), harassed and excluded from Jutland to maintain domestic pigs in a state of stabilized vulnerability.

The emotional disjuncture that calls attention to the normalization of death within the pig farming system, its precarious excess and hyper-exclusionary landscape making in relation to outside agents—farmworkers, viruses, wild boar—also calls our attention to other moments in other fields. Even though the hunting forest of BKL is a landscape arranged around death, a rich and uncontrollable life is shaping its own conditions of being and dying in the contested spaces between hunters, forest cover, and deer (Forssman and Root-Bernstein 2018). We might see killing as the act of greatest signification that structures every move, every land management method, undertaken by the hunters in the forest. But actually it is not the act of killing but the method of *hunting* (independently of the success or failure of killing; Forssman and Root-Bernstein 2018) that structures the forest and the hunters' knowledge of deer and forests. The act of killing piglets on Danish pig farms does not hold the same epistemological structuring role: there is no space (or time) between the production of an excess of piglets and humane killing methods, in which a complex landscape of desires and emotions, designs and counterdesigns of farmworkers, farm infrastructure, viruses, wild boar, and pigs can emerge.

There is, for sure, experimentation and methods development going on in the world of the pig farm, but our attention is drawn to how the possibilities and practices of their enactment are different than those in the hunting forest. Specifically, the enactment of methods in the field that do not leave a space of ambiguity between what we think we are doing (e.g., when we try to kill an animal), what the object of our method itself (the deer, the pig) thinks we are doing, and what some other actors (the trees, the wild boar) think we are doing is not generative of new understandings and ways of being. Our argument here is not specific to killing methods, and in any case, we can point to the close historical-epistemological links between methods of obtaining animals and plants for food and more general methods of learning about the world (Gell 1996; Liebenberg 2013). All methods that close down alternate possibilities through extreme efficacy do not generate the ambiguous, unresolved options and opportunities necessary for multispecies making and understanding of the world.

The sanitary-shoe-coverings phase of our methods tells us that methods that work perfectly on terrains that are stable and well defined merely reproduce and accumulate knowledge that we have already mastered and controlled, like so many pork chops. Thus what we describe doing in this chapter is methods, even if it looks like a failure of methods. It also answers our first question: to allow multispecies Anthropocene worlds to flourish, and to perceive them, we need to avoid highly effective, established, repeatable methods. The problematic ecological imprecision of the IDH is where we find the space of ambiguity to develop new interpretations. Furthermore, if we think of the pulse/disturbance concept introduced by Jentsch and White (2019), we should understand that for every upset, every failure, every boring abandoned mining pit—every discontinuity in whatever you thought was the local baseline of the world—there is a set of alternate affordances that can be reinterpreted as resources: resources for doing-otherwise and for understanding-otherwise. These are the ruderal resources that we were gathering with our undisciplinary and underdisciplined hodgepodge of methods.

CONCLUSION

In this chapter, we wanted to stay with the trouble of the rubber boots metaphor, to be troubled by its implication that methods and fieldwork are identical terms or that methods can be meaningful without reference to epistemology. We argue that disturbances to our expectations about how methods will yield insight do not just form failure or garbage but can be re-invented as intellectual resources for thinking differently. Like the impatient land manager for whom a big disturbance takes too long to return to high biodiversity, you may well ask, Where is the big intellectual payoff? What is the concrete result these vague and unconvincing methods have yielded? Overturning ecology and anthropology will not grow back as a new inter-disciplinary field within a mere five-year project. Yet BKL disturbed us on multiple levels. It disturbed us in the field, on the farm, and in the archive. It disturbed our expectations about understanding other species through observational field practices; it disturbed our perceptions of what is an inter-esting or a boring study object; and it disturbed our notions of what is upset-ting about killing animals. But we learned much more than how to observe, be interested, or be upset differently. We also learned about the relation-ship between methods that close down possibilities of being surprised and engaged by the study subjects and those that playfully open up such possi-bilities. Several years later, we can still make out the traces of these open methods in our epistemological landscapes.

NOTES

We thank all our interlocutors in the field as well as the other members of the BKL fieldwork team with whom we collaborated. Pierre du Plessis in particular assisted with fieldwork and helped shape some of our thinking. We also thank the editors for their engaging comments, which improved the chapter.

1. The red deer team included Pierre du Plessis, Natalie Forssman, Katy Over-street, and Meredith Root-Bernstein, with support from Mathilde Højrup.

2. All of the interlocutors' names in this chapter have been changed to preserve their anonymity.

3. My (rough) translation. Here is the original: "Herunder har et par feltkurser været henlagt til brunkulsgravene i Fasterholt; i A/S Carl Nielsens grav blev der således i 1970, umiddelbart før graven blev sat under vand, gennemført en opmåling, og i 1971 opmåltes lagserien i den delvis vandfyldte Damgaards sydlige grav."

REFERENCES

Barrios-Garcia, M. Noelia, and Sebastian A. Ballari. 2012. "Impact of Wild Boar *(Sus scrofa)* in Its Introduced and Native Range: A Review." *Biological Invasions* 14, no. 11: 2283–300.

Blanchette, Alex. 2015. "Herding Species: Biosecurity, Posthuman Labor, and the American Industrial Pig." *Cultural Anthropology* 30, no. 4: 640–69.

Brichet, Nathalia. 2016. "Mild Apocalypse: Feral Landscapes in Denmark." Exhibition, Moesgaard Museum, Aarhus, Denmark. https://anthropocene.au.dk/exhi bitions/mild-apocalypse-2016/.

Cardinale, Bradley J., Margaret A. Palmer, Anthony R. Ives, and S. S. Brooks. 2005. "Diversity–Productivity Relationships in Streams Vary as a Function of the Natural Disturbance Regime." *Ecology* 86, no. 3: 716–26.

Clark, Graeme F., & Emma L. Johnston. 2011. "Temporal Change in the Diversity–Invisibility Relationship in the Presence of a Disturbance Regime." *Ecology Letters* 14, no. 1: 52–57.

Clutton-Brock, Timothy Hugh, F. E. Guinness, and Steve D. Albon. 1982. *Red Deer: Behavior and Ecology of Two Sexes.* Chicago: University of Chicago Press.

Connell, Joseph H. 1978. "Diversity in Tropical Rain Forests and Coral Reefs." *Science* 199: 1302–10.

Despret, Vinciane. 2004. "The Body We Care For: Figures of Anthropo-Zoo-Genesis." *Body and Society* 10: 111–34.

Forssman, Natalie, and Meredith Root-Bernstein. 2018. "Landscapes of Anticipation of the Other: Ethno-ethology in a Deer Hunting Landscape." *Journal of Ethnobiology* 38, no. 1: 71–87.

Gell, Alfred. 1996. "Vogel's Net: Traps as Artworks and Artworks as Traps." *Journal of Material Culture* 1: 15–38.

Grime, John Philip. 1973. "Competitive Exclusion in Herbaceous Vegetation." *Nature* 242: 344–47.

Grist, Andrew, Jeff A. Lines, Toby G. Knowles, Charles W. Mason, and Stephen B. Wotton. 2018. "The Use of a Non-penetrating Captive Bolt for the Euthanasia of Neonate Piglets." *Animals* 8, no. 4: 48.

Hoag, Colin, Filippo Bertoni, and Nils Bubandt. 2018. "Wasteland Ecologies: Undomestication and Multispecies Gains on an Anthropocene Dumping Ground." *Journal of Ethnobiology* 38, no. 1: 88–104.

Horn, Henry S. 1975. "Markovian Properties of Forest Succession." In *Ecology and Evolution of Communities,* edited by M. L. Cody and J. M. Diamond, 196–211. Cambridge, Mass.: Belknap Press of Harvard University Press.

Hebblewhite, Mark, and Evelyn H. Merrill. 2009. "Trade-offs between Predation Risk and Forage Differ between Migrant Strategies in a Migratory Ungulate." *Ecology* 90: 3445–54.

Jentsch, Anke, and Peter White. 2019. "A Theory of Pulse Dynamics and Disturbance in Ecology." *Ecology* 100, no. 7: e02734.

Johnson, Edward A. 2017. "Disturbance." Oxford Bibliographies. https://doi.org/10.1093/OBO/9780199363445-0073.

Knudsen, Britta Timm, and Carsten Stage. 2015. *Affective Methodologies: Developing Cultural Research Strategies for the Study of Affect.* New York: Palgrave Macmillan.

Larsen, Gunnar, and Henrik Friis. 1973. "Sedimentologiske undersøgelser af det jydske ung-Tertiær." *Dansk geol. Foren., Årsskrift for 1972* 119: 128.

Laundré, John W., Lucina Hernández, and William J. Ripple, eds. 2010. "The Landscape of Fear: Ecological Implications of Being Afraid." Special issue. *Open Ecology Journal* 3, no. 1.

Liebenberg, Louis. 2013. *The Origin of Science.* Cape Town, South Africa: CyberTracker.

Mackey, Robin L., and David J. Currie. 2001. "The Diversity–Disturbance Relationship: Is It Generally Strong and Peaked?" *Ecology* 82: 3479–92.

Michalet, Richard, Robin W. Brooker, Lohengrin A. Cavieres, Zaal Kikvidze, Christopher J. Lortie, Franisco I. Pugnaire, Alfonso Valiente-Banuet, and Ragan M. Callaway. 2006. "Do Biotic Interactions Shape Both Sides of the Humped-Back Model of Species Richness in Plant Communities?" *Ecology Letters* 9: 767–73.

Miljøcenter Aarhus. 2009. "Afgørelse om Overgangsplan og revurdering for Affaldsselskabet Østdeponi's Affaldsbehandlingsanlæg" [Decision Regarding Transition Plan and Reassessment for Østdeponi Waste Treatment Plant]. http://afld.dk/UserFiles/file/Miljogodkendelser/Afgoerelse_om_overgangsplan_og_revuder ing_af_18_feb_2009.pdf.

Müller, Anke, Maria Dahm, Peder K. Bøcher, Meredith Root-Bernstein, and Jens-Christian Svenning. 2017. "Large Herbivores in Novel Ecosystems: Habitat Selection by Red Deer *(Cervus elaphus)* in a Former Brown-Coal Mining Area." *PLoS ONE* 12, no. 5: e0177431.

Pickett, Steward T. A., and P. S. White. 1985. *The Ecology of Natural Disturbance and Patch Dynamics.* New York: Academic Press.

Pulsford, Stephanie A., David B. Lindenmayer, and Don A. Driscoll. 2016. "A Succession of Theories: Purging Redundancy from Disturbance Theory." *Biological Reviews* 91, no. 1: 148–67.

Putman, Rory, and Werner T. Flueck. 2011. "Intraspecific Variation in Biology and Ecology of Deer: Magnitude and Causation." *Animal Production Science* 51: 277–91.

Root-Bernstein, Meredith, and Richard Ladle. 2019. "Ecology of a Widespread Large Omnivore, *Homo sapiens,* and Its Impacts on Ecosystem Processes." *Ecology and Evolution* 9, no. 19: 10874–94.

Rowell, Thelma E. 1987. "On the Significance of the Concept of the Harem When Applied to Animals." *Social Science Information* 26: 649–69.

Shea, Katriona, Stephen H. Roxburgh, and Emily S. J. Rauscher. 2004. "Moving from Pattern to Process: Coexistence Mechanisms under Intermediate Disturbance Regimes." *Ecology Letters* 7: 491–508.

Sorensen, Martin Selsoe. 2018. "Denmark Plans a Fence to Stop a Pig Disease. Will It Work?" *New York Times,* October 24.

Stodulka, Thomas, Nasima Selim, and Dominik Mattes. 2018. "Affective Scholarship: Doing Anthropology with Epistemic Affects." *Ethos* 46, no. 4: 519–36.

Swanson, Heather Anne. 2017. "The Banality of the Anthropocene." *Fieldsights,* February 22. https://culanth.org/fieldsights/the-banality-of-the-anthropocene.

Thompson, Jennifer A. 2014. "On Writing Notes in the Field: Interrogating Positionality, Emotion, Participation and Ethics." *McGill Journal of Education/Revue des sciences de l'éducation de McGill* 49, no. 1: 247–54.

Vellend, Mark. 2020. *The Theory of Ecological Communities.* MPB-57. Princeton, N.J.: Princeton University Press.

Waide, Robert B., Michael R. Willig, Chris F. Steiner, Gary Mittelbach, Laura Gough, Stanley I. Dodson, Glenn Patrick Juday, and Robert Parmenter. 1999. "The Relationship between Productivity and Species Richness." *Annual Review of Ecology and Systematics* 30: 257–300.

Willig, Michael R., and Steven J. Presley. 2018. "Biodiversity and Disturbance." *Encyclopedia of the Anthropocene* 3: 45–51.

Wilkinson, David M. 1999. "The Disturbing History of Intermediate Disturbance." *Oikos* 48, no. 1: 145–47.

CHAPTER 12

Cholera, Common Ground, and Project Drafts

Messages in a Bottle

NATHALIA BRICHET

"A cholera bottle," half-filled with a milky white fluid from the intestines of a cholera patient living in Scandinavia some 150 years ago, is the central character in this chapter. According to a group of researchers interested in the bottle, it most likely contains cholera bacteria, but possibly also phages, parasites, traces of gut bacteria, and other pathogen bacteria that have been transformed in the sealed bottle in unknown ways, creating a unique environment. As I will show, one might think of the bottle as a "Body without Organs" (BwO) (Deleuze and Guattari 2005), making up a matrix of potentials not (yet) organized.[1] It contains, as we shall see, possibilities for questioning entities like disease, nation, medical knowledge, cholera bacteria, pandemics, and transmission, thereby becoming "a milieu of experimentation" (Deleuze and Guattari 2005, 164). The overall point of this chapter is to engage the cholera bottle as such a "milieu of experimentation" and to chart anthropological methods for exploring relations between cholera, scientific disciplines, people, different parts of the world, ecologies, eras, and objects. In the following pages, I will make three interrelated propositions about rubber boots methods for the Anthropocene on the basis of my work with the cholera bottle. First, I will argue that any methodological choices are completely dependent on specific *analytical concerns* and thus are not a matter of independent and defined techniques. Second, I will suggest that methods for a vital anthropology should be tailored to the particular stories that one wants to tell and foreground; the issue is not just to notice and collect data from a field but to really consider what one wants to discuss with others. And third, I will argue that fields for rubber boots exploration are sites where the concerns of the fieldworker and the interlocutors of any kind must deliberately be brought together, even if across differences and contrasting points of view. I refer to this notion of the field as a "common ground" to highlight that it cannot be defined solely by the fieldworker, nor by reference to "ethnographic data" (cf. Brichet 2018). Even if these three

FIGURE 12.1. An image of the cholera bottle. Photograph from the Medical Museion; reprinted with permission.

propositions may not sound like matters of methods, my point is to argue for them as such and to show how they played out in my work with the cholera bottle.

My methodological forays were stimulated by the sudden interest of a group of researchers who approached the custodians of the bottle: the medical history museum in Denmark—Museion. Like many other museums, Museion is experiencing a growing interest from scientists wanting to use museum items for research and asking permission to compromise or even destroy material from old specimens to sequence DNA for further analyses (Yeates et al. 2016). So, too, in this case: a trained engineer specializing in environmental hygiene at the University of Copenhagen envisioned a research project in which he and a group of epidemiologists, microbiologists, and cholera researchers would break open the small, welded glass bottle from 1853—possibly destroying it—to map its content. Their ambition was to learn more about cholera's DNA and evolution, thereby enhancing knowledge about transmission ways. Museion eyed an opportunity to explore the potentials of the bottle from a social science and museological perspective, asking what could be gained or lost by allowing the bottle to be opened. Across disciplines, then, researchers and stakeholders agreed on the need to work together and find out what opening the cholera bottle might yield. By focusing on this process of developing an interdisciplinary project centered on cholera, my three points about anthropological methods for the Anthropocene became clear. Before substantiating these, I will lay out some conceptual terrain and present my "arrival story" that led to the cholera bottle becoming my field.

A Body without Organs:
Crafting a Half-Seen Project Draft

The "cholera bottle project" was first presented to me in 2018 by colleagues from Museion who employed me for six months to enter and develop the bottle's milieu of experimentation. The project PI was motivated by a belief that mapping the DNA of the found bacteria combined with detailed archival material from the 1853 outbreak in Copenhagen might generate knowledge about transmission that could revolutionize our understanding of cholera and pandemics more broadly. A new way of understanding cholera, the PI argued, might improve intervention strategies during present-day cholera outbreaks, particularly in Bangladesh, where he worked for years. However, staff members in the museum were not convinced about the link between nineteenth-century Denmark and contemporary Bangladesh, which is one of the reasons they asked me to chart research possibilities that the cholera bottle contains. Therefore my job became to explore the interdisciplinary knowledge potentials that might be implied in the sample bottle to find out as much as possible about the museum object, to suggest research opportunities for Museion, and to write a proposal for a common research project where all these issues were included.[2] Thus, in this particular moment in time, when ideas were up in the air, I began engaging actively with the cholera bottle. In terms of concrete activities, I did so by contacting people directly involved in the prospective research project and others who either showed a direct interest or might be interested in the cholera bottle project, and I read articles and followed lectures that were suggested to me or that I found interesting. This conjured up a spatially distributed network across Denmark, Bangladesh, North America, Norway, and Sweden, and consequently, I got involved in extensive email correspondences and meetings— a few online. All this to come up with suggestions of what to do with the bottle, unsealed or not. Even though it was not traditional anthropological fieldwork, I understand this intensely suggestive and relational engagement with the bottle as cognate in its situated, curious, personally invested and open-ended setup.

This interdisciplinary setting is the basis of my discussion of how anthropological methods can be practiced at the interface between an old museum object and a group of enthusiastic researchers, myself included. Deleuze and Guattari's figure of a BwO came to be an appropriate do-it-yourself guideline for me—an undogmatic, slightly tongue-in-cheek way of describing the practice of fieldwork—and ultimately as the method I have employed:

This is how it should be done: Lodge yourself on a stratum, experiment with the opportunities it offers, find an advantageous place on it, find potential movements of deterritorialization, possible lines of flight, experience them, produce flow conjunctions here and there, try out continuums of intensities segment by segment, have a small plot of new land at all times. It is through a meticulous relation with the strata that one succeeds in freeing lines of flight, causing conjugated flows to pass and escape and bringing forth continuous intensities for a BwO. Connect, conjugate, continue: a whole "diagram," as opposed to still signifying and subjective programs. We are in a social formation. (Deleuze and Guattari 2005, 161)

By stimulating a social formation through lines of flight, this thoroughly explorative and creative way of researching sidetracks ambitions to search for social representations 1:1. The idea that our methods—how we "lodge ourselves" and explore strata "through meticulous relation"—are part of the very formation that takes place is exciting not only for what happens *in* the field but also for what happens with anthropological analyses resulting from the methods employed. In a recent introduction, anthropologist Martin Holbraad (2018) identifies a shift in current anthropology. To him, methodological stances have increasingly become the dominant landmarks to situate one's anthropological work, supplanting earlier affiliations based mainly on (a) theory. Methods, then, encompass much more than a particular approach to "something else." Accordingly, he asks if theory, method, and analysis can be stabilized at all or if their internal indeterminacy is a sign of creativity within the field of anthropology (1). Attention to such changes in some of the fundamental academic elements is important. Indeed, the blurred distinctions between theory, methods, and analysis speak directly to my overall contribution to methods for the Anthropocene. My point here is that methods are not particular techniques "applied" to the world with the promise of certain yields but a way of engaging collaboratively with concerns that emerge within any one field.

In the same special issue that Holbraad introduces, anthropologist Alberto Jiménez argues that analysis makes empirical observations just as much as empirical observations make analysis—an interdependency that he calls symmetry (Jiménez 2018, 7ff.). Now, the symmetry that Jiménez points to enables "exercises in concurrency: how analysis and description partly anticipate, partly intuit, partly inscribe one another—how they half-see one another—in the material concurrency and environing of the social" (8). Jiménez refers to this half-seeing as a "draft"—a draft that (in our discipline) is situated and undertaken accompanied by others. As such, the draft is a continuous figure-ground exercise where what gets figured and backgrounded are results of "half-seeing" together with others (8). Paying attention to this process of drafting is my first point of what I take method to be: inherently analytic and powered by theory.

In the museum's protocols, there is no information about who sampled the bottle or why. But according to the Swedish handwriting on the label, the vessel contains pure cholera-infected gut juice—or rice water—from a patient who recovered from cholera in October 1853. Presumably, the bottle has never been opened since the day when someone poured rice water into it and sealed it. Instead, it somehow ended up in Museion, where it is now perceived as a unique museum object exhibited in a room with other pre-twentieth-century objects under the theme of humeral medicine. The museum is part of the Medical Faculty at the University of Copenhagen, and most of the objects stem from centuries of research done at the faculty. Given these ties, the museum is from the outset positive toward requests to use its objects for research but is also aware of the obligation to conserve the very same objects in the collections. There is no official policy for researchers' requests; Museion accepts or rejects them case by case. As for the cholera bottle, the staff is hesitant to compromise it because of its uniqueness and irreplaceability, but they are also willing to open it if it can help fight cholera in novel ways. In other words, they need good arguments and ideas to break it open—arguments that go beyond the gimmick of opening a unique bottle that contains a once highly virulent pathogen that, according to the project proposal's PI, might be dormant.

The request from the scientists and the ensuing discussions have made the bottle an opportunity for me, and eventually for Museion, to think about anthropological methods in an interdisciplinary field and which analyses to make and stories to tell in articles and in exhibitions in the twenty-first century. Choices about storytelling as an urgent methodological issue is my second point about methods (cf. Tsing 2015, 37). As I see it, being attentive to *which* stories to craft, draft, and make visible is also a matter of methods and not just of ensuing analysis. If noticing is an art (Tsing 2015), I take this art to be not just an intuitive and unrestricted mode of observation. From the outset, it is vested with concerns about which relations to nurture and foreground before, during, and after fieldwork—observation as half-seen, in Jiménez's words.

My fieldwork around the cholera bottle involved engaging the work of epidemiologists, microbiologists, global health workers, engineers, historians, glass specialists, medical doctors, conservators, and curators, among others. My third point here, then, is that this setting with a host of approaches and practices requires an analytical-cum-methodological approach that can craft—or draft—the cholera bottle as a "common ground." By this I mean a site intentionally made, nurtured, and used by the anthropologist to work across difference—a site where different concerns coalesce and can be discussed, where thoughts are shared, and where new relations are offered and others abandoned in an open-ended and joint, though not always smooth, process of collaboration. In this way, common ground is not an inversion of historian William Cronon's (1996) notion of "uncommon ground," pointing to disagreement and inequalities. Instead, common ground is a site held together by the fieldworker, where heterogeneity can be kept and presented to interlocutors to explore possibilities across differences through offering various drafts to people with whom one engages (cf. Brichet 2018). In this case, to draft such composite common ground requires an opening up to ever more questions about the ecologies of knowledge that cholera becomes part of and then a discussion and selection of which strata to embark on and which stories to cultivate.

Consequently, my point on methods so far is to engage fieldwork as a means to explore what the cholera bottle can be through an intersecting, situated, and incomplete host of concerns, encounters, and potential stories (Hastrup 2014). This entails a fieldwork in a network of different actors where cholera, in a sense, is the constantly fluid object, proliferating through ever more relations, touching on knowledge interests, academic politics, disturbed ecologies, and global connections. In terms of finishing a research grant proposal, this may be counterproductive and annoying. But in terms of developing anthropological methods for the Anthropocene, such creative and difficult nurturing of a common ground of stories and concerns seems vital to me. In this spirit, I thus work from Jiménez's idea of half-seeing in the company of others, drafting a proposal for Museion's future engagement with the bottle through stories that I consider important in an era of environmental and climatic crisis. In the chapter, I will dwell on fieldwork encounters and episodes that show how I worked to draft the cholera bottle project as a common ground—more or less successfully, and on different strata that mattered differently to the people involved, to borrow again from the concept of BwO.

Cholera, Common Ground, and Project Drafts 333

INITIAL OPENINGS: FIRST STRATUM OF
COLLECTIONS AND COMPARISON

Worried about the potential destruction of their unique object, staff in Museion encouraged me to investigate if other museums keep similar objects and how they manage requests to use and destroy specimen. A curator had already found that a group of researchers from McMaster University, in collaboration with the Mütter Museum in Philadelphia, had recently published an article based on a specimen of an intestine from a victim of cholera in 1849 kept in the museum collections (Devault et al. 2014). Despite old age, the researchers had succeeded in making the first genomic sequencing of the pathogen bacteria. This in itself was an accomplishment, because prior to this study, the DNA of cholera bacteria was only known from the twentieth century.

The comparison and possible collaboration between the cholera bottle project and the Philadelphia study seemed obvious. Given the interest of the prospective PI and Museion's wish to use the cholera bottle to fight present-day cholera, I was especially interested in how knowledge about the bacteria's ancient DNA could be or had been used in further research. On the home page about the research, it is stated, "Our wet specimens were used to crack a genetic code of cholera!"[3] foregrounding the mapping aspect of the research. Promisingly for the common ground that I was both working from and continuously establishing around Museion's bottle, the home page further read that "a better understanding of the genetics of this terrible disease may help efforts to control it in the future."[4] However, when I contacted the museum staff and group of researchers to hear more about their considerations, a focus on the present-day fight against cholera was beyond their research design. Instead, they aimed "investigations into the genomic origins of past pandemics" (Devault et al. 2014, 334), and as such their project differed from the one in Copenhagen. That said, the researchers welcomed if their study could be used to fight cholera or other pathogens today, but for them, they had succeeded by sequencing the oldest cholera bacteria to date.

Browsing the fifty-five articles quoting the Philadelphia study (autumn 2018), it was often listed among other early DNA sequenced pathogens, such as *Yersinia pestis, Shigella dysenteria,* and *Mycobacterium tuberculosis,* that have recently been described genetically and also mentioned as a reference for methods of measuring evolutionary timescale and mutation rates. In this way, my encounter with the Mütter Museum came to work as a comparison and an opportunity to inquire into the kinds of knowledge gained and research that it had further ignited. Although this was not necessarily along the same lines as the PI's project with its integral focus on fighting cholera outbreaks in the present, the comparison pointed to the difficulties—or work needed—to make that relation. Even though present-day challenges seem to be invoked as a legitimization in ancient DNA research, still no one has come out with results for medical use, as I was later told by a highly experienced researcher within the field. Instead, he added, "we are cataloging the presence of certain things"—which is exactly what the Philadelphia study had accomplished with its DNA sequencing.

This excursion to Philadelphia did not answer my question of how a better understanding of the ancient DNA of the bacteria could meet the PI's ambition in Denmark. Looking to a seemingly comparable experience yielded new ideas but did not substantiate the relation between ancient DNA and present-day interventions. Instead, paying attention to the processes of drafting by comparison made difficulties and potentials of the bottle visible. In the next section, I will lodge myself on the stratum that the PI's team was in the midst of creating—including the search for relations between nineteenth-century cholera and present-day outbreaks.

Drafting Biochemistry: Second Stratum across Time and Space

Encountering the PI and the group of researchers he had assembled was another important fieldwork experience. The questions guiding me were what their concerns were and what they envisioned could be achieved by opening the bottle. According to the PI, researchers and experts still do not know what exactly triggers a cholera outbreak and why it dies off again. Central to this lack of understanding is transmission ways. The bottle, the PI argued, might contain a unique possibility to shed light on the mysteries of cholera by making a comparative study between Copenhagen then (during the outbreak in 1853) and Arichpur, a suburb to Dhaka in Bangladesh, now (during one of the biannual outbreaks) because there are many similarities, such as population size and density, hygiene standards, and sanitary conditions, such as distance to wells. He hoped that a mapping of the old DNA of the cholera bacteria in combination with a mapping of population data, water infrastructure, and the spatial development of the outbreak in Copenhagen—archival material that is available—could unveil transmission ways to be used in research on contemporary Bangladesh. In this way, he envisioned the interdisciplinary project as a cross-temporal and cross-spatial comparative study.

The team had different fields of expertise: one was specializing in parasites, another in making images of bacteria, and a third in global health and coinfectious diseases. A handful were experts in recuperating damaged DNA, and finally, the project involved an anthropologist and a group of microbiologists from Bangladesh with a record of studying cholera and other infectious diseases. Apart from concurrently suggesting to look not only at the cholera bacteria but also into the microbiome of the sample, coinfections, diet, and how these presences might have shaped and been affected by the virulence of the cholera, this assembly of specialists made it clear that each person involved in the project saw themselves as contributing to a limited part of the project. Their different techniques and machinery would translate into what the PI organized as distinct working packages—or organs, to follow Deleuze and Guattari. A web of arrows connected the ever-expanding working packages that the PI thoughtfully organized in vibrant colors on A3 paper. This ordering made each working package able to participate in the overall organization while maintaining its own profiles. And even though everybody supported the idea of improving interventions during present-day outbreaks in Bangladesh, they had difficulties directly linking (their research on) the bottle to this ambition. Instead, when speaking about the potentials of the bottle, the involved scientists mentioned what we came to refer to as the "phylogenetic dot." In biology, phylogenetic dots are the units that make up a phylogenetic tree, which is used to understand evolutionary relationships among organisms. Each dot contains genetic information that can be compared to other dots, and in this relational way, the dots provide possibilities for better understanding changes among given organisms. Such an organization offers a history of changes in populations of, for example, the cholera bacteria. In the case of the cholera bottle, the phylogenetic

dot we talked about was based on DNA sequencing. It was exactly such a
"dot" that the Philadelphia specimen had resulted in, thereby expanding the
phylogenetic tree to the nineteenth century. In talking to the PI and his
team, I learned that sequencing DNA from various places and times makes
it possible for researchers to study the evolution of pathogen bacteria and
how they adapt and vary. Enhancing our understanding of pathogen bacteria in this way *might* be a key to fighting the disease in question, the PI
argued. And even though the cholera bacteria of the nineteenth century are
most likely not the same as those of the twenty-first century, it is relevant to
the story of species evolution.

But who, I asked the people involved, could or would commit to doing
this comparative microbiological work—and how was it part of the project
design? Clearly the lines of flight for the PI, to use Deleuze and Guattari's
vocabulary again, were directed toward the Bay of Bengal. In this way, the
potential phylogenetic dot of 1853 had to be managed for both its compara-
bility and its difference from a present-day Bangladesh setting—with all their
different ecologies. However, the hopes invested in making a link between
Copenhagen then and Bangladesh now seemed a challenge for everybody
involved (including the PI).

What we see here is a particular configuration—an organization of the
cholera bottle—drafting the biochemistry of the bottle in ways that are comparable to other pathogens and possibly other times and places. What is
interesting for this inquiry into methods are the ways in which the common
ground crafted through fieldwork around the cholera bottle kept spurring
new ideas while eluding others. Whereas the Philadelphia specimen resulted
in an article describing the cholera bacteria of the 1849 outbreak, the Museion
specimen is imagined to do that as well to improve intervention strategies in
Bangladesh. In a sense, the organization of the cholera bottle that the PI and
his team imagined and described could be organized into a body of working
packages where analysis is seen mostly as a matter of technique that could
either be accomplished or fail, depending on whether sequencing of ancient
DNA is successful. From these results, it is hoped, new lines of flight could set
off toward Dhaka. These germinating connections—"conjunctions," "levels," and "thresholds"—in this case along branches of the phylogenetic tree,
highlight that a common ground is a highly dynamic accomplishment that
feeds on the interests of all involved.

Pushing a Global Health Agenda:
A Bangladeshi–Danish Third Stratum

My field sprouted even more when I looked to present-day cholera outbreaks and the PI's research in Arichpur, an area on the outskirts of Dhaka City. Arichpur, I learned from one of the Bangladesh-based microbiologists involved in the project, is a low-income and "densely packed, rapidly urbanizing community . . . in close proximity to the nearby industrial area" (Hossain 2019, 40). Although far from Denmark, the industrial area, the PI elaborated, was primarily taken up by the garment industry, including Bestseller, a Danish company that happens to have more than sixty supplier factories around Dhaka.[5] This industry is labor intensive and notoriously in need of cheap workers and running water to dye, wash, and manufacture the garments, which is exactly what the region can offer. Unfortunately, neither Dhaka nor the industry can provide clean drinking water and sanitation to the expanding city, thereby making it a perfect breeding ground for the endemic cholera bacteria. In this expanding and constipated setting, the PI had conducted his research in households to see how pathogen bacteria, such as *E. coli* and cholera, were distributed and transmitted within the households. Taking samples from foodstuffs, doorknobs, and clothes, among other objects, his team had found that cholera (and *E. coli*) bacteria are transmitted in high numbers, and therefore they argued for "thinking beyond the water-supply in cholera epidemics" (Phelps et al. 2017, 1). As a result, his group emphasizes intervention strategies that focus on hygiene and sanitation, which are challenging in poor and densely populated areas. Cholera, he told me, is a disease that is preventable by good sanitary facilities and hygiene, and since the 1970s, it has also become easily curable with either oral rehydration solution or IV. Despite this knowledge, many people get cholera and do not make it to a medical center where they can be treated. In 2017, the World Health Organization (WHO; 2017a) had endorsed a call to action on ending cholera summarized in a publication called "Ending Cholera: A Global Roadmap to 2030." Here the task force wrote:

> Cholera is a disease of inequity—an ancient illness that today sickens and kills only the poorest and most vulnerable people. The map of cholera is essentially the same as a map of poverty. Every death from cholera is preventable

with the tools we have today, putting the goal of ending its public health impact within our reach.... The disease still affects at least 47 countries across the globe, resulting in an estimated 2.9 million cases and 95,000 deaths per year worldwide. (6–7)

Clearly, for the WHO, cholera is a disease of poverty. In the publication, it is further estimated that globally, "844 million people lack access to even a basic drinking water source . . . and 2.4 billion are without basic sanitation facilities, exposing them to a range of water-related diseases including cholera" (7). Just as it did in Europe in the nineteenth century, ending cholera basically demands a combination of clean water and better sanitation infrastructure. Even though cholera outbreaks can be kept at bay by investing in proper sanitation and water supply, the WHO recently stated that "global water and sanitation aid has declined in recent years, dropping from $10.4 billion in 2012 to $8.4 billion in 2015" (World Health Organization 2017b). Needless to say, this decline is scandalous, and I appreciate the PI's idea to relate the bottle to Bangladesh as a potential push to change things. However, according to the WHO, improving sanitation and water supply rather than getting a better understanding of the cholera bacteria and its phylogeny through DNA sequencing seem to be key. The PI is well aware of this but argues that with the current economic situation, every initiative matters.

I was further pushed along this line of exploring what he termed "global health" and the unequal distribution of health, care, and cures when I met with the anthropologist from Bangladesh who is involved in the project and who daily works for the International Centre for Diarrhoeal Disease Research, Bangladesh, where she identifies new outbreaks of all kinds of diseases in Dhaka. She reminded me that cholera was endemic and one among many deadly diseases in the region, where new pathogenic viruses and bacteria loom. Denmark, with its orderly infrastructure—at every level— seemed incomparable to the sense of urgency that prevailed in her daily life and work in Dhaka, where problems went far beyond the cholera bacteria's transmission ways. Obviously, talking to her, a narrow focus on cholera or ancient DNA seemed too provincial when the problem was much bigger. But on the other hand, and as we shall see, a focus on cholera transmission ways (combined with a global health perspective) is a place to start and embark on other flights.

This line of global health and sense of urgency intersected with my view of the exhibitions at Museion in Copenhagen. As it is now, this museum mainly explores local (Danish) concerns and celebrates Western medicine, then and now. However, part of the common ground of the cholera bottle, I suggest, could be to think about issues and concerns that we tend to neglect in our part of the world, where cholera is a disease of the past and where clean drinking water is used even to flush the toilet. The potential of the project, for me, was not just about transmission ways of pathogen bacteria in Denmark then and Bangladesh now but also about transmitting a message to museum visitors about living in a world where curable diseases, haunting Western societies in the nineteenth century, still flourish due to poverty and broader ecologies, as we shall explore in the next section. Could the opening of the cholera bottle and the succeeding microbiological research potentially published in international journals also be an opportunity to tell stories of inequality and (possibilities for) global health? Could the opening of the cholera bottle cure Museion of sticking to a narrow Western narrative about technological progress in medical care and thus broaden visitors' vision to also acknowledge the present-day relations and interdependencies between, say, Bangladesh and Denmark—think of the sixty garment factories that supply a Danish company operating and profiting from a cheap Bangladeshi labor force—and thus see the existence of cholera as a much more common problem than just "theirs"?

A World Wide Web Spilling into the Bay of Bengal: A Fourth Ecology Stratum

With these insights into the microbial and sanitary life of Dhaka, I suddenly found myself invited onto yet another stratum, which I immediately recognized and appreciated and thus lodged on to. I will call this stratum ecology. As already mentioned, I read articles and watched lectures explaining basic microbiology about bacteria, phages, and how the virulence of bacteria works. I read about the cholera outbreak in Copenhagen and explored the discussions of which it became part. I engaged with the Bangladeshi setting through academic articles, and I tried to relate all these newly gained insights to perspectives by which I had been inspired via my engagement with Anna Tsing's research group—the Aarhus University Research on the

Anthropocene (AURA) project (cf. the Introduction to this volume). In other words, I wanted to pour my curiosity toward landscapes, ecologies, and more-than-human relations into the cholera bottle. During my work to do so, I suddenly plunged into a talk available on YouTube by a professor in medicine from Stanford University, Gary Schoolnik (Schoolnik 2008; see also Schoolnik 2010). In *a single* talk, Schoolnik related the cholera bacteria to the Green Revolution in Southeast Asia, to the use of chemical fertilizer, and to deforestation, irrigation, heavy monsoon rains, algal bloom, and climate change. And after a fascinating explanation of the abilities of the bacteria and how they live with other aquatic species, he ended with a slide that urged listeners to cross disciplines, which he explained as the importance to understand and link the molecular and genetics to the regional and global scales by collaborating with people beyond one's own discipline. In forty-five minutes, this lecture had bridged broader ecologies to the microbiology of cholera and made it comprehensible to me by putting it into dialogue with perspectives that I knew well and cared about. This was a story that entailed scalar jumps and that had the kind of urgency in foregrounding ruined landscapes (Tsing 2015).

My work to create a common ground around the cholera bottle was further nurtured. A study had shown that the cholera bacteria crowd around the roots of a so-called water hyacinth *(Eichhornia crassipes)* (Spira et al. 1981; Takemura et al. 2014). When researching the hyacinth, sure enough, the water hyacinth opened into feral stories of an invasive species brought from Brazil by colonial officers, who somehow facilitated an infrastructure that led it into the Bay of Bengal in the beginning of the 1900s (Tsing et al. 2021). During the twentieth century, and in conjunction with modernization projects, it proliferated alarmingly quickly in the waters that also ran through Bangladesh. It blocked up vital waterways and paddy fields and reduced the number of fish dramatically. As described and argued by historian Iqbal Iftekhar, the water hyacinth became part of radical landscape changes in the region that ultimately led to the underproduction of food, culminating in the big famines of the 1940s (Iftekhar 2010, 140ff.). In this way, I was guided into a feral landscape of invasive species, hunger, and colonial administration instantiated by the hyacinth's relation to cholera.

I welcomed all these concerns for broader ecological stories that both Schoolnik and Iftekhar brought to the common ground that I was in the midst of drafting for Museion. As part of my fieldwork, I hoped to include them in the cholera bottle project. What fascinated me in particular were the ecologies and human-made landscapes to which the bottle could open up both on a microscopic level of DNA strains and on a macroscopic, global level. The crucial point for me became to see that cholera is not just bacteria acting on their own deep in our intestines—causing victims to lose up to one liter of liquid in an hour. Neither is it possible to maintain cholera as a local problem of poor countries with bad hygiene. As both Schoolnik and Iftekhar pointed to, cholera is part of a broader ecology—a particular landscape marked and transformed by all kinds of human interventions. I saw these kinds of relations as highly suitable to raise in a medical museum in the twenty-first century exactly because these relations made it possible to connect Bangladesh to Denmark—then and now, as was the project's ambition, but also its biggest challenge.

For the point of this chapter on methods, my challenge was exactly how I could introduce these perspectives that I nurture and treasure into a common ground mainly made of other concerns: the PI's focus on finding and sequencing ancient DNA to compare transmission ways of the cholera bacteria beyond the water medium in two settings with similar household setups (hygiene, sanitation, and population density). And second, is it at all the task of an anthropological fieldworker to "push" stories such as these into a common ground where nobody else involved in designing the research project talks about Green Revolution, hyacinth, copepod, algal blooms, or landscapes? In other words, should I try to include a focus beyond particular households on the wider ecology of the cholera bacteria as a shared concern? And if so, how? Again, analytical concerns and method issues become completely entwined. My point here is that what to include and exclude as a research focus is truly a methodological problem of any fieldworker. In the next section, I will dwell on these considerations of exclusion and inclusion of concerns while doing fieldwork.

CREATING FIELDS IN THE ANTHROPOCENE: FIFTH STRATUM OF BIOLOGICAL CONCEPTS

Considerations of inclusion and exclusion of themes are crucial for researchers doing fieldwork—a work that inevitably generates a surplus of material.

In this discussion on methods, I highlight how the cholera bottle mobilizes particular concerns and commitments—mine and those of the others involved—from the get-go. It foregrounds that fieldwork is not only a matter of listening, noticing, or observing but just as much a site of negotiation, where more or less successful sharing can take place. As argued earlier, as a fieldwork site, the cholera bottle could be seen as a BwO gesturing at different lines of flight to be explored. In the same playful vein, I will illustrate my considerations of inclusion and exclusion by way of two biological concepts. Using descriptions of biological processes as concepts to think with anthropologically might run the risk of reducing complex biological relations to metaphors (Tsing 2015, 144). This "reduction" (which is perhaps just as much an "amplification" of things half-seen), I will argue, goes for anything we do with our field—reducing and amplifying are at the heart of anthropological analyses-cum-methods; defining a field always entails foregrounding and backgrounding particular relations. Here, too, wanting actually to draft a story of the field of the cholera bottle, I go with the oversimplifications, or amplifications, and venture into the semantics of biological processes.

As a master in thinking relations across divides, Anna Tsing (2015) invites the reader to explore heterogeneity via fungi and interspecies relations. For her, the *hyphae*—the fungi's threadlike filaments—provide an apt image. Imagine, Tsing suggests, "if you could make the soil liquid and transparent and walk into the ground, you would find yourself surrounded by nets of fungal hyphae. Follow fungi into that underground city, and you will find the strange and varied pleasure of interspecies life" (137). Here hyphae are not just a metaphor but a particular biological way of studying life that acknowledges multispecies relations. Entailed in foregrounding the hyphae is a critique of species life as solo projects. Having us believe in such isolated solo projects—entities that can be scaled up and down—is to Tsing one of the biggest deceptions of our modern world. To counter such singular projects, she calls for a method—an art of noticing—where a focus on contingency and interspecies relations dominates and where interaction means something else than outcompeting each other in predator–prey relations. Mutualistic relations such as hyphae—or mycorrhizal networks—are a prime example (139). What I want to highlight by way of Tsing's hyphae is the way that our fieldwork methods include and exclude particular stories and how these processes entail critique. I treasure the focus on encounters across divides—subsoil as well as above—but instead of inviting the reader into a transparent underground city from which to tell stories across divides, I suggest thinking through yet another biological process that I encountered in the cholera bottle project and that speaks to the ways I envision anthropological methods.

In the online talks mentioned earlier, Schoolnik tells a breathtaking story about a new virulent strain of cholera called serotype 0139 that emerged in the Bay of Bengal in 1991–92. Until then, pathogenic cholera had been known as serotype 01, but immediately, 0139 caused a massive outbreak where not only young people died but also old people who should have acquired immunity to cholera—but only to serotype 01. In his talk, Schoolnik suggests how this new virulent strain was made possible because of some changes in the genetic setup of the cholera bacteria. These sudden changes, Schoolnik explains, happen under specific circumstances (particular associations between the bacteria and a polysaccharide called chitin). In this situated setting, the cholera bacteria can induce their competences for natural transformation—a process referred to as horizontal gene transfer. In this process, the bacteria can either grab or reject fragments of free-floating DNA released from other bacteria and in some cases incorporate fragments into their genome. In this way, horizontal gene transfer allows some bacteria "to take a quantum leap in evolution by acquiring big blocks of genes"— meaning that it is a much more rapid way to change and enhance one's abilities than organisms that evolve, for example, by mutations. In his lab, Schoolnik has tested that such a horizontal gene transfer might have been the instigator to the changes that suddenly resulted in serotype 0139. For my argument on method, I am inspired by the suggestive and transformative aspect of the transfer. I find this ability to reject or incorporate fragments of DNA a fruitful metaphor for the point I am making about fieldwork methods. Horizontal gene transfer points to the kind of anthropological method that I am proposing, in showing how analytical perspectives— strings of DNA—are suggested to a common ground in the making, to be rejected, acknowledged, or incorporated into the DNA of the partners engaged. *If* anthropological methods can take part in making *quantum leaps* and act exponentially in response to the constantly threatening ecological disasters and tipping points we face today, they are worth considering. The fundamentally suggestive and potentially transformative nature of the notion of horizontal gene transfer sums up my fieldwork around the cholera bottle.

If we return to the cholera bottle project, I navigated a landscape of concerns all ignited by the cholera bottle. Along the way, I grabbed some perspectives offered to me and explored and incorporated them, while questioning and suggesting yet others. Equally, at times, perspectives were rejected for well-argued reasons by some of the key persons engaged. For example, my suggestions to include the insights of scholars like Schoolnik and Iftekhar and broaden the ecologies of cholera toward the Bay of Bengal were not grabbed by the PI, who wanted to maintain a focus on the narrower unit of the household, where his team had done field studies and sampled interesting data over the years. As argued earlier, choosing which stories to tell is a methodological concern. If the field is not to be found out there but is something that we collectively and analytically create, then we depend on our collaborators and the shared concern for incorporating each other's concerns.

CONCLUSION: SHARED ECOLOGIES FOR BETTER OR WORSE

All this allows me to sum up my main points on methods for the Anthropocene. First, when thinking about fieldwork as a matter of horizontal gene transfer creating a common ground, there is no going solo. Nor really is the issue merely to notice multispecies worlds; surely this is a good place to start, but fieldwork, as I see it, is much more dependent on others to care for it to be captured in "noticing," however artful. What I propose is literally an unsettling approach; there are no solid techniques on which to lean. This is also why my fieldwork around the cholera bottle project is purposely presented as a series of concrete episodes where intersecting and diverging ideas and concerns emerged, rather than being specified according to distinct labels that we might otherwise use to name methods, that is, interviews, participant observation, deep hanging out, or what have you. Nonetheless, my points are methodological.

Second, the unsettled character of the common ground is an opportunity for all involved to bring a problem to the fore—in the hope that someone latches on to it and incorporates it enough for it to become part of the draft of concerns. In consequence, methods may not be a "starting point" for analyzing Anthropocene ecologies, as if analysis happens later. Seeing methods as a distinct starting point where different disciplines can meet seems to me to imply the field as too much of a given—which is surely not my experience from the cholera bottle. My dialogue with a large cast of researchers from different scholarly disciplines has taught me what I know about cholera, and I am grateful. But the joint making of a shared object of interest—shared enough, that is, to possibly break open an antique museum object—is the ongoing responsibility, joy, and privilege of all, none of us fully knowing what is in fact in the cholera bottle but all wanting to pour something into it. The concrete ways that this played out is what I have described and discussed herein.

Third, cholera and Museion's bottle provide a lesson about how the world is (also) connected and why we should not just say that cholera is a problem of "others." My fieldwork on the cholera bottle shows the collaboration between a Danish museum and a group of researchers as a site not primarily for expanding our understanding of cholera in Copenhagen in 1853 but as a shared probing of a global health and ecological problematic. My method, then, is also very much a theoretical stance, implying that anthropological objects of study—such as the cholera bottle—are figurations emerging in the uneven process of dialogue, as captured metaphorically in the notion of horizontal gene transfer. Put differently, the story of the cholera bottle told here shows how I have grappled methodologically with drafting a field, for better or worse, successfully pushing some stories, having other elements discarded, and ending up with anything but a transparent glass bottle. Indeed, during fieldwork, the cholera bottle has expanded and offered various lines of flight to be pursued. While this might seem disastrous in the genre of a research project application, it seems to me to be one way of inhabiting the precarity of the Anthropocene. It is only right for anthropological methods to change in response to fundamentally unsettling times.

Let me end with a vignette from my fieldwork. Some time ago, I attended a meeting on the cholera bottle project. The PI opened by nodding in my direction and exclaiming, "Well, inviting an anthropologist into the group is like opening Pandora's box!" Despite its connoting the unleashing of all sorts of challenges, I was flattered. In this article, I have revisited this Pandora's box, arguing for a fieldwork method that adds, complicates, discusses, and transforms the field in question.

NOTES

I thank all the people who engaged in and qualified the "cholera bottle project." In particular, I am grateful to Ken Arnold, Peter Mackie Kjær Jensen, Ion Meyer, Bente Vinge Pedersen, and Karin Tybjerg for inviting me into this intriguing project. Warm thanks to people in the AURA project, and especially to Frida Hastrup, for fruitful discussions along the way. I am grateful to Astrid Oberborbeck Andersen, Nils Bubandt, and Rachel Cypher for comments and suggestions on earlier drafts. This work was generously financed by Medical Museion, University of Copenhagen, Denmark.

    1. One might note that a BwO is not a concept but a practice that carefully and momentarily dismantles the organization of the organs to open up to new connections (Deleuze and Guattari 2005, 149–50, 161). Thus the BwO is not "before" the organism; it is adjacent to it and is continually in the process of constructing itself. Studying a BwO, therefore, always entails navigating both a body with and without organs.

    2. While working at Museion, I was introduced to other exciting projects, and therefore I have not been part of the further life of the bottle.

    3. http://muttermuseum.org/.

    4. http://memento.muttermuseum.org/.

    5. https://about.bestseller.com/sustainability/transparency-and-reporting/factory-list.

REFERENCES

Brichet, Nathalia. 2018. *An Anthropology of Common Ground*. Manchester, U.K.: Mattering Press.

Cronon, William. 1996. *Uncommon Ground: Rethinking the Human Place in Nature*. New York: W. W. Norton.

Deleuze, Gilles, and Félix Guattari. 2005. *A Thousand Plateaus: Capitalism and Schizophrenia*. Minneapolis: University of Minnesota Press.

Devault, Alison, Brian Golding, Nicholas Waglechner, Jacob M. Enk, Melanie Kuch, Joseph H. Tien, Mang Shi et al. 2014. "Second-Pandemic Strain of *Vibrio cholerae* from the Philadelphia Cholera Outbreak of 1849." *New England Journal of Medicine* 370, no. 4: 334–40.

Hastrup, Frida. 2014. "Analogue Analysis: Ethnography as Inventive Conversation." *Ethnologia Europaea* 44, no. 2: 48–60.

Holbraad, Martin. 2018. "The Analysis of Analysis." *Social Analysis* 62, no. 1: 1–2.

Hossain, Zenat. 2019. "Investigation of Household Transmission Pathways for *Vibrio cholerae* and *Escherichia coli* in Bangladesh." PhD diss., University of Copenhagen.

Iqbal, Iftekhar. 2010. *The Bengal Delta: Ecology, State and Social Change, 1840–1943*. London: Palgrave Macmillan.

Jiménez, Alberto Corsin. 2018. "Ethnography Half-Seen." *Social Analysis* 62, no. 1: 6–8.

Phelps, Matthew, Azman Andrew, Lewnard Joseph, Antillon Marina, Lone Simonsen, Viggo Andreasen, Peter K. M. Jensen, and Virginia E. Pitzer. 2017. "The Importance of Thinking beyond the Water-Supply in Cholera Epidemics: A Historical Urban Case-Study." *PLoS Neglected Tropical Diseases* 11, no. 11: 1–15.

Schoolnik, Gary. 2008. "Environment Degradation Begets Epidemics: Cholera in Bangladesh." YouTube video, 39:16. https://www.youtube.com/watch?v=ovUUs XoMfho.

Schoolnik, Gary. 2010. "The World Outside: A Changing Environment and How It Affects Us." YouTube video, 1:48:51. https://www.youtube.com/watch?v=VotBQ 8BNjx4.

Spira, William, Anwarul Huq, Shafi Ahmed Qazi, and Yusuf Saees. 1981. "Uptake of *Vibrio cholerae* Biotype eltor from Contaminated Water by Water Hyacinth *(Eichornia crassipes)*." *Applied and Environmental Microbiology* 42, no. 3: 550–53.

Takemura, Alison, Diana Chien, and Martin Polz. 2014. "Associations and Dynamics of Vibrionaceae in the Environment, from the Genus to the Population Level." *Frontiers in Microbiology* 5, no. 38: 1–26.

Tsing, Anna Lowenhaupt. 2015. *The Mushroom at the End of the World: On the Possibility of Life in Capitalist Ruins.* Princeton, N.J.: Princeton University Press.

Tsing, Anna, Jennifer Deger, Alder Keleman, and Feifei Zhou, eds. 2021. *Feral Atlas: The More-Than-Human Anthropocene.* Stanford, Calif.: Stanford University Press. http://www.feralatlas.org/.

World Health Organization. 2017a. "Ending Cholera: A Global Roadmap to 2030." http://www.who.int/cholera/.

World Health Organization. 2017b. "Prevention for a Cholera Free World." https:// www.who.int/news-room/feature-stories/detail/prevention-for-a-cholera-free -world.

Yeates, David, Andreas Zwick, and Alexander Mikheyev. 2016. "Museums Are Biobanks: Unlocking the Genetic Potential of the Three Billion Specimens in the World's Biological Collections." *Current Opinion in Insect Science* 18: 83–88.

CHAPTER 13

Rubber Boots Methods beyond the Field

Transformative Possibilities and Institutional Barriers in University Contexts

HEATHER ANNE SWANSON

R ubber boots methods are propositions for how to study and live on a damaged planet. Though heterogeneous, they are united in their commitment to fieldwork—to empirically grounded research attuned to the specificity of place. As described in the Introduction to this volume, rubber boots methods have emerged out of the Aarhus University Research on the Anthropocene (AURA) project's engagement with a former brown coal mining site in central Denmark (see also chapter 11). Both in and beyond this locale, rubber boots methods are a call to "stay with the site," in contrast to other popular methods in anthropology, such as "follow the thing" (Marcus 1995). One key argument of the rubber boots approach is that new forms of noticing and collaboration can best take shape through repeated contact with landscape patches at the scale of an embodied researcher, such as a stretch of river or patch of forest that one can walk in a day (see chapter 1). While the sensibilities of rubber boots methods are emergent from direct contact with multispecies landscapes, might they also be elaborated and transformed in other spaces? In dialogue with long-standing debates in anthropology about the construction and place of "the field" within research practice, I explore how one might expand the spatial and relational imaginaries of rubber boots methods. If rubber boots methods are aimed at better understanding and intervening in the world-making processes toward which the term *Anthropocene* gestures, might its techniques and approaches be further honed via experimentation in a broader range of academic sites?

This chapter engages these questions through a space that has been largely neglected within conversations about more-than-human methods: the university. It explores how the institution of the university enables rubber boots methods in the field at the same time that it is itself a location for rubber boots experimentation. My desire to focus on the university in this way is partially inspired by the work of Susan Wright, an educational anthropologist and feminist theorist who traces how universities have become increasingly oriented toward producing knowledge and skilled laborers for projects of capitalist expansion. In "Can the University Be a Livable Institution in the Anthropocene?," Wright (2017) lays bare how Danish universities in particular and Euro-American ones more generally are fundamentally intertwined with the logics and practices of economic growth that bring the Anthropocene into being. Most importantly, Wright issues a call to action: if we take seriously the ways that universities are structurally embedded in the making of planetary damage, how might we transform universities into institutions that foster more livable worlds? This is not a question about campus recycling bins or double-pane office windows—conversations that portray the ecological relations of universities in terms of their most direct waste or energy flows. In contrast, Wright's work (which I follow) attends most directly to the role that universities play in bolstering the politicoeconomic systems that reconfigure more-than-human landscapes in deeply problematic ways.

Maintaining the rubber boots commitment to empiricism, this chapter builds its insights from Aarhus University in Denmark, where the AURA project was based and where I continue to teach. Although it seeks to open questions that can resonate across a variety of contexts, it refuses to render "the university" in entirely abstract terms. It thus works outward from two concrete examples. First, it analyzes AURA project activities that took place within the monochromatic walls of university rooms but that coproduced the rubber boots methods that took place in more verdant locales. Second, it presents an attempt to bring rubber boots methods into a classroom that highlights the difficulties of expanding the scope of these methods. In both cases, I explore the relationships between field-based techniques and the academic practices that take place in classrooms, offices, and laboratories. Although such spaces are often viewed as separate domains—with the presumption that there is little overlap among the practices of field, pedagogical, and administrative work—I ask how they already shape each other and how they might do so in even more substantial ways. Like Klein and coauthors (chapter 9), I am interested in how rubber boots methods demand modes of collaboration that grate against current institutional structures. If we give further attention to the frictions produced by these new forms of collaboration, interdisciplinary field-based research, and their insights about more-than-human worlds, might they call out for substantial changes in universities' institutional practices? Or, put another way, what might rubber boots methods look like when one is trekking across the vinyl-tiled floors of a Danish university?[1]

I come to such questions in part via feminist anthropology and feminist science and technology studies, fields that have influenced rubber boots methods via shared commitments to generating research questions from observations of everyday relations and to crafting better and more just forms of scientific practice.[2] Yet rubber boots methods, I argue, also need to take up another part of feminist research practice: the undoing of categorical distinctions among field, classroom, conference, and office as spaces of analysis and experimentation.[3] Feminist scholars have long taken these spaces, especially classrooms, as sites of experimentation that are contiguous with other kinds of research, with critical attention to curricula, pedagogical dynamics, and analytical concepts (Geller and Stockett 2007). Because everyday life is a field study in gendered dynamics, feminist scholars frequently draw insights and seek to intervene in the gendered practices of the institutions of which they are a part. As the dynamics that produce economic growth and landscape destruction are as ubiquitous as those of gender, might rubber boots methods benefit from similarly blurring the boundaries between the field and other arenas of practice?

At the same time that I draw on feminist approaches, I am indebted to anthropological literatures that have questioned the construction of "the field" in diverse ways (Gupta and Ferguson 1997; Amit 2003; Coleman and Collins 2006; Candea 2007; Falzon 2016). While other chapters in this volume grapple with one issue raised by this line of work—the ongoing need to inherit and remake the colonial histories of field research (see the Introduction to this volume)—I take a somewhat different, albeit complementary, track as I explore how modes of delimiting the field may inadvertently constrain the abilities of rubber boots methods to intervene in some of the core dynamics of more-than-human world making.

Rubber Boots Methods: Key Features

As the overall volume illustrates, there is no singular definition of rubber boots methods: the term gestures toward a set of heterogeneous practices, albeit with meaningful overlap in their sensibilities. In line with the other authors in this collection, I take rubber boots to refer to a kind of "slow science" (Stengers 2016) where natural and social scientists, humanists, artists, and others meet *in* the field and discuss *what* we notice when conducting fieldwork (see also chapter 9). AURA strongly centered long-term grounded-site research at a moment when commitments to such practices have weakened within both biology and anthropology. AURA's stakes in foregrounding the field were a reaction against trends in cultural anthropology toward shorter periods of fieldwork, multisited studies that move too quickly to gain an in-depth feel for any one place, and the increasing valorization of abstract philosophical theorization over empirical detail (Marcus 2006). At the same time, the natural sciences have seen a sharp decline in the prestige of natural history and field biology (Anderson 2017) and a concomitant rise of "armchair ecology" or "keyboard ecology," especially big data methods that rely heavily on computer modeling (Noss 1996; Futuyma 1998; see also Custock 2020). To counteract these turns (while not rejecting them en toto), AURA has emphasized embodied, analog observation—a form of longer-duration research that does not fit well with normalized temporalities of scholarly production. Such commitments came to be enacted via repeated visits to the same field site, the former Søby brown coal fields, over a period of five years, in an attempt to develop analytical insights from the detailed particularities of an actual place, rather than from the abstract spaces of big data research or the Western philosophical canon (Bubandt and Tsing 2018b). One way to understand rubber boots methods is as a practice of *staying with a site*—of remaining committed to repeated fieldwork in a particular, spatially bounded terrain.[4] While the project examined the brown coal "patch" with an understanding that it is connected to other sites, rubber boots practices remained committed to the potential of staying put and looking closely.

Within these efforts to reassert the importance of grounded, long-term, empirical research, the field site became configured as a privileged and almost magical locale—one that not only would produce insights about the brown coal landscape but would reunite anthropologists and biologists in a novel and more generative way. Resonating with established tropes in both anthropology and natural history, treks to the field had a bit of an epic-journey aesthetic—departing from the university desk to travel to a privileged place of novelty, collection, and insight, as well as hardship and uncertainty. The AURA project was not blind to the fraught inheritances and critiques of fieldwork as gendered, racialized, and classed (e.g., Bell et al. 1993), and it sought to do fieldwork otherwise—to remake its practices and repurpose the tools of the colonial and scientific gaze for projects of more-than-human livability—in relation to its Danish site.[5]

Because the field was expected to be a place of transformation—where anthropologists and biologists would come to ask new questions that emerged from their encounters with each other and the field itself—the team intentionally started research with only an amorphous interest in developing collaborative, cross-disciplinary approaches for studying highly anthropogenic landscapes, rather than with predefined research questions. In the liminality of the field and the altered social mode of outdoor walking, disciplinary boundaries were expected to become more porous—with scholarly habitus and subjectivities transformed. The field itself was to make claims on its researchers, sparking new curiosities within them and new desires for collaboration with each other: humanists would quickly find that they need to know something about plants, or fungi, or Miocene coal formation to understand the region's deer-hunting cultures. Natural scientists would soon discover that highly particular human histories—such as practices of adding lime to brown coal lakes—are needed to understand the biological configurations and sediments of these water bodies. Rubber boots methods were a practice of trust in the transformative power of the field—the trust that it would yield fascinating and substantial insights, if only one remained curious enough and patient enough for them to emerge.

Again, such sensibilities draw on established commitments to the field in both anthropology and natural history as a site for not only answering research questions but also bringing new questions into being. As sociologist Paul Willis has phrased it, long-term, open-ended fieldwork contains "a profoundly important methodological possibility—that of being 'surprised,' of reaching knowledge not prefigured in one's starting paradigm" (as cited in Malkki 2007, 174). Embodying this promise, AURA experimental practices at the former brown coal mining site indeed generated new collaborations, practices of research, and modes of writing on the borders of the natural and social sciences (see Bubandt and Tsing 2018a; chapters 9 and 11).

However, AURA's commitments to a particular kind of field also separated fieldwork from the "everyday" of university life and limited the scope of the project's queries about the institutional implications of rubber boots methods. As we centered the important task of building new modes of field-based observation at the interface of biology and anthropology, we gave comparatively little attention to our conceptual and material separations of the field and the university, with its offices, labs, classrooms, and seminar locales. My goal in this chapter is to begin this work of expansion. To do so, I intentionally blur the line between *method* and *approach*. In the natural sciences, the term *methods* typically indexes concrete techniques and protocols. But such methods are always deployed within a particular framing—within specific ideas about what constitutes a research question, a field site, and even research itself. In this chapter, I interpret the "methods" of rubber boots methods in this broader sense, as an *approach* or *sensibility* (see also chapters 9 and 10).

I propose this expansion of method in relation to rubber boots practices because they explicitly seek broad, transformative effects on disciplinary divisions and modes of scholarly engagement *(approaches)* via shifts in everyday research modes and field techniques *(methods)*. This assumption that larger structures can be remade via shifts in seemingly routine practices is part of a broad sensibility inherited from feminist political and scholarly traditions. While feminist thinkers have asserted that seemingly mundane issues—such as who washes the dishes—are consequential political practices, so, too, do rubber boots methods assert that how one walks through a landscape matters to broader formations and enactments of research and scholarship.

Beyond the Field

In this vein, I want to briefly show how rubber boots methods have always been emergent from and enacted in spaces beyond the traditionally delineated field site, even though such spaces have received comparatively little attention within the AURA project's descriptions of its own practices. Alongside the brown coal fields, AURA's Slow Seminars—a monthly transdisciplinary reading and discussion series—has been a key site for cultivating rubber boots methods.[6] The Slow Seminars have been primarily textual (with a few forays into film and other forms of visual art), and they have generally taken place in mundane academic settings of off-white walls, florescent lighting, and badly brewed coffee. Yet Slow Seminars have nonetheless produced transformative modes of reading, talking, and listening across disciplines and genres. Like the AURA field excursions, the seminars gathered scholars from diverse disciplinary backgrounds and send them off to explore a messy terrain together—to see what kinds of new insights might arise at the junctures of their different modes of reading, thinking, and noticing (Swanson 2015). We read texts about the ecological dynamics of sea otters and urchins as well as the social behaviors of ants. We dove into the diagrammatic analytical practices of Bruno Latour and the future imaginaries of climate fiction. Like fieldwork, the Slow Seminars were a place where we learned to ask questions in new ways—always querying how and why it might matter to work across scientific and humanistic traditions in a sustained, integrative way, rather than one that is simplistically additive. In a similar spirit to the field practices that the AURA project explicitly considered to be rubber boots methods, the Slow Seminars enacted a kind of experimental practice—a tinkering with and challenge to "normal" reading groups that hewed more closely to disciplinary conventions or narrow, project-focused goals.

By briefly describing the Slow Seminars, my aim is to ask, what is the relationship between cross-disciplinary rubber boots fieldwork and *experimentations with other academic forms*? Is the field of rubber boots methods only "the field" in the conventional sense, or in light of its broader transformative aims, should it be understood in a more expansive way? My answers to the latter should be already apparent: I assert that it is methodologically essential not to overlook the ways that other crucial developments of AURA approaches occurred in seemingly ordinary university settings made less ordinary through their enactment of rubber boots sensibilities. I see this move—of noticing how rubber boots methods are neither limited to the field nor wholly generated within it—as one opening into the broad transformations in research, education, and institutional arrangements that are needed to remake the university as a force for livability in the Anthropocene. As long as rubber boots methods stay "in the field" and the field remains a domain apart from regular university life, their transformative potential is

constrained within the category of research, thus limiting their broader institutional and world-making implications.

In the next sections, I develop these arguments about the importance
and challenges of cultivating rubber boots methods beyond the conventional field site by drawing on concrete experiences with *classroom teaching* at Aarhus University that occurred concurrently with the brown coal
site experiments. Such work indeed shares many similarities with interdisciplinary field research in its experimental messiness, as well as its needs for
pragmatic co-creation and collaboration. Yet while the field is a valorized
site within anthropological and biological imaginaries, the classroom is often
portrayed as a space of content delivery and communication, rather than
of intellectual insight and novel discipline-transforming innovations. Even
when scholars seek to deconstruct the field–desk binary, they rarely invoke
the classroom in a substantial way. Analyses of teaching still tend to be rele-
gated to academic margins—assumed to be mere pedagogy—as other types
of thinking are allowed to claim the vaunted mantle of "original research."
Yet the classroom—like the field—is a relentlessly empirical space and thus
a useful one for experimenting with methods and practices. Holding to the
overall commitments of feminist approaches, I consider the everyday acts
of teaching, in this case, in a rubber boots–inspired mode, as practices that
are in no way lesser than those typically labeled research—and perhaps even
more important for reflecting on the dynamics that iteratively generate the
Anthropocene/Capitalocene, as described via Wright's work at the beginning of this chapter.[7] In doing so, I draw on a personal teaching experience
not as random anecdote but as part of an intentional commitment to long-
standing feminist approaches that are committed to thinking with biograph-
ical and autobiographical experience alongside other kinds of empirical data
(Cotterill and Letherby 1993).

Rubber Boots Teaching?

During the AURA project, two of our members (Pierre du Plessis and I)
were assigned to coteach a master's-level research methods course as part
of an interdisciplinary degree program in human security, a concept borrowed from United Nations (UN) efforts to foreground everyday precarities
rather than narrowly militarized notions of security.[8] The degree program is

fundamentally concerned with how to better analyze and intervene in intertwined social and ecological crises, such as those gathered under terms like climate change, water scarcity, food security, and ethnic violence. As it draws on UN terms, the program at once considers critiques of development and maintains an ongoing commitment to crafting better forms of development practice. Within the Danish educational system, the Human Security Program is somewhat of an outlier in that it is one of the very few programs that includes coursework from both natural and social sciences, with courses in agroecology and tropical ecosystems alongside conflict dynamics. Typically, students in Denmark take specific disciplinary degree programs with relatively rigid course plans and few electives, resulting in educations that tend to be relatively monodisciplinary in comparison to the liberal arts approaches of most U.S. institutions.

In light of the Human Security Program's heterodox approach, the methods course was conceptualized as a survey of common social science techniques, such as participant observation, semistructured interviews, and participatory assessments, with a brief gesture toward natural science practices that might aid students in assessing environmental and agroforestry issues, such as quantitative plant surveys. Although we were excited about teaching a course that sought to work and think across the social and natural sciences in a mode that partially overlapped with the AURA project's field-based experiments, it did not seem that past efforts had been very successful in promoting methodological interdisciplinarity among students. Course reviews from previous years indicated that students were widely dissatisfied with the natural science component, whose quantitative methods of classification and counting they rarely planned to integrate into their own research designs. Because the students were often considering projects that focused on everyday lived experiences of Anthropocene/Capitalocene problems, such as agricultural industrialization, water scarcity, and chemical exposure, we wondered if rubber boots approaches might be of use for their inquiries. We saw the overall AURA "slow science" approach as of likely use to those interested in improving development and aid: an arena where moving too quickly to "solutions"—without adequate attention to the complexities of more-than-human worlds—has caused profound harms, ranging from rural community displacement to soil loss.

Thus, alongside more classic social science practices—such as designing questionnaires and coding notes for discourse analysis—we designed a day of landscape-based noticing, for which we instructed students to bring outdoor footwear and rain gear. Because our teaching took place on a small campus in a rural area, we were able to quickly traverse a variety of landscape forms, from a gravel parking lot to farm fields to managed forests. It is an area whose temporal palimpsest is highly visible to a trained eye: trees bear marks of coppicing, Viking burial mounds dot distant hilltops, and the effects of a large eighteenth-century estate, whose central building is preserved on the campus, are visible in water, soil, and vegetation patterns. Through this site, we aimed to encourage students to consider whether and how noticing more-than-human landscapes might matter for their own upcoming MA research. Our assignment for the day was, in the spirit of AURA, focused on question generation in the field: at the end of the class, students were to propose a hypothetical research topic or question rooted in their curiosity about something they noticed during the outdoor landscape investigations and a plan for how they might further explore it.

Drawing on the landscape-walking and note-taking/sketching techniques of AURA, our own field practices, and a set of readings about arts of noticing,[9] we examined vegetation patterns, animal tracks, property boundary markers, buildings, discarded objects, and signs of landscape management. With attention not only to people but also to other species, we asked, Who lives here now? Who might have lived here in the past? How do different human activities seem to affect more-than-human worlds? As in the AURA project, we brought along natural science guidebooks and hand lenses to help in identifying plants, made sketches of marks on trees, examined maps of landownership and place-names, dipped nets into a pond to look for insects and amphibians, and jotted down information from interpretive signs. As part of the outing, we also emphasized the importance of walking with other scientists as method: archeologists and natural historians accompanied us,[10] pointing out the landscape features they noted, including an abandoned mill site, a planted grove of hardwood trees, and several bird species. Our focus was on how questions and techniques hang together: on how one might want to experiment with new tools to open up new questions at the same time that one may need to learn new techniques to better answer one's questions. Our goal was not to present some of the specific techniques we discussed (including such small things as noting the day's weather) as programmatically applicable for all contexts but to raise overall issues about the range of tools one could use to answer questions about more-than-human worlds and the possibility for given techniques to be used in diverse ways to better understand the relations and histories of a given place. Indeed, we couldn't do much more in the space of a single half-day seminar.

The students were good-natured and happily went along with the outing, but the effects of the course were transitory. When it came time for the end-of-the-course evaluations, many students thought the rubber boots methods of field sketching and journaling, species identification, and walking transects should be cut from the syllabus and replaced by more "practical" skills, such as the use of data analysis programs that they could list on their CVs to market themselves and increase their employability. They seemed to experience rubber boots methods as a diversion—while not offensive, such methods were "unproductive" and unnecessary. They didn't see themselves using such tools in future careers at the UN or development-oriented nongovernmental organizations. While part of the seeming irrelevance of rubber boots methods was undoubtedly due to our own pedagogical inadequacies, I think this case can nonetheless prompt some meaningful reflections about the challenges of "slow science" in the Anthropocene/Capitalocene.

Detailed, place-specific research is clearly necessary for addressing the complex social-political-ecological issues that Human Security students engage. But the students' experiences of rubber boots methods point to their temporal mismatch with currently dominant employment and educational logics. It takes time to learn and to use both rubber boots approaches and their specific techniques because they are dependent on excess, mistakes, and detours, which can prove transformative, but will not always be so. While this is true for all research and learning to some degree, the demands are even greater for field methods and greater still for those that involve open-ended observation for question generation. Furthermore, it takes time to learn techniques from a discipline that is not one's own with enough technical and epistemological understanding of them to redeploy them in new contexts. For example, only a small subset of the field biology techniques that we attempted to introduce to Human Security students will likely end up being relevant for their own research projects. While this very excess—of techniques, knowledge, and field journal notes that cannot be streamlined toward a clearly determined goal—sits at the core of the rubber boots approach, it was also precisely the source of the students' frustrations.

Students around the world have been increasingly encouraged to think about education in economic terms—that is, as an "investment" in their future employability—and such logics have been structurally incorporated into Danish universities in a particular way.[11] In contrast to U.S. universities, where such conversations work through configuring universities as revenue-generating corporations and students as consumers who take on debt to buy a product, Danish universities—which are government financed rather than funded by tuition—use performance metrics to monitor how well educational programs produce employable citizens. The primary criterion for evaluating degree programs, or majors, is the percentage of its graduates who finish their degrees quickly and are employed twenty-four months after completing their education. The system is structured in ways that incentivize both students and instructors to focus on swift progression and timely employment. To offer just one example, if the employed percentage is low, the number of students allowed to select that program in future years is reduced, along with the number of faculty members in the field. Such structures are contested by faculty in many ways—yet pressures remain.

We considered trying harder to convince the Human Security students that rubber boots methods could be useful job skills, but to adopt this course of action would, of course, implicitly reinforce conceptions of university education as preparation for employment within established, growth-oriented capitalist frames. This dilemma and its related temporal conflicts are in no way unique to rubber boots methods and pose profound challenges across wide swaths of the university. Yet they nonetheless manifest in particularly acute ways in relation to the open-ended, field-based practices of anthropology and natural history, which infuse the AURA approach, as well as in interdisciplinary contexts that demand serious understanding of multiple academic traditions.[12] To fully accept the rush toward standard, growth-oriented futures also runs counter to the broader insights derived from the AURA method itself: that such processes fundamentally transform everyday landscapes, often in damaging ways. Urgent more-than-human crises typically need approaches that depart from sameness of common sense, including the notion that current economic logics can be maintained as the planet continues to warm and unravel.

Slow science and its teaching correlates are not about delaying action in regard to such pressing concerns: instead, it is about rejecting the "fast" demands of contemporary capitalism and its institutions, including universities, to increase production and growth. Because rubber boots methods and other "slow science" techniques are intentionally disruptive of such dominant logics, it is not possible simply to add them on to a standard methods syllabus as an extra "bonus" day or quick skills package without reconsidering larger questions of course, program, and institutional structures.

This anecdote does not conclude with any kind of heroic resistance. Capitulating to the system, we removed rubber boots methods from the course altogether; we now keep students dry and out of the mud on their rush through the university and into a workplace. Thus, even in a model Scandinavian welfare democracy with free higher education, the commonsense principles of the university embody versions of the economic productivity rationales that undergird more violent and spectacular forms of planetary destruction.[13]

Lest this read as a trivial exposition, I want to emphasize that this failure to incorporate rubber boots methods into our teaching leads me back to the core concern of this chapter: *a worry about the ways that rubber boots, despite their inclusive and low-tech aesthetics, are often limited to particular spaces* (in this case, the former brown coal mining area). Although our effort to add rubber boots approaches to an existing class was an important experiment, it ultimately led us to reflect on the substantial tensions and barriers involved in pulling rubber boots methods into other university contexts and spaces.

A Danish University as Field Site

A Danish university is in many ways similar to the brown coal fields in terms of the nonspectacularity of its capitalist forms. Like the brown coal site, which pales in the severity of its damage when compared to landscapes of extraction in the Global South, Danish universities are a site of "mild apocalypse"—of banal capitalist damage tempered by Scandinavian social democracy. In contrast to the United States, where university students must typically take on large debts to attend college (Krabbe 2018), Danish higher education appears at first glance to be a socialist paradise: it is tuition-free, and most students are given a monthly stipend to cover their living expenses during their period of study. Yet, while Danish students are not burdened by the extreme student poverty, fears of bankruptcy, and decades of loan payments that are commonplace among young Americans, they are nonetheless thoroughly ensconced in a system that is similarly oriented toward employability and capitalist futures. Where U.S. students are encouraged to "invest wisely" in a university major that will "pay off" via solid career options and postgraduation earnings, Danish universities are more structurally linked to business worlds via state actions that orient education toward forms of normative economic productivity that continue to foster the logics within which Anthropocene/Capitalocene conditions are iteratively generated, even as the nation leads the way in the manufacture of wind turbines.

My aim is not to critique either the Danish higher education system or the Danish economy, as both are far less ruinous than their counterparts in most countries. Rather, it is to explore the limitations and possibilities of rubber boots methods where this approach congealed. AURA project members frequently invoke its Danish setting as one of the factors that enabled the development of its methodological approach—both intellectually and institutionally. While natural history observation is on the wane in Denmark, as it is in many places, its remnants remain relatively strong, for example, many Danish biologists (sometimes called "rubber boots biologists"; see the Introduction to this volume) are still capable of species identification in the field, even as they engage in big data research. Furthermore, Denmark has long valorized creative interdisciplinary thinking, particularly via public celebration of Niels Bohr, a Nobel Prize–winning Dane who developed quantum mechanics at the borders of physics and philosophy. The AURA project was also made possible through generous funding from the Danish National Research Foundation, whose grant programs reflect a willingness in Denmark and Europe more widely to sponsor large-group humanities- and social science–led collaborations in a way that is almost unheard of in the United States, where such funding structures are largely limited to the biosciences.

But while Denmark's research infrastructure provided the very conditions of possibility for rubber boots methods, *its educational and wider university forms have also curtailed their reach.* As grants create clusters of scholars who are encouraged to produce research that meshes with the conceptual frames and buzzwords of neoliberal universities—"innovative," "cutting-edge," "international," and "interdisciplinary"—they also locate this scholarship within centers and institutes. This structure generates wonderful spaces for creative play in part by separating them from the everyday business of educational, institutional, and economic replication. Although the research center as state of exception offers powerful affordances, including conditions for rubber boots fieldwork, it can also domesticate and politically limit research that questions and challenges wider university forms and logics.

RUBBER BOOTS FAILURES AS INFRASTRUCTURAL INVERSION

Universities are not mining companies or finance capital firms—the most eponymous Anthropocene/Capitalocene actors. Yet, over the past decades, they have indeed been remodeled into engines of economic growth, as producers of knowledge workers and capitalizable innovations. Thus it seems urgent that we consider the places—metaphorically and literally—of anti-Capitalocene scholarship in university worlds. Situated site-based research immediately foregrounds questions of temporality, as its uncertainties and slow pace run counter to the productivity demands of the corporate university. Because field research takes time, it resists easy capitalization, even as many field sciences are emergent from capitalist projects of forestry, rangeland management, agricultural development, and other forms of resource exploitation.

Perhaps counterintuitively, the failure of rubber boots methods to (as of yet) transform institutional structures or the research methods class may be part and parcel of its intellectual and political contributions. Regardless of discipline, research is often characterized by experimental tinkering and methodological failures, such as bugs in computer models and troubles with lab protocols. Yet rubber boots methods also fail in a more structural way—one akin to that described by cultural studies scholar Jack Halberstam (2011) in *The Queer Art of Failure.* Halberstam describes queer failure as the impossibility of living up to dominant heteronormative expectations. Though often violently punished, such failures can also be productive of other ways of living. For Halberstam, failure highlights the very structures and norms that foster social conformity, as well as how such norms marginalize those who do not fit their espoused ideals. Failure, here, is not a mark of individual insufficiency. It is, in Halberstam's words, "a way of refusing to acquiesce to dominant logics of power and discipline"—a form of critique that opens up already existing alternatives that dominant structures and ideologies deny (88). This kind of failure is not a necessary bump in the road on a journey to success but rather a process that brings the very definition of "success" into question.

Despite the different context, I propose that rubber boots methods and their relation to the university can be usefully understood in relation to Halberstam's work. Rubber boots methods and their failures can all too easily be contained within normative framings of research where high-risk creativity is celebrated and promoted as a strategy for innovation, university promotion/branding, and, ultimately, capitalization. This is the failure of Google and Silicon Valley—playful and necessary experimental mishaps in the service of economic growth and professional success. Rubber boots methods can tap into such discourses, presenting its methods as groundbreaking to university administrators and mainstream grant funders when they want to highlight novel modes of interdisciplinary research. Indeed, such slippage is strategically useful, as it helps garner resources for important work.

But its limitations become especially apparent when one seeks to move rubber boots methods from the spaces of the research group and the field project to the classroom. If one were to work for the expansion of rubber boots *within* existing university languages, teaching rubber boots methods would take the form of promoting transdisciplinarity as a broadly employable skill—as a capital-friendly, flexible-thinking technique that imagines it can solve environmental crises within existing political-economic structures. In certain contexts, this might be a reasonable tactic. If more university graduates were to value the skills of interdisciplinary field-based observation alongside computer programming and modeling as "knowledge," it would open up new possibilities—even if that "knowledge" continues to be rendered in the terms of the "knowledge economy." However, if we are serious that rubber boots methods are *methods for the Anthropocene/Capitalocene,* this is simply not enough. Such methods aim not only to describe and analyze Capitalocene conditions but also to offer skills and techniques that can serve as a foundation for fostering more livable worlds.

How, then, can we allow the failures of rubber boots—including those in the classroom—to do more of this transformative work, especially in relation to university institutions? The failure of rubber boots approaches to move beyond the field and into existing institutional and educational paradigms can be read as failures of the Halberstamian kind: as an inability of rubber boots approaches to conform to normative expectations for university practice beyond the domain of cutting-edge research. The reason to emphasize these failures is that they perform a version of what science and technology studies scholar Geoff Bowker has called "infrastructural inversion," a moment when underlying structures that are typically backgrounded are brought to the fore (Bowker 1994; see also Bowker and Star 1999). In the case of rubber boots approaches, it is their failures to find traction in wider university spaces that make visible how everyday university life is configured within infrastructures of capitalist production, including its demands for growth, speed, and linear productivity.

These kinds of institutional questions—and the ways we might intervene in them—are not external to questions about Anthropocene approaches and methodologies but central to them. In her contribution to the recent edited volume *Manifesto for Living in the Anthropocene,* Jenny Cameron (2015), an Australian geography professor, describes the importance of experimentation, including new modes of collaboration. She writes, "If we are now living in a planetary experiment, perhaps social research could also take a more experimental approach" (99). It is important to see this space of experimentation as extending beyond the traditional field site, not merely as an abstract principle, but because the tentacular enactments of economic growth—which are at the heart of most ecologic and geologic change—compel such moves. The reason for rethinking the spatially bounded field site (and expanding it in certain ways) is in no way a rejection or critique of the arts of field noticing or of natural history–inspired methods, which are fundamental to practices of critical description (Tsing 2013). Yet considering the limitations of the conventional field framing is also necessary to more fully reflect on the core stakes of rubber boots approaches, including their world-making aims. Explicitly putting the classroom into dialogue with the traditional field site forces one to confront urgent questions about how to torque the institutions and economic growth machines within which one is embedded. In this case, it spurs more robust inquiries and reflections about how commitments to field-based research raise questions about the need to more substantially experiment with university education and institutional practice.

The primary methodological technique foregrounded in this chapter is the more robust application of feminist practices and expanded notions of the field to rubber boots methods. It is also a call to stay with the rubber boots method of allowing research questions themselves to be emergent within encounters—at a brown coal site, in a classroom, or at an institutional meeting. The teaching experiments described in this chapter provoke wide-ranging questions about the implications and transformative potentials of rubber boots approaches, including the very question of what a truly more-than-human, or rubber boots, university would be. Based on research from biology, anthropology, and other fields, such a university would have to take seriously that economic growth may not be able to be rendered fully compatible with thriving ecologies. It would have to confront the likelihood that a green transition might require more substantial changes than technological innovation and fewer airline flights. If one were to take such calls for broad economic change seriously, how might one enact everyday teaching in different ways? How long and in what forms would education be undertaken? What kinds of coteaching and, indeed, degree programs and departments would be formed? What would a course plan attuned to a rubber boots approach look like? How would it move across spaces, disciplines, topics, and methods? Rubber boots education would certainly involve getting outside and engaging in arts of noticing more-than-human relations but also new modes of reading, studying, and collaborating. It would emphasize the importance of long-term research while also stressing the possibilities within forms of broad knowledge alongside areas of in-depth expertise. It would teach critical analysis of economic structures but also practices of pragmatic alliance and negotiation, and it would cultivate curiosity and appreciation of intricate beauty alongside techniques for mapping damage, inequality, and extraction.

Ongoing experiments—experiments that fail in substantial and frustrating ways—are integral to honing these questions and developing more precise propositions. Rubber boots methods need to fail more frequently within classrooms and other university spaces as we simultaneously develop further tools for allowing those failures to better perform the work of infrastructural inversion and thus to matter in more robust ways. Bringing university spaces, such as classrooms, into view requires abandoning hierarchies of field/desk and rethinking the institutional, economic, and world-making politics of bounded research spaces. One might argue that my insistence on enlarging the field risks rendering rubber boots meaningless by broadening them so much that they lose coherence—that importance of field noticing begins to slip from view. On the contrary, this chapter seeks to ask how universities—in the classroom and as institutions—would be different if they took the insights of more-than-human field-based noticing seriously. It is a plea to think more substantially about how world-making methods might, in tentative and provisional ways, begin to remake the worlds in which they are embedded—through their failures and inversions as well as through their constrained successes. To do so is to take seriously the challenge of rubber boots approaches: to experiment with other enactments of academia in response to Anthropocene/Capitalocene challenges.

Arts of noticing, as AURA enacts them as a field practice, include attention to morphology, form, and pattern. If one turns those arts toward an examination of the shape of contemporary higher education, what one sees is worrisome—universities that are ever more bound to the projects of neoliberal growth at the heart of the Anthropocene/Capitalocene and whose alignments with capital constrain the reach and scope of rubber boots approaches. At the Søby brown coal beds, which feature in several of the other descriptions of rubber boots methods in this volume,[14] we examined patterns of deer-browsing activity, geyser holes from postmining land subsidence, tomato plants at a former municipal dump site, and tree–fungi symbioses in eroding mounds of sandy overburden. In our work, we were committed to staying with the site—spending time to get to know the dynamics of a patch, in order to better understand how the Anthropocene is "patchy," that is, interconnected in difference. The patch of the university is *not* the same as that of the brown coal beds, but they can—and must—inform each other if we are to better diagnose and torque processes of more-than-human damage.

NOTES

1. This chapter thinks specifically with dilemmas and floor types that are frequent among northern European institutions. However, it is important to note that universities take many forms—with many different floor types. Furthermore, regardless of location, people cross floors in a wide range of footwear, such as boots, loafers, heels,

flip-flops, wheelchair wheels, and crutches, which leads to diverse relations with universities and their material forms.

2. This argument is developed in Swanson (2017) and Cova and Swanson (forthcoming), inspired by Haraway (2008) and Tsing (2015).

3. On "lateral theorization," see Brichet and Hastrup (2020) and Hastrup (2012).

4. My phrase "staying with the site" gestures to Haraway's (2008, 2016) concept of "staying with the trouble" in that the *site,* within the AURA context, is the "trouble" in Haraway-ian terms, that is, messy entanglements of bodies and histories.

5. Although rubber boots methods are alert to the colonial and racial histories of natural history, they could more explicitly consider how such continue to affect arts of noticing. Field-based noticing practices can be variously amplified and limited by state violence and other exclusions: as Tsing (2015) has noted, Southeast Asian war refugees cultivate in-depth forest knowledges and find spaces of partial possibility as they forage mushrooms along the U.S. West Coast. At the same time, more formalized natural history methods remain largely white and upper class, as outdoor recreation remains inhospitable to nonwhite communities. A recent case highlights the risks of "birding while black" (Lanham 2016): when Christian Cooper, an avid bird watcher and Audubon Society board member, asked a white woman to leash her dog, she threatened him with the possibility of racialized police violence by falsely reporting that a Black man was harassing her. Such issues are part and parcel of natural history and must be actively inherited in the process of recuperating and transforming its methods. The interface of natural history and disability studies also remains underexplored, as field methods—linked to field sports and the cultivation of masculine virility—are haunted by ideas of a normative body.

6. Other key events for developing rubber boots methods include two interdisciplinary conferences: "Anthropocene: Arts of Living on a Damaged Planet," held at the University of California, Santa Cruz in 2014, and "Anthropocene Woodlands," held at Aarhus University in 2018. For descriptions of the insights they sparked, see Tsing et al. (2017) and Swanson et al. (2021), respectively.

7. Feminist analyses of university practice have pointed toward gendered dimensions of university labor, including the lower status accorded to tasks like mentoring, teaching, and everyday administration. See Heijstra et al. (2017) for an analysis focused specifically on Iceland, but with citations to overall conversations. See also https://theconversation.com/why-higher-ed-needs-to-get-rid-of-the-gender-gap -for-academic-housekeeping-82135.

8. For information about the program's use of the human security concept, see https://kandidat.au.dk/en/humansecurity/ and https://humansecurity.dk.

9. Readings included excerpts from Wessels (1997), Rocheleau (1995), and William Cronon's website on landscape exploration: http://www.williamcronon.net/ researching/landscapes.htm.

10. We thank Peter Hambro Mikkelsen and Mette Løvshal, who took on this role at various iterations of the course.

11. My analysis of Danish universities is based not only on participant observation but also on the work of scholars including Nielsen (2015) and Susan Wright.

12. Anthropology education in Demark is especially affected by this squeeze, which makes it difficult to read the discipline's characteristic book-length ethnographic monographs and threatens the traditional extended fieldwork–based MA thesis, and many colleagues creatively challenge these dynamics.

13. For an account of a similar experiment in an American context, see Caple (2017).

14. See also a special section dedicated to the brown coal site research (Bubandt and Tsing 2018b).

REFERENCES

Amit, Vered, ed. 2003. *Constructing the Field: Ethnographic Fieldwork in the Contemporary World.* London: Routledge.

Anderson, John. 2017. "Why Ecology Needs Natural History." *American Scientist* 105, no. 5: 290–98.

Bell, Diane, Pat Caplan, and Wazir Jahan Karim, eds. 1993. *Gendered Fields: Women, Men and Ethnography.* London: Routledge.

Bowker, Geoffrey C. 1994. "Information Mythology: The World of/as Information." In *Information Acumen: The Understanding and Use of Knowledge in Modern Business,* edited by Lisa Bud-Frierman, 231–47. London: Routledge.

Bowker, Geoffrey C., and Susan Leigh Star. 1999. *Sorting Things Out: Classification and Its Consequences.* Cambridge, Mass.: MIT Press.

Brichet, Nathalia, and Frida Hastrup. 2020. "Ethnography, Exhibition Practices and Undisciplined Encounters: The Generative Work of Amulets in London." In *Exhibitions as Research,* edited by Peter Bjerregaard, 53–64. London: Routledge.

Bubandt, Nils, and Anna Tsing. 2018a. "Feral Dynamics of Post-industrial Ruin: An Introduction." *Journal of Ethnobiology* 38, no. 1: 1–7.

Bubandt, Nils, and Anna Tsing, eds. 2018b. "Feral Dynamics." Special issue. *Journal of Ethnobiology* 38, no. 1.

Cameron, Jenny. 2015. "On Experimentation." In *Manifesto for Living in the Anthropocene,* edited by Katherine Gibson, Deborah Bird Rose, and Ruth Fincher, 99–101. Santa Barbara, Calif.: Punctum Books.

Candea, Matei. 2007. "Arbitrary Locations: In Defence of the Bounded Field-Site." *Journal of the Royal Anthropological Institute* 13, no. 1: 167–84.

Caple, Zachary. 2017. "Holocene in Fragments: A Critical Landscape Ecology of Phosphorus in Florida." PhD diss., University of California, Santa Cruz.

Coleman, S., and P. Collins, eds. 2006. *Locating the Field: Space, Place and Context in Anthropology.* Oxford: Berg.

Cotterill, Pamela, and Gayle Letherby. 1993. "Weaving Stories: Personal Auto/biographies in Feminist Research." *Sociology* 27, no. 1: 67–79.

Cova, Victor Sacha, and Heather Anne Swanson. Forthcoming. "Reframing the Social, Rethinking the Body, Confronting Biologism." In *Cambridge Handbook on the Anthropology of Gender and Sexuality.* Cambridge: Cambridge University Press.

Custock, Adam. 2020. "How Rewilding Sciences Are Made and Matter." MA thesis, Aarhus University.

Falzon, Mark-Anthony, ed. 2016. *Multi-sited Ethnography: Theory, Praxis and Locality in Contemporary Research.* London: Routledge.

Futuyma, Douglas. 1998. "Wherefore and Whither the Naturalist?" *American Naturalist* 151, no. 1: 1–6.

Geller, Pamela L., and Miranda K. Stockett, eds. 2007. *Feminist Anthropology: Past, Present, and Future.* Philadelphia: University of Pennsylvania Press.

Gupta, Akhil, and James Ferguson, eds. 1997. *Anthropological Locations: Boundaries and Grounds of a Field Science.* Berkeley: University of California Press.

Halberstam, Jack. 2011. *The Queer Art of Failure.* Durham, N.C.: Duke University Press.

Haraway, Donna J. 2008. *When Species Meet.* Minneapolis: University of Minnesota Press.

Haraway, Donna J. 2016. *Staying with the Trouble: Making Kin in the Chthulucene.* Durham, N.C.: Duke University Press.

Hastrup, Frida. 2012. "Shady Plantations: Theorizing Coastal Shelter in Tamil Nadu." *Anthropological Theory* 11, no. 4: 425–39.

Heijstra, Thamar Melanie, Finnborg Salome Steinthorsdóttir, and Thorgerdur Einarsdóttir. 2017. "Academic Career Making and the Double-Edged Role of Academic Housework." *Gender and Education* 29, no. 6: 764–80.

Krabbe, Mathias. 2018. "A Means to an End: Aspirations and Realities of American Student Loan Borrowers." MA thesis, Aarhus University.

Lanham, J. Drew. 2016. *The Home Place: Memoirs of a Colored Man's Love Affair with Nature.* Minneapolis, Minn.: Milkweed Editions.

Malkki, Liisa. 2007. "Tradition and Improvisation in Ethnographic Fieldwork." In *Improvising Theory: Process and Temporality in Ethnographic Fieldwork,* edited by Allaine Cerwonka and Liisa Malkki, 167–88. Berkeley: University of California Press.

Marcus, George. 1995. "Ethnography in/of the World System: The Emergence of Multi-sited Ethnography." *Annual Review of Anthropology* 24, no. 1: 95–117.

Marcus, George. 2006. "Where Have All the Tales of Fieldwork Gone?" *Ethnos* 71, no. 1: 113–22.

Nielsen, Gritt. 2015. *Figuration Work: Student Participation, Democracy and University Reform in a Global Knowledge Economy.* Oxford: Berghahn Books.

Noss, Reed. 1996. "The Naturalists Are Dying Off." *Conservation Biology* 10, no. 1: 1–3.

Rocheleau, Dianne. 1995. "Maps, Numbers, Text, and Context: Mixing Methods in Feminist Political Ecology." *Professional Geographer* 47, no. 4: 458–66.

Stengers, Isabelle. 2016. "'Another Science Is Possible!': A Plea for Slow Science." In *Demo(s),* edited by Hugo Letiche, Geoffrey Lightfoot, and Jean-Luc Moriceau, 53–70. Boston: Brill Sense.

Swanson, Heather Anne. 2015. "Who Is in the Room? The Importance of Multi-disciplinary Spaces for Anthropology and STS." *HAU: Journal of Ethnographic Theory* 5, no. 1: 445–48.

Swanson, Heather Anne. 2017. "Methods for Multispecies Anthropology: Thinking with Salmon Otoliths and Scales." *Social Analysis* 61, no. 2: 81–99.

Swanson, Heather Anne, Jens-Christian Svenning, Alark Saxena, Robert Muscarella, Janet Franklin, Matteo Garbelotto, Andrew S. Mathews et al. 2021. "History as Grounds for Interdisciplinarity: Promoting Sustainable Woodlands via an Integrative Ecological and Socio-cultural Perspective." *One Earth* 4, no. 2: 226–37.

Tsing, Anna Lowenhaupt. 2013. "More-Than-Human Sociality: A Call for Critical Description." In *Anthropology and Nature,* edited by Kirsten Hastrup, 27–42. New York: Routledge.

Tsing, Anna Lowenhaupt. 2015. *The Mushroom at the End of the World: On the Possibility of Life in Capitalist Ruins.* Princeton, N.J.: Princeton University Press.

Tsing, Anna Lowenhaupt, Nils Bubandt, Elaine Gan, and Heather Anne Swanson, eds. 2017. *Arts of Living on a Damaged Planet: Ghosts and Monsters of the Anthropocene.* Minneapolis: University of Minnesota Press.

Wessels, Tom. 1997. *Reading the Forested Landscape.* Woodstock, Vt.: Countryman Press.

Wright, Susan. 2017. "Can the University Be a Liveable Institution in the Anthropocene?" In *The University as a Critical Institution?,* edited by Rosemary Deem, 15–37. Boston: Brill.

Troubling Methods in the Anthropocene

A Roundtable Discussion

KIRSTEN HASTRUP, URSULA MÜNSTER,
ANNA TSING, *and* NILS BUBANDT

This is an edited transcription of a conversation about some of the challenges we face in answering the question of how to include other-than-humans, their lives and their traces, in social science accounts of life on a planet dominated by humans. The conversation took place on November 26, 2018, in Aarhus.

NILS: We modern humans have made our mark on the world. Our colonial boot prints, our carbon footprints, our toxic traces, our boreholes, and our oil spills are evident in the soil of the earth, in the water of the ocean, in the atmosphere. We have stamped out numerous forms of life and decimated the number of most species. Forty percent of the number of the world's wild animals has disappeared in the last forty years. I am not talking species here: I am talking number of individual animals. As recently as in the early 1970s, there were almost twice as many wild animals in the world as there are today. We and our four main species of domesticated mammals (cows, pigs, sheep, goats) weigh more than twenty times as much as all wild mammals put together. However, there are other footprints besides those of humans and their domesticated mammals. The Great Acceleration notwithstanding, human prints, tracks, and traces are not the only ones one earth. Is it not time that we—we all in general and we social scientists in particular—begin to pay attention to them? Indeed, this is the question that the multispecies turn in social science and multispecies ethnography have asked and affirmed. The question that drives this anthology has been the logically next one: *how* do we go about doing that, not just in theory, but in our scientific practice? The Anthropocene is a multispecies event, a multispecies crisis. It is also, arguably, an invitation to multispecies analysis—to experiments with methods that span the social and natural sciences, the arts and the academy, in the recognition that we need a variegated toolbox to better notice the complex relations in time and space that link many kinds of humans to many kinds of nonhumans.

Rubber boots methods is a tongue-in-cheek term to describe the diverse set of methods from biology and beyond that researchers associated with AURA—Aarhus University Research on the Anthropocene—have been experimenting with in the last five years: methods for how we might empirically trace and critically describe co-species relations in disturbed landscapes. This set of methods is based on a simple claim that also sparked the AURA project initially: in order to study the marks, the tracks, and the imprints that humans, along with their companion species and companion technologies, are making on the planet in their conjunction and disjunction with those marks and traces left by other forms of life, anthropologists and their disciplinary allies need to try out new methodological rubber boots—allow these co-species relations and changing landscapes to challenge our methods as we follow the tracks of multiple species across historical landscapes. You might object that this is nothing new. Anthropologists have for a long time done fieldwork in all kinds of methodological footwear attuned to the terrain in which they have found themselves, be it in boots, flip-flops, or bare feet. This objection is of course true. Equally true, however, is that the tracks in the terrains followed by anthropologists and other social scientists have rarely been other-than-human. Social methods for tracing these tracks and their intersections with human tracks are therefore poorly developed. If we are interested in following also those more-than-human forms of world-making in their disjointed relation to human world-making, how does this challenge the conventional methods—the methodological footwear, as it were —of social science? What equipment, in the broad sense, does the charting of such terrain require? What might we learn from other field-based disciplines—biology, environmental science, political ecology, landscape history—that already have long-established methods for following the tracks of more-than-human sociality? How might this trouble conventional social science, and how will we need to trouble those methods themselves? The way in which one answers these questions has implications for what kind of transdisciplinary methods one can imagine and build. I am hoping that this conversation could touch on some of the challenges and possibilities of transdisciplinary experiments with methods in the Anthropocene along four lines of inquiry:

Critical description: how do we take Anthropocene theory beyond concepts into situated analysis and critical description of ruined, patchy landscapes?

Blending methods: how do we collaborate across social and natural sciences after the science wars?

Decolonizing methods: how do we push multispecies methods beyond a Western and male Anthropocene and bring in multiple Anthropocene realities?

Desecularizing methods: how might multispecies methods be retooled to notice also the many uncanny dimensions of a nonsecular Anthropocene?

CRITICAL DESCRIPTION

NILS: Let us jump right into the first set of issues. A lot of the responses from the human and social sciences to the concept of the Anthropocene have been conceptual: they have begun and ended in theory. They have engaged in conceptual critique, in critical discussion of the concept itself, in epistemological reflection. Is *Anthropocene* the right word, or should it be another (Haraway 2015; Moore 2016)? What is the genealogy of the *anthropos* in *Anthropocene* (Grusin 2017)? What does *Anthropocene* leave unsaid (Malm and Hornborg 2014)? What are its politics, enchantments, blind spots (Baskin 2015; Cook and Balayannis 2015; Vetlesen 2019)? Why did it emerge when it did, and what kinds of futures does it envision and block (Yusoff 2018)? These are important critical and conceptual questions. But critical questions of this kind also raise their own counterquestions: Is conceptual critique enough? Is there not more to an anthropological response to the Anthropocene than conceptual analysis and philosophical critique?

Anna: The way you ask the question makes me think about what happened in anthropology, and cultural theory more generally, that stimulated the initial discussions of the Anthropocene as questions of concepts and philosophies—and how AURA tried to do something different than this. There was a history behind the original impulse to abstraction, a history in which culturally oriented social scientists felt that the most important contributions were to interrupt unquestioned dogma through conceptual rethinking. In development studies, for example, Arturo Escobar (1995) reopened the question of what development means. In science studies, Bruno Latour (1988) dissected scientific empiricism. When discussion of the Anthropocene came along, cultural theorists jumped immediately to examine its contours as a philosophical concept—and as a word (e.g., Malm and Hornborg 2014; Haraway 2015; Gilroy 2015).

AURA has been an attempt to push anthropological responses to the Anthropocene in a different direction: away from conceptual critique toward new forms of empirical engagement with the world, yet with a critical theory eye, that is, what we've called "critical description." Of course, this is not to say that critical descriptions are so new. In fact, anthropologists have been doing theoretically engaged empirical fieldwork since the beginning of the modern field. But, more recently, empirical description passed out of the attention of ambitious scholars. AURA was an attempt to revitalize and bring back for a new time the kinds of fieldwork empiricism for which anthropologists have been known. But, of course, to add this element: that social relations, the stuff that was at the center of the field of anthropology, involves nonhumans as well as humans. This was an integral dimension of what we have called "critical description" (Bubandt and Tsing 2018; Tsing 2013). With this in mind, there are so many new problems to examine, and they challenge us both empirically and theoretically. Far from being boring and uninspired, this kind of description allows a new moment of engagement with the world, in all its messy pleasures and dangers.

Kirsten: What Anna is saying here is very important, I think, and I would like to quote her view of what we should mean by "critical description" not being self-evident. She suggests that it is "critical, because it asks urgent questions; and description, because it extends and disciplines curiosity about life" (Tsing 2013, 28). The extension implies that we should not break away from our own tradition of going out there, into the field, to see what actually happens there, as we have always done. But we must do so with a broader perspective in mind, one that incorporates more-than-human socialities and requires more openness toward other disciplines. This may gradually refashion anthropology, but then it has always been refashioned in the wake of new fieldworks, changing global relations, and emerging theoretical vocabularies.

In other words, I believe that critical description is not so much a conceptual issue as it is a way of approaching the field (of whatever making) with a sense of humility that will allow the field to reveal itself, before it is classified and boxed. This is possibly the only way to deal with the Anthropocene in a truly anthropological spirit, being both localized and analytical. What I struggle with in Nils's question is the idea of an Anthropocene *theory*. The Anthropocene is a concept that has been invented by humans and has been used in different ways, and by different people from different sciences, often stopping the discussion at the fate of the planet rather than opening it up for new insights into the details of lived life in multispecies communities. In anthropology, it has equally been used in many different ways, and we must of course pay heed to its implications, yet in practice, anthropology has never really moved beyond situated analysis, being its hallmark since the beginning of the discipline in the early twentieth century. So, the Anthropocene may play a prominent conceptual role, but anthropology has many details to add to it—some of them emerging out of new multispecies fields—which may undermine the essentialist discussion of what the Anthropocene *is*. With Donna Haraway (2015, 160), I believe that our job is to make the Anthropocene as short/thin as possible. The term may have outrun itself in the process of seeking a definition and become more of stumbling block for interdisciplinary discussion of global multiplicities and interdisciplinary challenges.

Going back to the notion of "method" as the main issue in this part of our discussion, I want to stress that methods are just practices; they are not fixed notions, concepts, or relations. They are practices by which we attempt to understand whatever world we are studying from our different vantage points. Sometimes the methods direct us in particular directions; at other times, it is the inverse: particular questions and concerns prompt new methods, new collaborations, new ways of seeing the world. I think the recent urge to collaborate with natural scientists, notably biologists, is a sign that we are taking situated analysis seriously; the notion of the Anthropocene (for all its shortcomings) has made it impossible to overlook how deeply the landscape is marked by human footprints. Humans are all over the place not just as persons but also in the traces left over thousands of years, in having affected other species, for instance, in the global megafauna development at the Pleistocene–Holocene transition (Sandom et al. 2014). Instead of discarding the past as "history" or even "prehistory," we should take these histories seriously as "contemporary" and see what we can learn from them as partially human footprints. In the process, we might reassess the notion of critical description as a way of getting beyond what we can immediately *see* in any landscape. In other words, there are many different layers of concern in this opening discussion of the notion of methods. What are they, apart from motley knowledge practices?

URSULA: AURA's call for a situated analysis and critical description of more-than-human constellations in the ruined landscapes of the Anthropocene resonates with me. However, I also think that a conceptual critique of the term *Anthropocene* coming from the social sciences and humanities has been very useful. Especially those interventions that have shown us that not all humans are equally responsible for causing our current moment of ecological and climatic crisis and that a critical description of multispecies relations in the Anthropocene needs to be situated in the political ecologies of capitalist destruction, Indigenous dispossession, and colonial violence (Moore 2017; Haraway 2015; Tsing et al. 2017; Chakrabarty 2009). Other notions, such as the "Capitalocene," "Plantationocene," or "Chthulucene" (Bonneuil and Fressoz 2015; Haraway 2015; Moore 2015, 2016), might be more analytically adequate for describing the current planetary moment. In my own work, I have used the notion of Anthropocene as an invitation to "undiscipline" (Armiero et al. 2019, 2) myself in a time of ecocide and climate crisis. The Anthropocene has taught us that the traditional separation between disciplines concerned with "nature" and those that study "culture" no longer makes sense. The term demands a bold extension of our methods beyond the confines of one discipline. Many scholars in the environmental humanities, committed to understanding our contemporary crisis, are creatively borrowing methods from many disciplines and working collaboratively across boundaries, together with scientists and artists, to situate the Anthropocene in time and space and to trace its histories. To address the Anthropocene as a multispecies crisis, and to come to a closer understanding of nonhuman lifeworlds, I agree with Bruno Latour (2016, vii) that "all the resources of science and the humanities have to be put to work."

I can give an example of my ethnographic work at a conflicted natural–cultural contact zone in South India. While doing ethnographic field-work at the forest edges in Kerala, I realized that for understanding the troubled relations between people and other species in this postcolonial landscape, my training in cultural anthropology soon reached its limits. To grasp, for example, how the forest had become a violent space of clashes between elephants and people, insights from behavioral biology, forest ecology, environmental history, disease ecology, and botany were essen-tial and helped me to ask better questions. To tell the stories of inter-species conflicts, I needed to engage with the simplified ecologies of a timber plantation and the complexities of social and psychological ele-phant lives, where traumas caused by histories of violent capturing and habitat loss are passed down over generations, turning some animals into violent killers. Violent interspecies clashes are situated in an anthropo-genic landscape where histories of colonialism and capitalist extraction play out within a broader "web of life" (Moore 2015).

Going beyond a conceptual critique of the term *Anthropocene* and mov-ing toward a "situated analysis and critical description of patchy land-scapes" can also be very productive when experimenting with new ways of interdisciplinary environmental teaching. The term *Anthropocene* gives us the chance to learn and creatively apply the skills and methods of dif-ferent disciplines to address the multiple planetary crises we are cur-rently facing—of climate, extinction, health, agriculture, economy, and so on. AURA served as a hopeful example of transdisciplinary collaboration while teaching environmental studies to a group of interdisciplinary stu-dents at the Rachel Carson Center for Environment and Society in Munich (and soon at the Oslo School of Environmental Humanities). My stu-dents and I have been very inspired, Anna, by your notion of "cultivat-ing arts of noticing" (Tsing 2015). My favorite teaching takes place out of the classroom, going out on field trips and place-based workshops with groups of interdisciplinary students. For example, we explore a national park, a river, a mountain, or a postindustrial landscape together, combining perspectives of geology, ecology, environmental history, geography, or anthropology. On these excursions, it becomes obvious that for a place-based, historically situated analysis of anthropogenic times, we needed to cultivate diverse and interdisciplinary "arts of noticing." While walking together in a high Alpine national park, for example, a botanist can help us to notice the complex symbiotic connections between plants, insects, and birds. You soon realize that some plants and animals will not be able to adjust to climate change because they can't migrate beyond the mountaintop to survive in the conditions they need, causing chains of extinction. Exploring a Bavarian natural park with the methods and eyes of an environmental historian can teach you how a mountain is a political entity and the site's contested history is deeply entangled with Germany's National Socialism. At the same time, walking with a geologist and observing folds and thrusts in the rocks takes you into a story of deep time, when the rocks were still ocean floor and the Alps were formed.

It has been interesting to note for me that many of my students with a background in the natural sciences, in biology, physics, or chemistry, have been the clearest voices in insisting that we need to include nonhuman others in our research. Yet, for many of them, a first contact with the theories and critical methods of the social sciences and humanities was a great revelation. For example, engaging with the conceptual critique of the Anthropocene and introducing terms like *Capitalocene* (Moore 2016) to science students provided them with a new conceptual language to voice their long-felt critique against capitalist destruction, environmental injustice, and ecocide and a naive belief in a technofix.

Because you have asked us, Nils, to reflect on the question of "decolonizing methods," I would like to share some thoughts on the metaphor of "rubber boots methods." When I heard first heard the term, it carried for me an overly masculine connotation, not so well suited in times when scholars (Indigenous and others) call for a "decolonization" of methods and scholarly practices and to uncover the colonial and racist roots of our discipline (TallBear 2011; Smith 2012; Parreñas 2018). The metaphor "rubber boots" evokes the image of the "muddy boots"—anthropologist, the figure of the lone, heroic, white, colonial researcher-explorer: these men who enter the field in a tough, hypermasculine way, wearing muddy boots, to collect data on "exotic" and "alien" worlds. Maybe "rubber boots methods" is not an ideal metaphor for reimagining anthropological methods in the Anthropocene. We live in a time that requires us to be *more,* not less, sensitive. I am not sure, for example, if rubber boots are the best footwear for walking with heighted sensitivity. Maybe sometimes it would be better to take off your shoes and go barefoot—in a forest, for example—so you can feel the moss and the ground and become more sensitive to the life below your feet. Walking barefoot, you would also, for example, feel the difference to walking on concreted or sealed soils, another growing characteristic of the Anthropocene. My call would be to invent methods for the Anthropocene that are more attentive to how other species make their worlds, not less. It is not just humans who build their worlds; other life-forms do so as well. I would encourage researchers to take their shoes off sometimes and become more sensitive in the Anthropocene, to cultivate skills and practices that help us to feel more, see more, hear more, and to learn, for example, how nonhuman others communicate. I argue that we need more "barefoot methods" in the Anthropocene, to draw a parallel to "barefoot doctors," rural health care workers, including traditional healers, who are sensitive to needs of people living in rural and marginalized areas. Perhaps, rather than putting on rubber boots that desensitize ourselves to the world, we need to learn to use more "barefoot methods" in multispecies anthropology that are more sensitive to the needs of other-than-human life-forms suffering from anthropogenic destruction.

Becoming more attuned to nonhuman ways of being became a major requirement during my fieldwork on human–elephant relations in Wayanad, Kerala. One way to better understand why elephants become aggressive killers when living in close contact with people at the forest boundary was to investigate how they have become what I call "anthropogenic wildlife": how elephants' ways of life have changed through living in the close contact zone with people. Long histories of colonial and postcolonial timber extraction, elephant capturing, ecological simplification, and habitat fragmentation have traumatized elephants over generations. Human infrastructure hinders them following their long migration routes, in search of food, rainfall, and medicine. Even elephants' ways of communicating have been deeply disturbed in the Anthropocene. During fieldwork, scientists, forest workers, and Indigenous elephant mahouts (handlers) pointed out to me that elephants are very sensitive to sound and can communicate over long distances. This required a heightened attentiveness to the ways elephants sense their worlds. Elephants can produce infrasonic (low-frequency) rumbles that travel via seismic underground vibrations over many kilometers. Elephants can hear or sense these seismic waves through their feet (O'Connell-Rodwell, Hart, and Arnason 2001; Payne, Langbauer, and Thomas 1986). In Wayanad, Indigenous forest guards reported that elephants felt increasingly irritated by human sounds. Especially the vibration caused by hundreds of tourist Jeeps running daily through the core area of the Wildlife Sanctuary turn some elephants angry and agressive. To come closer to an understanding of how elephants perceive their worlds, sound ecologists and bioacousticians use sensitive microphones to detect the ultrasonic sounds of elephants (see also https://www.birds.cornell.edu/ccb/). This way, science can provide "speech prostheses" (Kirksey 2012, 24; Latour 2004, 67) to better communicate with other-than-humans and to imagine what other species need to lead meaningful lives, what makes them happy, and what makes them suffer. Simultaneously, we can also learn about this by listening to people who have very intimate relationships with other species, such as Indigenous forest guards, farmers, hunters, or some experimental field biologists. Experimental and immersive embodied and participatory fieldwork is of course not unique to social and cultural anthropologists. Rather, the method of intimately knowing an animal by closely living together with it has been borrowed from experimental behavioral biologists. Primatologists in particular have used methods of participatory observation and creative interspecies communication, as the work of scientists like Barbara Smuts (1985), Shirley Strum (1987), Robert Sapolsky (2001), and Thelma Rowell (1973) shows. I think that the Anthropocene demands a serious search for methods that teach us to become more sensitive, empathetic, and attuned to the needs and desires of nonhuman companion species.

KIRSTEN: I would like to continue from where Ursula let off, with the issue of learning from other species. I do not think this is new in anthropology.

If you go back and read Evans-Pritchard (1940), for instance, it is all about cattle. If you read Malinowski (1935), it is all about plants and sea animals. So, animals have always been there in anthropology. What is new, of course, is that we by way of the notion of the Anthropocene have become more observant of the fact that humans and animals (of all kinds) have been more deeply involved with each other than we would have thought. With the legacy of the early anthropologists in mind, I believe we could still learn a lot from the classically trained field ethnographers who certainly mastered the art of noticing. And they took their time to do so. They were also more readily transdisciplinary; the founding fathers naturally came from other disciplines, like Boas (physics), Rivers (medicine), and Malinowski (philosophy), to mention but a few. The fieldworks of the founders were long term, and often lonesome, and even if the tempo and the means of communication have changed dramatically over the past one hundred years, our forebears sat the scene for fieldwork as based in an art of noticing. With the AURA project, it has become vastly refined and conceptualized, of course.

I am not so critical about anthropology these years, as many others seem to be, because I do think that we generally do our best in the fields where we work, paying attention to detail in multispecies relations and learning a lot from studying with people who are dealing with various kinds of natural "others" in their own landscapes—as they see it. From this basic foothold in knowing other realities, and given our analytical powers, we can contribute to general anthropological vocabulary that may affect global thinking. This just to stress that there is nothing parochial about fieldwork in small places, as long as we know how to make the results applicable to larger issues.

This leads me to another question that Nils also asked us, namely, how we could go from conceptual issues into "situated analysis and critical description of ruined, patchy landscapes"? I think that is a very important question these days, as we are now admitting to working in patchy landscapes, where the earlier anthropologists would probably focus more explicitly on a social entity. I work in the Arctic, where the landscapes of water and rock and ice make such a huge impression that it is hard to imagine that they are more than marginally affected by Anthropocene realities when you look at them. But, of course, they are deeply shaped and altered by these realities and inevitably marked by environmental changes, rapidly altering the landscape with which people have lived for ages (Hastrup 2016). The Anthropocene is "all over the place," even if the fewness of people in the High Arctic makes it hard to see how humans can affect the magnificent surroundings. This makes it critical to anthropological analysis to notice how fieldwork also *takes place.* It takes place in landscapes that are marked in many different ways. Because of the radical diversity of and in the Anthropocene, there cannot be just one recipe of how we should deal with the concept; we have, first, to understand it in situ, so to speak. And let me add, perhaps controversially in this room, that I think we should still follow our anthropological instinct to communicate mainly with humans, both in the field and in our dissemination of knowledge. It is not least by sharing the landscape with which (other) people live that we may get to know them and their concerns—and their companions. In my High Arctic field, this sharing was to be taken literally; conversation was impossible, if only abstract and general. When moving about in the landscape, what was significant showed itself in practice, and questions only became possible when they concerned lived realities; philosophizing might follow later but could not be called up by the guest.

ANNA: I would like to follow up on what Kirsten, kindly, made me remember, namely, the issue of patchy landscapes. I think that if there is one place where AURA really has pushed Anthropocene discussions, it has been in trying to argue that the Anthropocene deserves a spatial analysis as well as a temporal one (Tsing, Mathews, and Bubandt 2019; Tsing and Bubandt 2018). To foreshadow a little the next question about collaborations with natural scientists, we argue that in spatial analysis of anthropogenic landscapes we need all the old skills of the anthropologist as well as some of what geologists, biologists, and ecologists can show us about these landscapes (Tsing et al. 2017). Crucially, it is also in spatial analysis of the patchy landscapes of the Anthropocene that we have the possibility and the challenge to bring in the kinds of things that anthropologists have traditionally cared about—human social inequalities and forms of violence and injustice (Bubandt, Mathews, and Tsing 2019). Spatial and situated analysis is the place where we can bring these agendas into our understanding of the ecological problems of the world.

Through this attention to how landscapes themselves are heterogeneous, we may be able to see how Anthropocene phenomena are not "just" planetary. They can be planetary too, of course, but they also have particular histories that relate to places across space.

Anthropocene thinkers need to be in much better dialogue with the environmental justice movement (e.g., Mascarenhas 2020; Sze 2020; Taylor 2014). Working through patchy landscapes is one important way to push forward this dialogue. When toxic wastes, for example, are situated in the communities made vulnerable by racism and poverty, patches of toxins are created (Bullard and Wright 2012). Such patches should be at the heart of Anthropocene—and anthropological—analysis.

KIRSTEN: If I may add a short note to what Anna just said here, it made us remember that patchy landscapes, all landscapes, are not simply places. They are also histories; there is a deep historical component in all landscapes. There has been an upsurge of attention, from people in AURA and from others, to the historical depth that all landscapes have, but in many cases, this recedes from view in the analysis; we need a more consistent "anthropology of deep time," as suggested by Irvine (2020). What is really interesting when you see it from the point of view of the (ill-defined) Anthropocene is that there are many temporalities and previous histories in all landscapes, whether we are aware of them or not. And I think that is one thing of which AURA has reminded us forcefully. In my work in both sub-Arctic Iceland and High Arctic Greenland, these histories have been relatively conspicuous, partly because these landscapes have not been plastered over with massive modernity until recently, if one may say so; now both these landscapes are changing dramatically, owing to global climate changes, challenging the fabric of the surface upon which people live. What came to the fore in my work in Iceland in the 1980s was the ways in which the landscape was changing even from one generation to the next, as my hosts in the countryside were keen to tell (Hastrup 1998). Seagulls fertilized the mountain slopes and made them green, in turn attracting the free-roaming sheep, which were increasingly stuck on the mountain ledges, from where they had to be roped and hauled up again. Such landscape histories, of course, went down into my notebooks. Later visits have more than confirmed that such benign "generational" changes caused by seagulls are speeding up and literally shattering the entire island, as even the underground protests against modern energy ventures—extracting power from deep thermal repositories (Maguire 2020). It is physically frightening to sense the potentially destructive earthquakes, speeding up and roaring for a better future (that is my interpretation). There has always been volcanic activity, but now human-made technologies are adding to a sense of an uncertain future of the old landscape. As Nils has phrased it so aptly on the basis of his work in Indonesia, an unsettling feature of the Anthropocene is "the increasing impossibility of distinguishing human from nonhuman forces, the *anthropos* from *geos*" (Bubandt 2017, G122).

A comparable drama unfolds in High Arctic Greenland, where changes in the landscape are no less threatening. Just a good generation ago, a renowned biologist wrote about climate change and animal fluctuations in the region over a couple of centuries, seeing them as nothing but benign, more or less cyclical events (Vibe 1967). Today there is no way to overlook the fact that the glaciers are melting, the sea ice dwindling, and whole habitats for animals and humans changing dramatically, from one year to the next, and with no recourse to earlier conditions; this deeply affects "politics." The inhabitants are amazed by the changes but also see new potentialities—bigger boats, possibly new mining adventures, and easier communication; they are used to living in an anticipatory mode, as it were (Hastrup 2013; Nuttall 2017).

In short, I think it is critical that we, as anthropologists, continue to find new ways to pay attention to the complex historicity of landscapes, which goes far beyond the Anthropocene as commonly understood.

BLENDING METHODS

NILS: This is a good segue for the next set of questions. I very much appreciate what you said here, Kirsten, namely, that to trace the histories of landscapes, we need to listen to people who have lived in and helped shape them for long time. Oral history is a critical method for paying attention to landscape historicity. At the same time, the patchiness that Anna talked about also entails forms of temporal and spatial connections that exceed what local oral history may capture—connections that require other methods, broader perspectives, longer trajectories of history, of power, of slow violence. I would agree also entirely with everything that has been said about learning from the ethnographic attention to animals, ecology, and landscapes that early modern anthropologists like E. E. Evans-Pritchard (1940), Bronisław Malinowski (1935), or Henry Lewis Morgan (1868) displayed in their writings. At the same time, it is clear that this attention was itself circumscribed: the landscape histories they noticed tended to omit that they were colonial landscapes, for instance (Asad 1973; Crosby 2003). One might say that their attention to animals was similarly circumscribed: anthropological attention to animals tended to focus on the animal-for-humans, the animal-as-function, and the animal-as-meaning (Kirksey and Helmreich 2010). Add to this the challenge of the anthropological "field" in the Anthropocene. Anthropologists, of course, know that their field is analytically defined, following two decades of critical debate about what the "field" in anthropology is (Clifford 1997; Faubion and Marcus 2009; Gupta and Ferguson 1997). But in the Anthropocene, such analytical situatedness may itself be problematic when animal extinction, climate change, toxic pollution, warming oceans, and acidification are causes and have effects at scales across any given "place." Relying on classical ethnographic situatedness alone may therefore be problematic in the Anthropocene. Lots of problems—methodological, political, epistemological, ontological—wrangle for attention here. But one key methodological question that I was hoping we could zoom in on is this: how do we methodologically draw in the multispecies and historical relationships that take place in and across specific landscapes and "fields"? Through oral (and other critical) history? Definitely. Through ethnographic observation? For sure. Through archival work? Yes, of course. But in addition to these and other social science methods, might we also benefit from approaches and methods of landscape study employed by natural science in order to get to multispecies forms of patchy connection across time and space (Bubandt, Mathews, and Tsing 2019)? And if so, how do we best do it? What are the pitfalls and challenges of employing or learning from natural science methods? Another way to ask this question is, how can we establish concrete human–social science collaboration in the wake of the so-called science wars, which cemented a seemingly unbridgeable gap between how we come to know the world in the human and the natural sciences? Is there an "after" the science wars (Latour and

Weibel 2002)? For people in AURA, a key starting point and constant
challenge has been to begin collaboration with the natural sciences, not
with questions of epistemological difference, but with methodological
commonalities. Biologists and anthropologists, after all, have a shared
history of doing fieldwork. It is on this shared history that the term rubber boots methods seeks to put a spotlight. I agree completely with you,
Ursula, that this shared history is also a masculinist, colonialist, and white
history. Rubber boots methods are, as we discuss in the Introduction,
exactly about being critical and staying with the trouble of this history
in the attempt to imagine collaboration anew. They are not a celebration
of the colonial and masculinist legacies of fieldwork but, on the contrary,
an acknowledgment of them, by inquiring into the methodological footwear that made fieldwork both problematic and possible. All fieldwork
methods, even those of a barefoot ethnography, will have to work with
the challenges of this history. I worry that claims to "bare-foot methods"
would ignore the footwear we actually use in the field. My Indonesian
friends and informants can walk barefooted in a sago swamp, but I cannot, because the soles of my feet are not hardened by many years of hard
labor. Ignoring that difference in order to make a claim to methodological sensitivity runs the risk of forgetting the colonial history of this difference. Colonial legacies are one set of challenges of this history that we
need to bring forward. But another set of challenges that I hope we could
discuss are the epistemological differences between natural sciences and
social and human sciences. These differences do not disappear easily, and
rubber boots methods have to work with the troubled legacy and presence of epistemological difference too. When John Law, the brilliant and
generous scholar of STS, visited AURA, he challenged us by saying that
he feared AURA set back science studies thirty-five years. What are your
thoughts on the epistemological possibilities and dangers of transdisciplinary methods collaboration?

URSULA: One of the great inspirations of AURA for my work was indeed
the experimentation with a combination and blending of methods from
the social sciences and natural sciences, the arts of doing shared interdisciplinary field research for a critical description of a patchy landscape in
the Anthropocene. I think that we can learn a lot from AURA, and you
have inspired my own work at the Wayanad Wildlife Sanctuary in Kerala

in South India. I learned a lot about how to take nonhumans seriously in my ethnography from people who are "species-whisperers" in one way or the other. A wildlife veterinarian, for example, whose task it was to conduct autopsies of the corpses of the charismatic mammals, such as elephants or tigers, who had had died in the forest taught me a lot about nonhuman life and death in the Anthropocene. He provided me with insights into the invisible world of the forest, of microbes, viruses, and bacterial spread between animals. At the same time, I also engaged with forest workers at the sanctuary. They knew the forest in a different way, as laborers extracting teak. They had been used as cheap workers for the forest department, working side by side with trained elephants, since colonial times. Different human perspectives gave me access to very different ways of experiencing and living with nonhuman life in the forest. But sometimes these perspectives and experiences overlapped and merged in unexpected ways. For instance, the forest workers knew elephants as remarkably intelligent beings with astonishingly long-term memories, complex social lives, the ability to use tools and medicines, individual personalities, self-awareness, and the skills to plan strategically into the future. The forest workers recognized that elephants share many traits with humans, including culture and a complex social organization. But this so-called Indigenous perspective is, in fact, analogous to what scientists at the sanctuary were also finding out about elephants. Behavioral biologists and wildlife veterinarians considered elephants' smartness and cognitive abilities as the main reason why they are so difficult to control and to govern in conservation. Elephants often outsmart people and find creative ways of crossing electric fences and elephant-proof walls built by the forest department (Münster 2016a). In other words, quite different perspectives and ways of knowing overlapped in a surprising way. In the same way, it is definitely worth trying and continuing to work across disciplines. Different disciplines might complement each other and agree on many issues in surprising ways when focusing on a joint problem.

KIRSTEN: In my view, anthropological methods are inherently "blended," whatever you are doing. They are blended in the sense that they are always mixed up with the concrete place where we are working and with its inhabitants—or, indeed, *places* in the plural, as they gradually show themselves. We have to be flexible and "go with the field," so to speak. This is one of the challenges we have when we engage in interdisciplinary work. Over more than three years (2014–17), I worked with biologists, among others, in Northwest Greenland on an interdisciplinary project. I saw collaboration with biologists as an extension of a general anthropological bent toward collaboration with each and every person that could teach me something new about the field within which I was working (Hastrup 2018). We always collaborate in the field: with local inhabitants, farmers, hunters, miners, cleaning ladies, policemen, teachers, and whoever is there in the field that we are studying. Anthropology *is* collaboration in that very basic sense. We cannot do anything from a bird's-eye perspective, setting ourselves above and beyond the field we are studying. That kind of involved or "mixed-up" methodology is a challenge to biologists. In my experience of working together with biologists, we have succeeded in publishing lots of papers together (e.g., Hastrup, Grønnow, and Mosbech 2018; Hastrup et al. 2018; Jeppesen et al. 2018). But it was *so* complicated. Not because we had "seen" different worlds, because we had not. In fact, we had taken a lot of trouble agreeing to see the landscape through each other's eyes. But when it came to *describing* what we had seen and to disseminating and publishing that description, we were worlds apart (see also chapter 10). Many of the primarily biological papers that we coauthored had fifteen and sometimes more authors (of which maybe only two had taken part in the fieldwork). An average paper might be twelve pages long, implying that most authors contributed only small "parcels of facts" to somebody, who was then in charge of gathering together those parcels. Natural scientific results are often presented in that way: a compilation of different kinds of expertise generating different kinds of facts. In contrast, anthropologists work by integrating everything they know into one coherent argument. I read a lot of scientific papers, and I often find that they are too technical to be linked up with my own human interest, because "data" are fixated as such and can never be questioned. A dialogue across genres becomes difficult when one party offers "facts" that cannot be questioned. The anthropological group took some grave verbal bashing for sloppy thinking from some of the colleagues in the biological group. This is more than just a battle between quantitative and qualitative analysis. Anthropologists also often work quantitatively. Conversely, biologists studying the behavior of the little auk *(Alle*

alle) in the High Arctic have to work qualitatively. They have to think about how the little auk might think about their young and their prey, and they have to resort to human analogies for a suitable vocabulary (see, e.g., Mosbech et al. 2018). In that sense, the differences between natural sciences and social or human sciences are not necessarily a dramatic distinction between quantitative and qualitative science. It is primarily a difference of genre and jargon in the final output and of an inherited perception of hard "science" as radically different from soft "humanities." It takes some energy to convince collaborators from natural science that anthropology is also serious scholarship, but we need to try. If anything, the current times call for collaboration across received categories.

ANNA: Nils recently challenged me to think of a new name for the kind of work that AURA has tried to offer, in which field biologists and ecologists work together with fieldwork-oriented anthropologists (see chapter 9). Nils's point was that this is not just "transdisciplinary" or "blended methods" but perhaps something with a new kind of traction. For the moment, I'm thinking of it as "critical empiricism," that is, a form of empirical inquiry in which critical reflection remains central to the process of formulating research objects—at the heart of the collaboration—as well as the research and analysis that follow. In contrast to the fashion in anthropology for "theory for theory's sake," critical empiricism revitalizes attention to what's happening in the world. It shows us social relations as more-than-human. It remains radically engaged with questions of justice, violence, and dispossession. And it does this, in part, through finding out what's going on, for both humans and nonhumans.

This kind of collaboration is not easy. Let me continue Kirsten's reflections on how difficult collaborations can be. In hindsight, it was one of the most brave and crazy aspirations of AURA to stage collaborations that were not just about adding up separate facts from natural and social sciences but instead allowed natural scientists and humanists to try to figure out "what mattered": what research we were going to do in common. It turned out to be completely impossible in many ways. And yet, in understanding the Anthropocene, it is still really the most important and the most urgent thing of all!

Just as Kirsten has said, identifying a potential joint research object across human and natural sciences is really hard. One key reason is that from a natural science point of view, the kinds of questions that anthropologists ask are so huge that no natural science method can possibly answer those questions in a responsible way. Anthropological questions are, from a natural science perspective, too big, crazy big. I have been working with a mycologist from Copenhagen, and in response to what I was telling him interested me about fungi, he said, "You are asking the kinds of questions that the UN might ask." Anthropological questions cannot be handled easily by biological methods, which only count as science when they offer precise results within the severely limiting constraints of the research setup. In contrast, and as Kirsten just described, we anthropologists often find the questions that natural scientists ask to be too small. We may think, "Can't we get something more out of it than that?" I am not saying this to judge on either side but rather to reflect on why the process is so difficult. The miracle, then, is when it starts to work.

Let me offer a few examples from AURA projects where we gained some traction. In June 2018, AURA anthropologists worked together with biologist Jens-Christian Svenning to organize a conference on "Woodlands in the Anthropocene." The conference gathered biologists, anthropologists, ecologists, and landscape historians, and as we began inviting this very disparate group, we were terrified that we would hardly be able to speak to each other. Yet the conference was great. We produced a review paper to survey what we know about anthropogenic forests in a changing world (Swanson et al. 2021). When the question of "what matters" came up, the natural scientists and the humanists turned out to have a lot of common commitments. Still, the review paper only came into being because biologist Svenning and anthropologist Heather Swanson were generous enough to engage the work across the great divide of disciplinary difference. And we spent several years trying to convince professional journals that this kind of work across disciplines is important.

A second example is represented in this volume (see chapter 9). Through AURA, biologists Peter Funch and Stine Vestbo collaborated with anthropologist Joe Klein and me to plan an investigation of "marine hitchhikers," that is, organisms that travel without human design in human-sponsored global supply chains. As we describe in the chapter, the work of collaboration was long and painstaking, if also fun and rewarding. We had to learn to navigate the difference between "methods" in biology and anthropology—and questions of humans versus nonhumans was hardly the issue at all. After several years of working together, we have a good research proposal; in collaborations of this sort, taking time is key. It entails patience, perseverance, slowing down: skills that, indeed, are familiar in anthropological fieldwork.

One more AURA project that has required patience and negotiation to cross genre and disciplinary borders is *Feral Atlas: The More-Than-Human Anthropocene* (Tsing et al. 2021). What began as a way to document AURA research expanded into a flowing cascade of stories of the Anthropocene, told through the nondesigned consequences of imperial and industrial infrastructure. The expansion occurred because scholars and scientists kept volunteering to add their stories. Even if such collaboration is painstaking, even mad, everyone recognizes that we cannot do without it. If you want to see what "critical empiricism" might look like in the cross-talk between the natural and social sciences, take a look at *Feral Atlas,* where we argue that we cannot know the Anthropocene without working through patches and across radical difference.

KIRSTEN: I subscribe wholeheartedly to Anna's point about the need to extend collaboration. And I should add to what I said before about the challenges implied in collaborating across received scientific categories, that I have certainly learned a lot from working with natural scientists. It has been very rewarding, even if it remained difficult to make a synthesis out of these different strands of thinking (chapter 10; see also Flora and Andersen 2017).

URSULA: Hearing about these difficulties is a little disheartening, because for me, AURA seemed such a promising example of how transdisciplinary collaboration could work. Many scholars in the vibrant field of environmental humanities are experimenting with cross-disciplinary collaborations, engaging the sciences and the arts to find creative and experimental ways of addressing environmental problems and telling new stories that engage a broader public (see, e.g., "The Living Archive," http://www.extinctionstories.org/, or, more recently, Anna Tsing's collaboration on the amazing *Feral Atlas*). Such crossings of disciplinary boundaries are not new, of course. I have scholars like Rachel Carson in mind, for example, a marine biologist who touched people well beyond science with her writing and kicked off the environmental movement with the publication of her book *Silent Spring* (1962). It is exciting to work with students who often cross disciplinary divides with ease and support them to communicate their own environmental stories through creative methods like film, art projects, photography, or exhibitions. In many environmental humanities programs, including our newly established MA certificate program at the Oslo School of Environmental Humanities, students learn to experiment with experimental ways of environmental storytelling. They can, for example, experiment with a poem, a utopian science fiction story, or an environmental audioguide to convey their environmental research. As many of us have experienced, sometimes a picture or a story has a greater impact than mere numbers or a graph. Many of my students with a background in the hard sciences are very aware that science alone will not save us but that we need new narratives and a radical social, political, economic, and cultural shift as well.

ANNA: I would like to add one other small place for hope. Throughout its life span, we have in AURA held monthly seminars (http://anthropocene .au.dk/activities/seminars/). Altogether, we have now held more than forty of these "slow seminars," as we called them. We use the term *slow* to echo philosopher Isabelle Stengers's idea of "slow science," that is, science in which we take the time to figure out what matters (Stengers and Muecke 2018). The slow seminars have a place for biologists; for the occasional geologist; and for artists, philosophers, anthropologists, ecologists, STS scholars, and people from other kinds of backgrounds to come together. And they have been extremely useful. Here I am going back to Ursula's and Kirsten's considerations about the importance of genre experiments. The slow seminar itself was a genre experiment. It also allowed us to consider what kinds of writing could stimulate good discussion. One such kind of book was that written by natural scientists who reflected on their research practices in books intent on both introducing their research and speaking to nonspecialists about it. This genre of books by people as diverse as Fredrik Sjöberg (2016), Chris Thomas (2017), David Foster (2014), Oliver Rackham (2012), Deborah Gordon (2010), and James Estes (2016) has been very successful in starting conversations between people from across the natural and the human and social sciences.

Decolonizing Methods

Nils: One key challenge that anthropologists and others from the human sciences have when they engage with natural science is what to do with the "One-World world" claim upon which natural science is often built (Law 2011). I say "often" because not all natural science is based on the claim about a One-World world. Take astrophysics, which is sometimes, as an anthropologist might note with interest, referred to as "cosmology." Debates within astrophysics, among other things, center on the possibility of a *multiverse,* a term that has recently become popular in anthropology as well as part of an attempt to move beyond a One-World world perspective (Barad 2007; de la Cadena and Blaser 2018; Wallace 2012). Still, the One-World world perspective dominates in natural sciences. It is a perspective that tends to ban other perspectives than those of Western rationalism and evidence-based science. Reality is a playing field onto which only certain ideas and players are allowed—a game in which only certain conceptual "moves" are possible. Among other things, a One-World world entails a particular kind of knowing, namely, one that comes from a "bird's-eye view" and where data are "fixated" in certain ways, as you put it earlier, Kirsten. That type of knowing has a particular history that is also Western, modern, and predominantly male. In this context, the question that I am here putting under the heading of "decolonizing methods" becomes, how do we collaborate with natural science, "take natural science data seriously," and learn from its methods without also implicitly adopting a One-World world epistemology? For anthropology, which sees itself as the comparative study of common sense (Herzfeld 2001), a discipline that regards itself as the voice of other worlds muted by the idea of a One-World world, there is a particular challenge here. How do we take natural science seriously—because we have to in the Anthropocene (Chakrabarty 2009)—without forfeiting other worlds, in particular, the worlds that are constituted by the perspectives that Indigenous people, nonexperts, and others bring to the Anthropocene (Danowski and Viveiros de Castro 2017)? One might call this the "Chakrabarty–de Castro paradox." On one hand, there is an urgent need to retool anthropology and the human sciences by employing also natural science insights and methods to understand the planetary scale of the environmental, ecological, and climatic crises of our time. On the other hand, it is crucial to be epistemologically critical of the knowledge practices and colonial histories through which "the planetary" comes into view in the first place (see Coen 2018). This is really a question about the global hierarchy of knowledge that centers some people and some forms of knowledge while marginalizing other people and knowledges. Is there a place in the study of the Anthropocene from which to take the perspectives of natural science and Amazonian Indians equally seriously at the same time? Is it possible to align the natural science methods we need for an anthropology of *the* Anthropocene (Steffen, Crutzen, and McNeill 2007) with the critical methods of an anthropology of a billion Black Anthropocenes (Yusoff 2018)?

KIRSTEN: Well, I have some trouble with this question. Of course, it seems "right" to decolonize our approaches and vocabularies as far as we can do so from our standpoint in the first place. But I think it is not really up to us to define what should be done and how we should decolonize the world, because that would tendentially continue the very process that it is criticized in this word. Another issue I have with this is in what sense we can define the Anthropocene, as an object of study. In my thinking, the Anthropocene is a concept beyond both colonial and postcolonial, even beyond definition, and therefore beyond anthropological analysis as an entity.

NILS: My question was an attempt to paraphrase—and thereby to begin to address for our discussions about multispecies methods—an amalgam of critiques from decolonial, feminist, and neo-Marxist scholars of the concept of the Anthropocene (Crist 2013; Davis and Todd 2017; Haraway 2016; Haraway et al. 2016; Malm and Hornborg 2014; Yusoff 2018). For instance, the critique that the etymology of the ancient Greek word *anthropos* in *Anthropocene* comes with the strong gender and class bias of a slave-based ancient Greek society where the *anthropos* was associated primarily with free and adult men. The naturalization of all humanity into the term *anthropos,* in other words, obscures an androcentric and occidental bias—as well as an anthropocentric one. Such critiques are clearly of the conceptual kind that we talked about earlier, and we could say "never mind those; let's just get on with our study of the Anthropocene." But the critique points to a broader issue: on one hand, the concept of the Anthropocene brings us into a species and planetary kind of

conversation. It is an antiparochial argument that suggests a "world crisis" in which we are all in the same boat. But in so many ways, we are
not in the same boat. Across divides of geography, economy, gender, eth-
nicity, and species, the Anthropocene is plural. The argument that the
Anthropocene incorporates a legacy of colonialism and cultural imperialism—and that the One-World world perspective is the universalization
of one parochial perspective, a predominantly human, male, and Western
one—has, it seems to me, some important implications for what methods
we employ to study the Anthropocene. How do we invent methods that
capture the planetary scale of our shared problems *and* stay critical of
who exactly this "our" is, remaining alert to that in "our shared problems"
which is not shared at all?

KIRSTEN: Yes, there is no doubt that the term *Anthropocene* is problematic,
except as an indication of human interference with the globe, where we
all live. We are subject to the same destructive processes, although some,
evidently, are more hit by them than others: we are not living in the same
places, we have not had the same opportunities, and so on. I think anthropologists are very much aware of that. But I don't think we can blame the
term. I think doing so would be a problem. We have to be aware of these
critical notions and just see the world as it is now. The endless debates
about when and where the Anthropocene hits whom are a dead end. The
strength of the concept of the Anthropocene is that it is a way of con-
ceptualizing the plights that affect humans and nonhumans all over the
globe today, even in the most faraway places. It is already plural. What we
can do to counter its consequences is the real question for me. Whether
the Anthropocene *as a concept* is "ours" or "theirs" is, in my view, less
important than the all-pervasive consequence of human activities across
the globe. In a sense, it brings us back to Alexander von Humboldt's
([1814–29] 1995, 7) "great problem" of "how to determine the laws that
relate the phenomena of life with inanimate nature"—and to his acknowledgment that he has "not always succeeded in separating the observations of detail from the general results that interest all educated minds"
(10). Humboldt then continues,

These results should bring together the influence of climate on organized
beings, the look of the landscape, the variety of soil and plants, the mountains

and rivers that separate tribes as much as plants. I do not regret lingering on these interesting objects for modern civilization can be characterized by how it broadens our ideas, making us perceive the connections between the physical and the intellectual worlds. (10)

This quote is from Humboldt's personal narrative of his journey to the New World, 1799–1804. My reason for quoting him here is to stress the intellectual rather than the conceptual challenge that we are facing when dealing with the embroilment of different landscapes, species, and temporalities. What are "*our*" shared problems—and indeed, who are "*we*"? How can we generalize our results to make them work elsewhere, without stepping out of the meshwork of detail?

ANNA: One importance of Nils's question is that it asks what anthropologists can learn with our privilege of talking to local people. What role should vernacular natural histories, for example, play in knowledge of the Anthropocene? Local people often have insights that natural scientists might not have thought to inquire into as they are each trying to understand a particular ecology.

I am a firm advocate of the idea that we just cannot know the Anthropocene without tapping a diversity of perspectives. For example, in my work on mushrooms (Tsing 2015), I found that the kinds of empirical natural history questions that mushroom pickers ask are completely different from the ones that mycologists ask. Mycologists are responding to an academic literature on evolution and ecology. Pickers want to know "Where are the mushrooms? When and under what conditions do they emerge?" In trying to answer these kinds of questions, pickers use lots of observations and even do their own experiments. In many cases, these do not contradict professional scientific knowledge; they just explore questions that professionals have not thought to ask. In a similar way, I found that American and Japanese scientists, both trained in international scientific conventions, asked quite different questions about the mushrooms I was studying. We ignore these varied streams of empirical knowledge to our detriment.

Where things get tricky and interesting is where local people tell you things that are not part of or immediately coherent with what a natural scientist would say. In the research that I am beginning in West Papua,

Indonesia, in collaboration with a Western-trained ornithologist, we often come across such instances (Tsing 2022). For instance, one knowledge-able local man alerted me to a bird's song as follows: "Did you hear that kingfisher? It just announced the tide is coming in [*air naik*]." How should I interpret this statement within my ornithological collaboration? And what is its similarity to or difference from my ornithologist's statement, alerting me to a killdeer (i.e., another coastal bird) call in the eastern United States by saying, "It's saying 'kill a deer'"?

Anthropologists too often think that what we get from working with our human interlocutors is some beautifully organized cosmology of the structure and origin of the universe. But what we often get is something like "that bird just said the tide is coming in." It is possible that those birds sing when the tides come in; it is also possible that this is wordplay, mak-ing the bird call legible through our attempts to give it language. And even if the latter, perhaps this attempt to put language in the mouths of birds might be ontologically important; I don't yet know. The challenge of work-ing across different kinds of disciplinary methods in the Anthropocene is that we are obliged to be open to what Eduardo Viveiros de Castro (2019) calls "ontological anarchism"—the multiplicity and unpredictable juxta-positions of very different world-making projects (see also Viveiros de Castro 2004)—because we are dealing with real differences in how we will understand landscapes. But, at the same time, we are looking at where these world-making projects meet and overlap to create new histories with each other. I think it is our job to *both* really listen seriously to vernacular under-standings of what is going on *and* work with them as part of an intersect-ing set of histories, including the histories of ecology and natural science.

URSULA: I could not agree more. That is what continues to make me excited to have been trained in the field of anthropology. I have learned from environmental historians to engage with the layered histories of land-scapes and their more-than human inhabitants, to understand life in the Anthropocene. What anthropologists can add to these histories is an engagement with subaltern and marginalized perspectives and voices. Anthropology has much to contribute, as it offers access to alternative ways of experiencing an ecology through an alternative kind of noticing. Sometimes science is fairly slow to arrive at insights that Indigenous observers and practitioners had seen long before. Let me give another elephant example here. For the Indigenous mahouts in India, it is obvious that elephants have culture. They know very well that elephants teach their young, educate their offspring, and transfer knowledge about medicinal plants, food sources, and migration routes from one generation to the next (Münster 2016b). But natural scientists have only quite recently begun to accept the contested notion of nonhuman culture and social learning (de Waal 1999; Plotnik, de Waal, and Reiss 2006). For this reason, I find it important for multispecies studies to combine listening to both science and other ways of knowing. In addition, we live in a time when we can draw on and learn from Indigenous philosophers and Indigenous theories to rethink more-than-human relations in the Anthropocene (Kimmerer 2015; TallBear 2011). Anthropology is no longer the study *of* people, a discipline that claims to speak *for* local people. We study and talk *with* people (Ingold 2018).

Can I go back to the question of the word *Anthropocene* here? Because I was surprised to see that you in AURA stick to the term *Anthropocene*, rather than embracing one or several of the many alternative terms that have been suggested, such as *Capitalocene* or *Chthulucene* (Bonneuil and Fressoz 2015; Haraway 2015; Moore 2016).

ANNA: The reason why I stick with that term, although I am perfectly happy to use all the others also, is related exactly to the question of collaboration with the natural sciences. If you want to be in dialogue across disciplines, then *Anthropocene* is the term that allows for transdisciplinary dialogue. That possibility of dialogue makes using the term worth it, despite all its pitfalls.

URSULA: I very much agree with that in fact. In teaching, the term *Anthropocene* is the most productive, because we can speak to scholars across many disciplines. Scientists can relate to the Anthropocene because for them it is *there* in the sense that you can pinpoint it in the layers of the earth, it manifests materially. For that reason, it is a helpful term, even if this "there-ness," of course, is deeply problematic (Yusoff 2018).

KIRSTEN: We should add that the term *Anthropocene* is still a very much debated term in natural science (Finney and Edwards 2016; Walker, Gibbard, and Lowe 2015; Zalasiewicz et al. 2017). So it is not as if the Anthropocene is just one thing that you have to accept.

URSULA: That is correct.

ANNA: I want to address one dimension of this natural science disagreement, which is the different proposals for when the Anthropocene began,
because for me that is a really productive disagreement. Disagreement,
after all, is the anthropologist's bread and butter, and these disagreements
have to do partly with different kinds of processes happening at different
historical periods. Did the Anthropocene begin with landscape modifi-
cation through fire (Glikson 2013), the advent of agriculture (Ruddiman
2003), colonialism (Lewis and Maslin 2015), or capitalism and its acceleration after World War II (Zalasiewicz et al. 2014)? These are disagreements about periods that have always mattered to anthropology and
that we can work on. In that sense, I am delighted with the many delays
to the attempt of the Anthropocene Working Group under the International Commission on Stratigraphy to settle on a date, because the delays
demonstrate how hard geologists have had to work to declare an official
answer to this question. This, in turn, allows anthropologists to reexam-
ine history with the Anthropocene in mind. We can rethink the consequences of the European invasion of the New World and the onset
of industrialization and, of course, to investigate the many post–World
War II developments in the world. It is the very disagreement about the
Anthropocene, the still molten form of the concept (Swanson, Bubandt,
and Tsing 2015), that allows anthropologists to be robust interlocutors with
natural science—exactly because issues of capitalism and colonialism as
well as unequal patterns of production and consumption now matter to
natural science in a new way. (See also *Feral Atlas* on "detonation" instead
of periodization as a better analytic frame for Anthropocene timelines:
https://feralatlas.supdigital.org/index?text=ad-anthropocene-detonators).

DESECULARIZING METHOD

NILS: The last question I would like to spend a little time on is less a question than a sense of unease or doubt that is stirred, for instance, by Anna's
story of the kingfisher that announces the rising tide. As we endeavor to
align natural science and human sciences methods and perspectives in a
study of life in the Anthropocene, how do we do so without losing sight
of many forms of animation and vitality that fall outside a secular natural
science worldview? I am thinking of the phenomena and beings of the

unseen world that anthropologists have conventionally studied: ghosts, ancestors, magic, spirits, witchcraft, and sorcery, for instance. It is the haunting and the sense of the uncanny that accompanies everyday life everywhere and which we have learned in anthropology to take seriously, not just as a remnant of a superstitious past. The uncanny is part of everyday life; it determines our relation to the past and helps shape the future, particularly in a time of crisis. Postsocialism and the financial crisis a decade ago each had its own ghosts and monsters. So, too, with the current environmental crisis (Tsing et al. 2017). The Anthropocene is replete with examples of the uncanny (Bubandt 2018). Human industry and agriculture are the cause of ongoing mass extinctions among amphibians and insects while we raise fifty billion chickens for slaughter every year. Our domestic chickens weigh three times as much as all wild birds put together (Bar-On, Phillips, and Miloa 2018). Antibiotic medication is creating multiresistant superbugs, viruses are mutating faster than our vaccines can keep up with, and we all consume perhaps fifty thousand pieces of microplastics a year. Death and uncanny proliferation, unseen agents that may harm us in ways that cannot be predicted and against which we struggle to protect ourselves: this, it seems to me, is the uncanny realm of ghosts and spirits (Bubandt 2022).

Let me give you a concrete example from West Papua in Indonesia, where Anna and I have been doing fieldwork. In this region, crocodiles have in the last ten years increasingly begun to attack and kill people. People in the region say that crocodiles ordinarily do not attack people. Crocodiles only do so when they are possessed by a human spirit. As evidence for this, they provide the kind of natural history observation perspective that Anna talked about earlier: people observe that when crocodiles attack and prepare to do the death roll, they appear to hug their victims. This may (or may not) be an adequate description of what scientists call a "death roll." But the anthropomorphism of crocodile attacks does not stop there. When the victim is dead, people say the crocodile will carry the body on its back in the same manner that locals carry produce back from the field in baskets on their backs. And finally, the crocodile will place the body under a log in the water, much as humans put their foodstuff in a food cupboard until it is dinnertime. All of this is natural history observation that supports a "more-than-natural" proposition:

when crocodiles attack, they are really humans in disguise, commandeered by ancestors or a sorcerer's spirit (Tsing and Bubandt 2020). This instance is similar to Anna's example earlier about the kingfisher that announces the tides, except here the local and nonsecular account directly clashes with that of secular, evolutionary biology. Biologists say that the legs of modern crocodiles are vestigial and are not used in their attacks, which are anything but human. So, what do we do when our observations of vernacular natural history go in directions not supported by biological insights? How do we stay true to the anthropological kind of craftsmanship of wanting to insist on the reality of these things and, at the same time, insist on the reality of what is biologically true? Crocodiles in West Papua provide one example of this clash. Another example is how we study mountains that are geological formations and places that people mine as well as spirits and ancestors (Bubandt 2017; de la Cadena 2015; Povinelli 2016). It is here that the term *Anthropocene* might also be helpful because it points to a new kind of convergence where the natural science and vernacular knowledge do not need to clash. After all, the advent of the Anthropocene, as Bruno Latour (2017) suggests, points us "back" to an animist perspective, not *despite* the natural sciences, but exactly *because* of them. The reason is that the natural sciences no longer recognize their classical objects of analysis either. The Anthropocene is possibly a world where it is possible to have our secular cake and eat it too. For the natural science of the Anthropocene is full of animated nonhuman agents that secular modernity and science would have denied not too long ago: a world of nonhuman agency stirred to life by human intervention. This is, for me, also worth keeping in mind as we work to establish new forms of collaboration with the natural sciences. The method question in this is: can we combine our methods for studying the Anthropocene with methods for studying the Anthropo-not-seen (de la Cadena 2014)? Is it possible to play with the methods and epistemologies of the natural sciences while also insisting on the need for a study of a nonsecular Anthropocene?

URSULA: I am also very inspired by Latour's (2016, xi) idea of "additive realism" and his suggestion of "accumulating knowledge, instead of subtracting it." And I agree that there is a real double challenge in how we both decenter the human and decenter secular accounts of it.

KIRSTEN: We have to acknowledge that *Anthropocene* is a human term. It was devised by humans who saw some disturbing developments in the world and then called it the Anthropocene for very good reasons, given the human footprint all over the place. But, of course, the term calls for as many questions as it gives answers. Therefore, I also think, as both Anna (Tsing 2015) and Donna Haraway (2016) have suggested, that we should work toward making the Anthropocene as short and as thin as possible. This means, among other things, making room for the analysis of other species that may "fight back" against what the notion tends to take for granted. Your recent chapter on sustainability, Anna, is a very important one here, because it posits some responsibility of humans (Tsing 2017). Just as humanity as a whole has some responsibility for creating this very disturbing situation in which everybody is now living, we also have an obligation to fight back. We can do so by working to create refuges for marginalized species. This is one way for us to fight back against the Anthropocene and make it less self-evident. I think anthropologists have something to add to this because we work in all these patchy landscapes with so many specific histories and stories, while also being part of a modern world of knowledge. It may be that they talk about ancestors and crocodiles in a particular way, but they are also very much part of the same world as the rest of us. We should never forget that, and we should not lock them into "Indigenous spaces."

URSULA: When you say, Kirsten, that we need to create more refuges for other species, do you think that this is the site for a new activism?

KIRSTEN: Possibly, yes. It is really Anna's term, and I am just echoing her here. But I do think that it is very important to make sure that there are still such spaces and that we cherish them. Personally, I am not so much calling for activism as I am pledging my allegiance to anthropology as a singularly important intellectual space, making room for multiplicity— and for making new modes of thinking about the earth possible.

ANNA: I would like to try to respond to Nils's really difficult question about alternatives to secular assumptions, even though I do not have an answer to it. It is a question with which I have been struggling. One of the key insights on this issue to which we have come in AURA, with Nils's help, is that it is a modernist conceit that spirits are features of a definitive cosmology, which in turn is a matter of belief. Making spirits a matter of belief is, as Kirsten just said, a way to lock spirit up into "Indigenous spaces," which we can then comfortably say have nothing to do with our reality. Instead, as Nils has argued, spirits are sometimes features of the kind of doubt and questioning and indeterminacy of our world (Bubandt 2014). If we start from there, then an anthropological description of the world has to include spirits if that is what people tell you is there, or possibly there. This means working with doubt, theirs and ours. In my experience, I have thought things were spirits that were not spirits. When I was working in Kalimantan, everybody said that the pigs were only in that place for a particular season. When they come to have their babies, this is their special place. I thought this was mythology. The pigs I knew do not migrate long distances. But it turned out this was a different species of pig, and their explanation was empirically correct, according to observers and natural scientists. Some of the time, we as researchers cannot tell the difference between spirits and natural histories. So that's a piece of it. And that is also how I feel about the bird that calls the tide in. I do not know if that is about spirits or ecology. So for that very reason, when we are committed to critical description, we have to include ghosts and spirits and all that stuff, because we do not know what is going on, and they are conduits there. The second element of this situation is doubt. Doubt is a piece of figuring out what is going on in a particular place. If the Anthropocene is the condition of nondesigned effects of human actions, then doubt is central to its description. Perhaps working with cosmological uncertainties, rather than certainties, can help us with the kinds of overlapping and juxtaposed ontological situations in which we find ourselves. To know the Anthropocene also may require unknowing the taken-for-granted contours of the world.

References

Armiero, Marco, Thanos Andritsos, Stefania Barca, Rita Brás, Sergio Ruiz Cauyela, Çağdaş Dedeoğlu, Marica Di Pierri et al. 2019. "Toxic Bios: Toxic Autobiographies—a Public Environmental Humanities Project." *Environmental Justice* 12, no. 1: 7–11.

Asad, Talal. 1973. *Anthropology and the Colonial Encounter.* London: Ithaca Press.

Barad, Karen. 2007. *Meeting the Universe Halfway: Quantum Physics and the Entanglement of Matter and Meaning.* Durham, N.C.: Duke University Press.

Bar-On, Yinon M., Rob Phillips, and Ron Miloa. 2018. "The Biomass Distribution on Earth." *Proceedings of the National Academy of Sciences of the United States of America* 115, no. 25: 6506–11.

Baskin, Jeremy. 2015. "Paradigm Dressed as Epoch: The Ideology of the Anthropocene." *Environmental Values* 24: 9–29.

Bonneuil, Christophe, and Jean-Baptiste Fressoz. 2015. *The Shock of the Anthropocene: The Earth, History, and Us.* London: Verso.

Bubandt, Nils. 2014. *The Empty Seashell: Witchcraft and Doubt on an Indonesian Island.* Ithaca, N.Y.: Cornell University Press.

Bubandt, Nils. 2017. "Haunted Geologies: Spirits, Stones, and the Necropolitics of the Anthropocene." In *Arts of Living on a Damaged Planet: Ghosts and Monsters of the Anthropocene,* edited by Anna Lowenhaupt Tsing, Heather Anne Swanson, Elaine Gan, and Nils Bubandt, G121–41. Minneapolis: University of Minnesota Press.

Bubandt, Nils. 2018. "Anthropocene Uncanny: Nonsecular Approaches to Environmental Change." In *A Non-secular Anthropocene: Spirits, Specters and Other Non-humans in a Time of Environmental Change,* edited by Nils Bubandt, 2–19. More-Than-Human Working Paper 3. Aarhus, Denmark: Aarhus University Research on the Anthropocene (AURA). http://anthropocene.au.dk/working-papers-series/.

Bubandt, Nils. 2022. "The Uncanny Valley of the Anthropocene: Short Stories about the Undead under the Brightest of Lights." In *The Anthropocene and the Undead: Cultural Anxieties in Contemporary Narratives,* edited by Simon Bacon, 67–84. Washington, D.C.: Lexington Books.

Bubandt, Nils, Andrew Mathews, and Anna Tsing, eds. 2019. "Patchy Anthropocene: The Frenzies and Afterlives of Violent Simplifications." Special issue of *Current Anthropology* 60, no. S20.

Bubandt, Nils, and Anna Tsing. 2018. "An Ethnoecology for the Anthropocene: How a Former Brown-Coal Mine in Denmark Shows Us the Feral Dynamics of Post-industrial Ruin." *Journal of Ethnobiology* 38, no. 1 (online suppl.): 1–13.

Bullard, Robert, and Beverly Wright. 2012. *The Wrong Complexion for Protection: How the Government Response to Disaster Endangers African-American Communities.* New York: New York University Press.

Carson, Rachel. 1962. *Silent Spring.* Boston: Houghton Mifflin.

Chakrabarty, Dipesh. 2009. "The Climate of History: Four Theses." *Critical Inquiry* 35, no. 2: 197–222.

Clifford, James. 1997. "Spatial Practices: Fieldwork, Travel, and the Disciplining of Anthropology." In *Routes: Travel and Translation in the Late Twentieth Century,* edited by J. Clifford, 52–76. Cambridge, Mass.: Harvard University Press.

Coen, Deborah R. 2018. *Climate in Motion: Science, Empire, and the Problem of Scale.* Chicago: University of Chicago Press.

Cook, Brian, and Angeliki Balayannis. 2015. "Co-producing (a Fearful) Anthropocene." *Geographical Research* 53, no. 3: 270–79.

Crist, Eileen. 2013. "On the Poverty of Our Nomenclature." *Environmental Humanities* 3, no. 1: 129–47.

Crosby, Alfred W. 2003. *The Columbian Exchange: Biological and Cultural Consequences of 1492.* 30th Anniversary ed. Westport, Conn.: Praeger.

Danowski, Déborah, and Eduardo Viveiros de Castro. 2017. *The Ends of the World.* Cambridge: Polity Press.

Davis, Heather, and Zoe Todd. 2017. "On the Importance of a Date; or, Decolonizing the Anthropocene." *ACME: An International Journal for Critical Geographies* 16, no. 4: 761–80.

de la Cadena, Marisol. 2014. "Runa: Human but Not Only." *HAU: Journal of Ethnographic Theory* 4, no. 2: 253–59.

de la Cadena, Marisol. 2015. *Earth Beings: Ecologies of Practice across Andean Worlds.* Durham, N.C.: Duke University Press.

de la Cadena, Marisol, and Mario Blaser, eds. 2018. *A World of Many Worlds.* Durham, N.C.: Duke University Press.

de Waal, Frans. 1999. "Anthropomorphism and Anthropodenial." *Philosophical Topics* 27, no. 1: 255–80.

Escobar, Arturo. 1995. *Encountering Development: The Making and Unmaking of the Third World.* Princeton, N.J.: Princeton University Press.

Estes, James. 2016. *Serendipity: An Ecologist's Quest to Understand Nature.* Berkeley: University of California Press.

Evans-Pritchard, E. E. 1940. *The Nuer: A Description of the Modes of Livelihood and Political Institutions of a Nilotic People.* Oxford: Clarendon Press.

Faubion, James D., and George Marcus. 2009. *Fieldwork Is Not What It Used to Be: Learning Anthropology's Method in a Time of Transition.* Ithaca, N.Y.: Cornell University Press.

Finney, S. C., and L. E. Edwards. 2016. "The 'Anthropocene' Epoch: Scientific Decision or Political Statement?" *GSA Today* 26, no. 2: 4–10.

Flora, Janne, and Astrid Oberborbeck Andersen. 2017. "Whose Track Is It Anyway? An Anthropological Perspective on Collaboration with Biologists and Hunters in Thule, Northwest Greenland." *Collaborative Anthropologies* 9, no. 1–2: 79–116.

Foster, David, ed. 2014. *Hemlock: A Forest Giant on the Edge.* New Haven, Conn.: Yale University Press.

Gilroy, Paul. 2015. "'Where Every Breeze Speaks of Courage and Liberty': Offshore Humanism and Marine Xenology; or, Racism and the Problem of Critique at Sea Level." *Antipode* 50, no. 1: 3–22.

Glikson, Andrew. 2013. "Fire and Human Evolution: The Deep-Time Blue-prints of the Anthropocene." *Anthropocene* 3: 89–92.

Gordon, Deborah. 2010. *Ant Encounters: Interaction Networks and Colony Behavior.* Princeton, N.J.: Princeton University Press.

Grusin, Richard, ed. 2017. *Anthropocene Feminism.* Minneapolis: University of Minnesota Press.

Gupta, Akhil, and James Ferguson. 1997. *Anthropological Locations: Boundaries and the Grounds of a Field Science.* Berkeley: University of California Press.

Haraway, Donna. 2015. "Anthropocene, Capitalocene, Plantationocene, Chthulucene: Making Kin." *Environmental Humanities* 6: 159–65.

Haraway, Donna. 2016. *Staying with the Trouble: Making Kin in the Chthulucene.* Durham, N.C.: Duke University Press.

Haraway, Donna, Noboru Ishikawa, Scott F. Gilbert, Kenneth Olwig, Anna Tsing, and Nils Bubandt. 2016. "Anthropologists Are Talking—about the Anthropocene." *Ethnos* 81, no. 3: 535–64.

Hastrup, Kirsten. 1998. *A Place Apart: An Anthropological Study of the Icelandic World.* Oxford: Clarendon Press.

Hastrup, Kirsten. 2013. "Anticipation on Thin Ice: Diagrammatic Reasoning in the High Arctic." In *The Social Life of Climate Change Models: Anticipating Nature,* edited by Kirsten Hastrup and Martin Skrydstrup, 77–99. London: Routledge.

Hastrup, Kirsten. 2016. "A History of Climate Change: Inughuit Responses to Changing Ice Conditions in North-West Greenland." *Climatic Change* 2016: 1–12.

Hastrup, Kirsten. 2018. "Collaborative Moments: Expanding the Anthropological Field through Cross-Disciplinary Practice." *Ethnos* 83, no. 2: 316–34.

Hastrup, Kirsten, Bjarne Grønnow, and Anders Mosbech. 2018. "Introducing the North Water: Histories of Exploration, Ice Dynamics, Living Resources, and Human Settlement in the Thule Region." *Ambio* 47: 162–74.

Hastrup, Kirsten, Astrid Oberborbeck Andersen, Bjarne Grønnow, and Mads Peter Heide-Jørgensen. 2018. "Life around the North Water Ecosystem: Natural and Social Drivers of Change over a Millennium." *Ambio* 47 (suppl. 2): 213–25.

Herzfeld, Michael. 2001. *Anthropology: Theoretical Practice in Culture and Society.* Oxford: Blackwell.

Humboldt, Alexander von. (1814–29) 1995. *Personal Narrative of a Journey to the Equinoctical Regions of the New Continent.* London: Penguin.

Ingold, Tim. 2018. *Anthropology: Why It Matters.* Cambridge: Polity Press.

Irvine, Richard D. G. 2020. *An Anthropology of Deep Time: Geological Temporality and Social Life.* Cambridge: Cambridge University Press.

Jeppesen, Erik, Martin Appelt, Kirsten Hastrup, Bjarne Grønnow, Anders Mosbech, John P. Smol, and Thomas A. Davidson. 2018. "Living in an Oasis: Rapid Transformations, Resilience and Resistance in the North Water Area Societies and Ecosystems." *Ambio* 47 (suppl. 2): 296–309.

Kimmerer, Robin Wall. 2015. *Braiding Sweetgrass: Indigenous Wisdom, Scientific Knowledge and the Teachings of Plants.* Minneapolis, Minn.: Milkweed.

Kirksey, Eben. 2012. "Living with Parasites in Palo Verde National Park." *Environmental Humanities* 1: 23–55.

Kirksey, Eben S., and Stefan Helmreich. 2010. "The Emergence of Multispecies Ethnography." *Cultural Anthropology* 25, no. 4: 545–76.

Latour, Bruno. 1988. *Science in Action.* Cambridge, Mass.: Harvard University Press.

Latour, Bruno. 2004. *Politics of Nature: How to Bring the Sciences into Democracy.* Cambridge, Mass.: Harvard University Press.

Latour, Bruno. 2016. "The Scientific Fables of an Empirical La Fontaine." In *What Would Animals Say If We Asked the Right Questions?,* edited by Vinciane Despret, vii–xiv. Minneapolis: University of Minnesota Press.

Latour, Bruno. 2017. *Facing Gaia: Eight Lectures on the New Climatic Regime.* London: Polity Press.

Latour, Bruno, and Peter Weibel, eds. 2002. *Iconoclash: Beyond the Image—Wars in Science, Religion and Art.* Cambridge, Mass.: MIT Press.

Law, John. 2011. "What's Wrong with a One-World World." *Heterogeneities,* September 25.

Lewis, Simon, and Mark Maslin. 2015. "Defining the Anthropocene." *Nature* 519: 171–80.

Maguire, James. 2020. "Icelandic Resource Landscapes and the State." *Anthropological Journal of European Cultures* 29, no. 1: 20–41.

Malinowski, Bronisław. 1935. *Coral Gardens and Their Magic: A Study of the Methods of Tilling the Soil and of Agricultural Rites in the Trobriand Islands.* 2 vols. New York: Routledge.

Malm, Andreas, and Alf Hornborg. 2014. "The Geology of Mankind? A Critique of the Anthropocene Narrative." *Anthropocene Review* 1, no. 1: 62–69.

Mascarenhas, Michael. 2020. *Lessons in Environmental Justice: From Civil Rights to Black Lives Matter and Idle No More.* London: SAGE.

Moore, Jason W. 2015. *Capitalism in the Web of Life: Ecology and the Accumulation of Capital.* London: Verso.

Moore, Jason W. 2016. *Anthropocene or Capitalocene? Nature, History, and the Crisis of Capitalism.* Oakland, Calif.: PM Press.

Moore, Jason W. 2017. "The Capitalocene, Part I: On the Nature and Origins of Our Ecological Crisis." *Journal of Peasant Studies* 44, no. 3: 594–630.

Morgan, Henry Lewis. 1868. *The American Beaver and His Works.* Philadelphia: Lippincott.

Mosbech, Anders, Kasper L. Johansen, Thomas A. Davidson, Martin Appelt, Bjarne Grønnow, Christine Cuyler, Peter Lyngs, and Janne Flora. 2018. "On the Crucial Importance of a Small Bird: The Ecosystem-Services of the Little Auk *(Alle alle).*" *Ambio* 47 (suppl. 2): 226–43.

Münster, Ursula. 2016a. "Challenges of Coexistence: Human–Elephant Conflicts in Wayanad, Kerala, South India." In *Conflict, Negotiation, and Coexistence: Rethinking Human–Elephant Relations in South Asia,* edited by Piers Locke and Jane Buckingham, 272–99. New Delhi: Oxford University Press.

Münster, Ursula. 2016b. "Working for the Forest: The Ambivalent Intimacies of Human–Elephant Collaboration in South Indian Wildlife Conservation." *Ethnos: Journal of Anthropology* 81, no. 3: 425–47.

Nuttall, Mark. 2017. *Under the Great Ice: Climate, Society and Subsurface Politics in Greenland.* London: Routledge.

O'Connell-Rodwell, C., L. Hart, and B. Arnason. 2001. "Exploring the Potential Use of Seismic Waves as a Communication Channel by Elephants and Other Large Mammals." *American Zoologist* 41, no. 5: 1157–70.

Parreñas, Juno Salazar. 2018. *Decolonizing Extinction: The Work of Care in Orangutan Rehabilitation.* Durham, N.C.: Duke University Press.

Payne, Katharine, William Langbauer, and Elizabeth Thomas. 1986. "Infrasonic Calls of the Asian Elephant *(Elephas maximus).*" *Behavioral Ecology and Sociobiology* 18, no. 4: 297–301.

Plotnik, J. M., F. B. M. de Waal, and D. Reiss. 2006. "Self-Recognition in Asian Elephant." *Proceedings of the National Academy of Sciences of the United States of America* 103: 17053–57.

Povinelli, Elizabeth. 2016. *Geontologies: A Requiem to Late Liberalism*. Durham, N.C.: Duke University Press.

Rackham, Oliver. 2012. *Woodlands*. London: HarperCollins.

Rowell, Thelma. 1973. *The Social Behaviour of Monkeys*. Harmondsworth, U.K.: Penguin.

Ruddiman, William F. 2003. "The Anthropogenic Greenhouse Era Began Thousands of Years Ago." *Climatic Change* 61, no. 3: 261–93.

Sandom, Christopher, Søren Faurby, Brody Sandel, and Jens-Christian Svenning. 2014. "Global Late Quaternary Megafauna Extinctions Linked to Humans, Not Climate Change." *Proceedings of the Royal Society, Series B* 281, no. 1787: 20133254.

Sapolsky, Robert M. 2001. *A Primate's Memoir*. New York: Scribner.

Sjöberg, Fredrik. 2016. *The Fly Trap*. New York: Vintage.

Smith, Linda Tuhiwai. 2012. *Decolonizing Methodologies: Research and Indigenous Peoples*. London: Zed Books.

Smuts, Barbara B. 1985. *Sex and Friendship in Baboons*. New York: Aldine.

Steffen, Will, Paul Crutzen, and John McNeill. 2007. "The Anthropocene: Are Humans Now Overwhelming the Great Forces of Nature?" *Ambio* 36, no. 8: 614–21.

Stengers, Isabelle, and Stephen Muecke. 2018. *Another Science Is Possible: A Manifesto for Slow Science*. Newark, N.J.: Polity Press.

Strum, Shirley C. 1987. *Almost Human: A Journey into the World of Baboons*. New York: Random House.

Swanson, Heather, Nils Bubandt, and Anna Tsing. 2015. "Less Than One but More Than Many: Anthropocene as Science Fiction and Scholarship-in-the-Making." *Environment and Society: Advances in Research* 6: 149–66.

Swanson, Heather, Jens-Christian Svenning, Alark Saxena, Robert Muscarella, Janet Franklin, Matteo Garbelotto, Andrew Mathews et al. 2021. "History as Grounds for Interdisciplinarity: Promoting Sustainable Woodlands through an Integrative Ecological and Socio-cultural Historical Perspective." *One Earth* 4, no. 2: 226–37.

Sze, Julie. 2020. *Environmental Justice in a Moment of Danger*. Berkeley: University of California Press.

TallBear, Kim. 2011. "Why Interspecies Thinking Needs Indigenous Standpoints: Theorizing the Contemporary." https://culanth.org/fieldsights/260-why-interspecies-thinking-needs-indigenous-standpoints.

Taylor, Dorceta. 2014. *Toxic Communities: Environmental Racism, Industrial Pollution, and Residential Mobility*. New York: New York University Press.

Thomas, Chris. 2017. *Inheritors of the Earth: How Nature Is Thriving in an Age of Extinction*. New York: Penguin.

Tsing, Anna Lowenhaupt. 2013. "More-Than-Human Sociality: A Call for Critical Description." In *Anthropology and Nature*, edited by Kirsten Hastrup, 27–42. New York: Routledge.

Tsing, Anna Lowenhaupt. 2015. *The Mushroom at the End of the World: On the Possibility of Life in Capitalist Ruins*. Princeton, N.J.: Princeton University Press.

Tsing, Anna Lowenhaupt. 2017. "A Threat to Holocene Resurgence Is a Threat to Livability." In *The Anthropology of Sustainability: Beyond Development and Progress,* edited by Marc Brightman and Jerome Lewis, 51–65. New York: Palgrave Macmillan.

Tsing, Anna Lowenhaupt. 2022. "On the Sociality of Birds: Reflections on Ontological Edge Effects." In *KIN: Thinking with Deborah Bird Rose,* edited by Thom van Dooren and Mathew Chrulew, 15–32. Durham, N.C.: Duke University Press.

Tsing, Anna, and Nils Bubandt, eds. 2018. "Feral Dynamics of Post-industrial Ruin." Special issue. *Journal of Ethnobiology* 38, no. 1.

Tsing, Anna, and Nils Bubandt. 2020. "Swimming with Crocodiles." *Orion Magazine,* Spring, 35–41. https://orionmagazine.org/article/swimming-with-crocodiles/.

Tsing, Anna, Jennifer Deger, Alder Keleman Saxena, and Feifei Zhou, eds. 2021. *Feral Atlas: The More-Than-Human Anthropocene.* Stanford, Calif.: Stanford University Press. http://www.feralatlas.org/.

Tsing, Anna, Andrew Mathews, and Nils Bubandt. 2019. "Patchy Anthropocene: Landscape Morphology, Multispecies History and the Retooling of Anthropology." *Current Anthropology* 60, no. 20: S186–97.

Tsing, Anna Lowenhaupt, Heather Anne Swanson, Elaine Gan, and Nils Bubandt, eds. 2017. *Arts of Living on a Damaged Planet: Ghosts and Monsters of the Anthropocene.* Minneapolis: University of Minnesota Press.

Vetlesen, Arne Johan. 2019. *Cosmologies of the Anthropocene: Panpsychism, Animism, and the Limits of Posthumanism.* London: Routledge.

Vibe, Christian. 1967. "Arctic Animals in Relation to Climatic Fluctuations." *Meddelelser om Grønland* 170, no. 5.

Viveiros de Castro, Eduardo. 2004. "Perspectival Anthropology and the Method of Controlled Equivocation." *Tipití* 2, no. 1: 3–22.

Viveiros de Castro, Eduardo. 2019. "On Models and Examples: Engineers and Bricoleurs in the Anthropocene." *Current Anthropology* 60, no. S20: S296–308.

Walker, Mike, Phil Gibbard, and John Lowe. 2015. "Comment on 'When Did the Anthropocene Begin? A Mid-Twentieth Century Boundary Is Stratigraphically Optimal' by Jan Zalasiewicz et al. (2015)." *Quaternary International* 383: 204–7.

Wallace, David. 2012. *The Emergent Multiverse: Quantum Theory According to the Everett Interpretation.* Oxford: Oxford University Press.

Yusoff, Kathryn. 2018. *A Billion Black Anthropocenes or None.* Minneapolis: University of Minnesota Press.

Zalasiewicz, Jan, Colin N. Waters, Mark Williams, Anthony D. Barnosky, Alejandro Cearreta, Paul Crutzen et al. 2014. "When Did the Anthropocene Begin? A Mid-Twentieth Century Boundary Level Is Stratospherically Optimal." *Quaternary International* 383: 1–8.

Zalasiewicz, Jan A., Colin N. Waters, Alexander P. Wolfe, Anthony D. Barnosky, A. Cearreta, M. Edgeworth, E. C. Ellis et al. 2017. "Making the Case for a Formal Anthropocene Epoch: An Analysis of Ongoing Critiques." *Newsletters on Stratigraphy* 50, no. 2: 205–26.

Contributors

ASTRID OBERBORBECK ANDERSEN is associate professor of techno-anthropology at Aalborg University, Denmark. She is coeditor of *Anthropology Inside Out: Fieldworkers Taking Notes.*

FILIPPO BERTONI is a freelancing transdisciplinary developmental editing specialist based in Amsterdam, Netherlands.

HARSHAVARDHAN BHAT is a researcher and writer interested in the social study of monsoonal futures.

NATHALIA BRICHET is associate professor of environmental anthropology at the University of Copenhagen, Denmark. She is the author of *An Anthropology of Common Ground: Awkward Encounters in Heritage Work* and curator of several exhibitions.

NILS BUBANDT is professor of anthropology at Aarhus University, Denmark. He is author of *The Empty Seashell: Witchcraft and Doubt on an Indonesian Island* and coeditor of *Arts of Living on a Damaged Planet* (Minnesota, 2017) and *Philosophy on Fieldwork: Case Studies in Anthropological Analysis.*

RACHEL CYPHER is a postdoctoral fellow at the University of Pennsylvania's Penn Program in Environmental Humanities.

JANNE FLORA is associate professor in anthropology at Aarhus University, Denmark. She is author of *Wandering Spirits: Loneliness and Longing in Greenland.*

NATALIE FORSSMAN is lecturer of technical communications at the University of British Columbia, Canada.

PETER FUNCH is associate professor in the Department of Biology and in the Arctic Research Center at Aarhus University, Denmark.

KIRSTEN HASTRUP is professor emeritus of anthropology at the University of Copenhagen, Denmark. She is author of *A Passage to Anthropology: Between Experience and Theory* and numerous other books.

COLIN HOAG is assistant professor of anthropology at Smith College in Massachusetts. He is author of *The Fluvial Imagination: On Lesotho's Water-Export Economy.*

JOSEPH KLEIN is PhD candidate in anthropology at the University of California, Santa Cruz.

ANDREW S. MATHEWS is professor of anthropology at the University of California, Santa Cruz. He is author of *Instituting Nature: Authority, Expertise, and Power in Mexican Forests* and *Trees Are Shape Shifters: How Cultivation, Climate Change, and Disaster Create Landscapes.*

DANIEL MÜNSTER is associate professor of medical anthropology at the University of Oslo, Norway. He is author of *Postkoloniale Traditionen: Eine Ethnografie über Dorf, Kaste, und Ritual in Südindien* and coeditor of *Suicide and Agency: Anthropological Perspectives on Self-Destruction, Personhood, and Power.*

URSULA MÜNSTER is associate professor of environmental humanities and director of the Oslo School of Environmental Humanities at the University of Oslo, Norway.

JON RASMUS NYQUIST is lecturer in social anthropology at the University of Oslo, Norway.

KATY OVERSTREET is assistant professor at the Centre for Sustainable Futures at the University of Copenhagen, Denmark.

PIERRE DU PLESSIS is a researcher at the Oslo School of Environmental Humanities, Department of Culture Studies and Oriental Languages, University of Oslo.

MEREDITH ROOT-BERNSTEIN is research scientist at the Centre national de la recherche scientifique (CNRS), based in the National Museum of Natural History in Paris, France.

HEATHER ANNE SWANSON is professor of anthropology and director of the Centre for Environmental Humanities at Aarhus University, Denmark. She is coeditor of *Arts of Living on a Damaged Planet* (Minnesota, 2017) and *Domestication Gone Wild: Politics and Practices of Multispecies Relations* and author of *Spawning Modern Fish: Transnational Comparison in the Making of Japanese Salmon*.

ANNA TSING is professor of anthropology at the University of California, Santa Cruz and at Aarhus University, Denmark. She is author or editor of numerous books, was codirector of Aarhus University Research on the Anthropocene, and co-curated the recently released interactive digital project "Feral Atlas: The More-Than-Human Anthropocene" (http://www.feralatlas.org/).

STINE VESTBO received her PhD on species redistributions in the Anthropocene from the Department of Biology at Aarhus University, Denmark.

Index